Space Science and Astronomy

THE MACMILLAN SKY AND TELESCOPE LIBRARY OF ASTRONOMY

With the advent of space exploration the science of astronomy enters a new phase—that of practical applications significant not only to scientists but also to the public at large. The rapid and critical developments in astronomy that preceded this new phase are presented in the unique library of astronomy made up of articles that first appeared in the prominent journals *Sky and Telescope, The Sky,* and *The Telescope.*

Space Science and Astronomy

ESCAPE FROM EARTH

EDITED BY THORNTON PAGE AND LOU WILLIAMS PAGE

VOLUME 9   *Sky and Telescope* Library of Astronomy

Illustrated with over 150 photographs and diagrams

MACMILLAN PUBLISHING CO., INC.
NEW YORK

COLLIER MACMILLAN PUBLISHERS
LONDON

043007

Macmillan Publishing Co., Inc.
866 Third Avenue, New York, N.Y. 10022
Collier Macmillan Canada, Ltd.

Library of Congress Cataloging in Publication Data
Main entry under title:
Space science and astronomy.
  (The Macmillan Sky and telescope library of astronomy; v. 9)
  "Articles that first appeared in . . . Sky and telescope."
  Bibliography: p. 450
  Includes index.
  1. Astronomy—Addresses, essays, lectures.
2. Astronautics—Addresses, essays, lectures. I. Page, Thornton. II. Page, Lou Williams.
QB51.S68   520'.8   76-5879
ISBN 0-02-594310-3

FIRST PRINTING 1976

Printed in the United States of America

Contents

3. Exploring the Moon and Its History

4. Trips to Venus and Mercury

5. Preparations for Landing on Mars

6. Close-Up Views of Jupiter

7. Spacecraft Design and Workshops in Space

8. Optical Observations of the Sun, Earth, and Stars

9. X-Ray and Gamma-Ray Astronomy

10. The Frontier in Space

Illustrations

Figure

Figure

Figure

Figure

Tables

xvi   ILLUSTRATIONS

Table

About the Editors

Thornton Page is Research Astrophysicist for the Naval Research Laboratory at NASA Johnson Space Center in Houston. He got his B.S. from Yale in 1934, Ph.D. from Oxford, where he was a Rhodes Scholar, in 1938, and has taught astronomy at Wesleyan University, the University of Chicago, the University of Colorado, U.C.L.A., and Yale. His early research was on nebulae and galaxies, using the 82-inch Struve Telescope at the McDonald Observatory in western Texas and the 60-inch Bosque Alegre Telescope of Córdoba University, Argentina. Recently, he has been involved in two NASA missions with the S201 Far-Ultraviolet Camera: Apollo 16 to the Descartes landing on the Moon, and the 85-day Skylab-4 Earth-orbiter, studying, among other things, Comet Kohoutek.

Mrs. Page received a Ph.D. in geology from the University of Chicago. She is the author of A Dipper Full of Stars for young readers; Ideas from Astronomy, Ideas from Geology, and The Earth and Its Story for high-school readers; and a text for an introductory course for college nonscience majors, Astronomy: How Man Learned About the Universe. She has taught at Stephens College, in the College of the University of Chicago, and at Wesleyan University, was managing editor of the Journal of Geology, and is now editor of the Connecticut Geological and Natural History Survey.

Space Science and Astronomy

1

Concepts of
the Space Age

What is the "space age"? and when did it start? Astronomers can rightly say that they have been studying planets, stars, and other things in space for more than 5000 years. Various aspects of these studies have been covered in earlier volumes of this series: the solar system in Volumes 1–3, telescopes in Volume 4, stars in Volumes 5 and 6, the Milky Way and other galaxies in Volumes 7 and 8.

Before discussing this broad topic of space science—which has vastly expanded the activities and horizons of astronomy—we consider it worthwhile to establish just what the space age means to various groups of people. As the following articles will show, it means everything from getting outside the Earth's atmosphere (among astronomers) to visiting or communicating with other civilizations. Earth scientists want to get material samples from other parts of the solar system, and some physicists want to experiment with accurate clocks moving at high speed or in high gravitational fields. All scientists welcome the observations possible under conditions very different from those on the Earth's surface. People have long been intrigued with the notion of space travel, and books were written about a trip to the Moon well before the Apollo-11 astronauts made *it. See Jules Verne's* From the Earth to the Moon *(1865) or H. G. Wells'* The First Men in the Moon *(1901).*

*We might date the space age from when the first men actually stepped on the lunar surface (Armstrong and Aldrin in 1969), or when a man first orbited the Earth above the atmosphere (Gagarin in 1961). To many it started on Oct. 4, 1957, when the Soviet Union orbited the first satellite, Sputnik I, about 370 mi. above the Earth's surface. Well before that, astronomers at the Naval Research Laboratory shot rockets carrying instruments above the atmosphere, starting in 1946, and by early 1957 space astronomy was well thought out.—*TLP

An Astronomer Looks at Space Travel

VICTOR M. BLANCO

(*Sky and Telescope,* May 1957)

Man has always dreamed of traveling away from the Earth to visit the Moon, the other planets, and the stars.

Why should we wish to venture away from the safety of mother Earth out into the vastness of space?

Apart from the possible military uses of space travel, we will gain valuable new knowledge in many fields of science. Curiously enough, space travel will help geophysicists to learn more about the interior of the Earth. The accelerations experienced by spacecraft in the vicinity of the Earth will yield new information about the internal structure of our planet.

Meteorology will also benefit by observation from outer space. In spite of our global network of weather stations, there are at present huge ocean areas where a hurricane or typhoon may lurk unobserved. A watch post far out in space would help us to track the course of major storms with accuracy.

The nature of cosmic rays may be studied in space without the marked changes caused by the Earth's atmosphere and magnetic field. Such observations could well make the analysis of cosmic rays a common part of astronomical research, adding the Geiger-counter telescope to optical and radio telescopes as major tools of the astronomer of the future.

For astronomers, the ability to observe the heavenly bodies from outside the atmosphere will mean a revolutionary improvement in the accuracy of their observations. The variations in telescopic-image quality collectively known as "seeing," which are caused by the refraction and diffraction (bending) of light rays as they pass through the Earth's atmosphere, will be eliminated entirely.

Photographic plates will not be fogged as badly in long exposures by the over-all sky brightness, which is due partly to the glowing air in the ionospheric regions, and partly to the random scattering of light rays in the atmosphere. Not all the sky brightness will be eliminated in observations made outside the atmosphere, however, for the dust found between the planets also scatters light (the zodiacal light—see p. 334), and the background of the heavens will always be illuminated by the total aggregate of distant stars and galaxies. Indeed, the study of this residual illumination will help in determining how both the interplanetary dust and the distant galaxies are distributed in space.

The ultraviolet and infrared regions of the spectrum will no longer be blocked out by the atmosphere. This blocking is serious, for stars as blue or bluer than Vega have their maximum intensity of radiation in parts of the spectrum too far in the violet to shine through our atmosphere, whereas the reddest stars known radiate chiefly in infrared colors that are at least partly absorbed by the atmosphere.

Atmospheric extinction, the amount by which the intensity of light rays is cut down by our atmosphere, would be eliminated in an outer-space observing station. Very short radio waves are affected by water vapor, whereas waves longer than 30 m. are not passed by the Earth's ionosphere. But observations from a lunar observatory could cover the

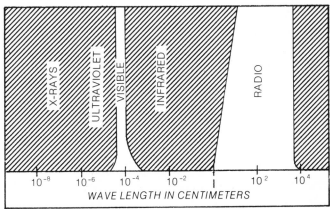

FIG. 1. Optical and radio telescopes can study astronomical bodies only through two restricted atmospheric windows. From a space station above the atmosphere, observations can be made of the blocked-out wavelengths as well.

entire electromagnetic spectrum—x-rays, ultraviolet, infrared, and radio waves.

The bolometric magnitude of a star is a measure of the total rate of radiant-energy output at all wavelengths. A knowledge of accurate bolometric magnitudes is fundamental in understanding the details of energy production, internal structure, and ageing effects of the stars. Our present information on bolometric magnitudes is based upon the crude measurements that we can make through the atmosphere.

Most of the spectroscopic features originating from interstellar atoms and molecules in their lowest energy states are in the ultraviolet, not observable at present. The observation of these features will permit a vastly more thorough analysis of the gases present in the space between the stars.

In view of the rapid engineering development of large rockets in the last few years, it appears that the age of spaceflight is not too distant. However, many of us overlook the serious problems that must be solved before men travel away from the Earth. Great advances in rocket design have been achieved since the first V-2 rockets were devised by the Germans. Current secret work on guided missiles is no doubt advancing the day of spaceflight but all present rockets have one serious shortcoming. They are entirely too small for manned space flights.

Large quantities of rocket fuel will be required to overcome the gravitational attraction of the Earth. Furthermore, to provide for the comfort of the crew, vast increases in the size of rocket vehicles and in their fuel capacities will be necessary. We must find economical ways of building very large spaceships and, if possible, devise more efficient fuels to propel these ships. The acceleration of a rocket depends not only on the speed with which burned fuel is ejected but also on its quantity. During take-off, the more rapidly the rocket is accelerated, the more efficient it will be, and the less fuel will be needed to overcome the Earth's attraction.

But the presence of a human crew will limit the acceleration to the level of human endurance. This means that a compromise will have to be made in the efficiency of the rocket, and fuel requirements must be increased correspondingly.

So far we have no answer to the question, "How fast can a person be accelerated and stay not only alive but in good health?" Acceleration is measured in g units, where one g is the amount of acceleration of a freely falling body near the Earth's surface. Human beings can stand up to about five g's for limited periods. As yet, however, we do not know what accelerations can be tolerated for longer intervals than a few seconds.

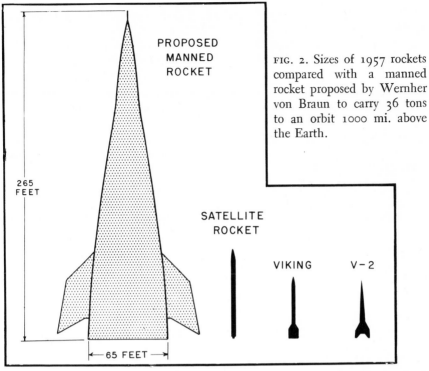

PROPOSED MANNED ROCKET

265 FEET

SATELLITE ROCKET

VIKING

V-2

← 65 FEET →

FIG. 2. Sizes of 1957 rockets compared with a manned rocket proposed by Wernher von Braun to carry 36 tons to an orbit 1000 mi. above the Earth.

The effect of weightlessness on a space traveler is another medical problem of great interest (p. 41). There are reasons for believing that serious physiological and psychological maladjustments could result from long weightlessness, because such a condition would render ineffectual the balance and orientation mechanism of the human body as well as its gravity-dependent muscular responses.

Collisions with meteoroids (see Glossary) will be a hazard to space-flight. The most frequent collisions will be with dust-grain-size meteoroids. According to Harvard astronomer Fred L. Whipple, a meteor bumper—a thin outer skin surrounding the spacecraft—will offer sufficient protection. The effectiveness of such a bumper will depend on how often meteoritic particles of various sizes are encountered.

This information could be gathered in advance by means of artificial satellites. If space travelers stay away from regions of high particle concentration, such as exist in the orbits of the meteor swarms, the chances of collisions with meteorites large enough to penetrate the walls of the vehicle will be small. It should be possible to avoid collisions with meteorites sufficiently large for early radar detection, as well as to repair minor collision damage while in flight.

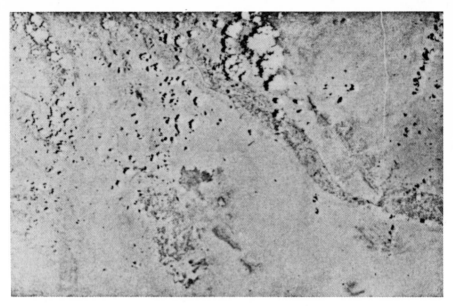

FIG. 3. View of the Earth from a Viking rocket at the record altitude of 158 mi. in May 1954. The Rio Grande runs diagonally across the picture; north is at the top. White puffs are cumulus clouds. (U. S. Navy photo.)

Considering the problems that must be solved before space travel becomes a reality, one may hazard a guess as to the sequence of developments in the near future. After the successful completion of present artificial-satellite experiments, we will probably increase the instrument-carrying capacity of the satellites. The next step will be to send relatively small unmanned rockets to the Moon, possibly on round-trip journeys.

By means of automatic observing and recording instruments, we will study the Earth from a distance, measure the magnetic field of the Moon, and record the lunar surface, including the never-seen far side, in detail. These experiments appear so near at hand that one may venture to say that in the next decade or so we will be living through the excitement of such explorations. Similar studies of some of the other planets and of the space near the sun will follow. The scientific rewards and the desire for international prestige should stimulate continuous government-supported efforts along these lines.

Many of Blanco's predictions came true in the next two decades, and large rockets became the mainstay of space sciences. But some of the astronomers' objectives can be met without the use of rockets, as demonstrated by Stratoscope.—TLP

Stratoscope Observation
of a Seyfert Galaxy

(*Sky and Telescope,* December 1968)

Princeton University scientists have been developing the techniques for high-resolution celestial photography from stabilized unmanned balloons. Their current program uses a 3½-ton, 36-in. reflecting telescope known as Stratoscope II, which can be operated by remote control while floating above 95 percent of the Earth's atmosphere.

Martin Schwarzschild reported some results from the latest Stratoscope-II ascent, in a paper with Robert E. Danielson and Blair D. Savage. The launch took place on the evening of May 18, 1968, from the Scientific Balloon Flight Station near Palestine, Texas. Nearly 14 hrs later the Stratoscope telescope was safely landed 2 mi. northeast of Cushing, Texas. Six hours were required for ascent and descent, and the remainer of the time was spent observing at the 80,000-ft level.

The telescope worked reliably during the whole night. It was aimed by radio command, using data from television cameras aboard the balloon. More than 100 frames of 35-mm film were exposed.

In-flight measurements of focus showed that air turbulence inside the telescope tube caused some image degradation, so that the diffraction-resolution limit of 0.1 arc-sec was not reached. Stellar images recorded during the flight were elongated, generally smaller than 0.3 by 0.8 arc-sec. The average tracking jitter was as low as 0.03 arc-sec on guide stars of magnitudes 6 to 8.

A main object of observation was the Seyfert galaxy NGC 4151 (see Glossary). The bright, compact nucleus, which is variable and has a bright-line spectrum, somewhat resembles a quasar.

The Princeton astronomers concluded that the size of the nucleus (out to half its peak brightness) is no larger than 0.18 arc-sec. At the distance of NGC 4151, this corresponds to about 30 light years (1.7×10^{14} mi.—see App. V).

Many other balloon flights have obtained valuable astronomical data, but rockets are required to get entirely outside the atmosphere. An excellent review is the History of Rocketry and Space Travel *by Wernher von Braun and F. I. Ordway III (3d ed., Crowell, 1975). It covers the ancient*

development of rockets for military use and the later shift to space probes by Tsiolkovsky (a Russian), Goddard (an American), and Oberth (a German). One important difference between a gun shell and a rocket is the length of time it can be accelerated—the gun-barrel-length vs. the rocket's burn time. Another is the accuracy of flight—initially poor for rockets. It is clear that space-age applications required highly accurate guidance and control of rockets—achievements started by the Germans during World War II. But the accuracy required for shooting instruments to the Moon is much higher than for dropping V-2 bombs on London, and it took ten post-war years to develop such accuracy.

The astronomers' early interest in space centered on making observations outside the atmosphere, but the remainder of this book will show that a great deal more was achieved, and that some current plans outdo the science fiction of the last decade. Space science has bridged astronomy, physics, geology, and several other sciences, so the limits of space astronomy are difficult to define. The chronology given in Appendix II shows major steps that may give some perspective to the following chapters—particularly to the early accomplishments cited in Chapter 2. Note that the steps that benefited science would never have been taken without the politicians' interest in the "space race" and could not have been taken without the remarkable engineering developments of the last twenty years. Many of these developments (rockets, radio, radar, lasers, etc.) were sponsored by the military, and recent NASA activities even show some influence of diplomats.

So the emergence of space astronomy—space science in general—depended on an interlocking group of special interests: military, political, engineering (industrial), and science. The following articles give some sample accounts, showing what was popular at times from 1958 to 1975. Note that, as time went on, more and more specialized terms became necessary in the reporting of space activities. Words basic to the reasoning of space science are discussed in the Editors' Comments, and others of importance are defined in the Glossary, Appendix IV.—TLP

Space Age: Year One

MARSHALL MELIN

(*Sky and Telescope,* October 1958)

Oct. 4, 1958, marks the anniversary of the greatest technological achievement since atomic energy was unleashed. On that date last year the first artificial satellite was successfully placed in orbit.

Ever since Newton's day, an artificial Earth satellite moving in a stable orbit was theoretically conceivable, while early in our century H. Oberth's analyses and R. H. Goddard's experiments made it clear that multiple-stage rockets (p. 274) could be used to put such bodies aloft.

Even today, the general public scarcely realizes the intricate engineering problems that were mastered—the high-speed fuel pumps for unstable liquids; the reaction chambers to work at extreme temperatures; the elaborate systems to control the brief moments of guided flight with millisecond precision. . . .

As in few other undertakings, the launching of a satellite demands the successful co-ordination of hundreds of mechanical and electrical systems. Even when all the components perform well, an improbable chance can mean disaster—as on Aug. 24, 1958, when the first-stage Jupiter C of Explorer V separated properly, but later collided with the remaining stages of the package!

There have been seven successful launchings during the first 11 months of the space age, putting 10 separate objects into orbits. Six of these are still circling the Earth, three of them with radios transmitting observations.

Looking back over the space age's first year, we can trace its deep impact on America, aside from international politics. The Sputniks gave a sharp spur to scientific education in the United States, leading Congress to pass on Aug. 23, 1958, the Science Aid Act, which provides $900 million to help train scientists. Soon after Sputnik I was launched, J. R. Killian was appointed science adviser to the President, and the Advanced Research Projects Agency was created within the Department of Defense.

More recently, the scope of the National Advisory Committee for Aeronautics was broadened to include a space program, and the NACA was renamed the National Aeronautics and Space Administration (NASA—see p. 10). In another development, the National Academy of Sciences and the National Research Council created a Science Space Board, for a joint survey of all aspects of man's advance into space.

Satellite research is, of course, a part of the International Geophysical Year, which was scheduled to end in December 1958. However, in August the IGY governing committee decided at Moscow for an extension to be known as the International Geophysical Co-operation-1959.

Initial satellite results are exciting. The upper atmosphere turns out to be appreciably denser than we had thought, and temperature measurements in those regions now replace uncertain estimates. Satellite observations are allowing a redetermination of the oblateness of the Earth and of the distribution of mass inside it; eventually they will tie together the geodetic networks covering Europe and the United States more accurately than ever before.

Counts of micrometeoroids (see Glossary) in the neighborhood of the Earth have been announced. Other programs measure the intensity of solar radiation and the magnetic and electrical properties of the Earth's surroundings. Biological studies have begun, with indications that future space voyagers can re-enter the Earth's atmosphere without undergoing lethal temperatures. But a new hazard to space travel is the unexpectedly high intensity of cosmic radiation, which doubles for each 60 mi. of altitude above 250 mi.

As the first year of the space age ends, we can foresee some other advances in the near future. One possibility is an artificial satellite that will release an inflated balloon perhaps 12 ft in diameter; because of its sensitivity to air resistance, it should allow accurate determinations of upper-atmosphere densities. A telescope mounted on a satellite platform is being planned and the first successful rocket to the Moon may soon be history.

For historical completeness we cite here a few of the beginning sections of Public Law 85–568, passed by the U.S. Congress on July 29, 1958. President Eisenhower established NASA in accordance with this Act in October 1958, and appointed James E. Webb as Administrator.— TLP

NATIONAL AERONAUTICS AND SPACE ACT OF 1958, AS AMENDED SEPTEMBER 3, 1974

SEC. 102. (a) The Congress hereby declares that it is the policy of the United States that activities in space should be devoted to peaceful purposes for the benefit of all mankind.

. . .

(c) The aeronautical and space activities of the United States shall be conducted so as to contribute materially to one or more of the following objectives:

(1) The expansion of human knowledge of phenomena in the atmosphere and space;

(2) The improvement of the usefulness, performance, speed, safety, and efficiency of aeronautical and space vehicles;

(3) The development and operation of vehicles capable of carrying instruments, equipment, supplies, and living organisms through space;

(4) The establishment of long-range studies of the potential benefits to be gained from, the opportunities for, and the problems involved in the utilization of aeronautical and space activities for peaceful and scientific purposes;

(5) The preservation of the role of the United States as a leader in aeronautical and space science and technology and in the application thereof to the conduct of peaceful activities within and outside the atmosphere;

. . .

(7) Cooperation by the United States with other nations and groups of nations in work done pursuant to this Act and in the peaceful application of the results thereof; . . .

Sec. 201. (a) There is hereby established, in the Executive Office of the President, the National Aeronautics and Space Council (hereinafter called the "Council") which shall be composed of—

(1) the Vice President, who shall be Chairman of the Council;

(2) the Secretary of State;

(3) the Secretary of Defense;

(4) the Administrator of the National Aeronautics and Space Administration; and

(5) the Chairman of the Atomic Energy Commission.

. . .

(e) It shall be the function of the Council to advise and assist the President, as he may request, with respect to the performance of functions in the aeronautics and space field, including the following functions:

(1) survey all significant aeronautical and space activities, including the policies, plans, programs, and accomplishments of all departments and agencies of the United States engaged in such activities;

(2) develop a comprehensive program of aeronautical and space activities to be conducted by departments and agencies of the United States;

(3) designate and fix responsibility for the direction of major aeronautical and space activities;

(4) provide for effective cooperation among all departments and agencies of the United States engaged in aeronautical and space activities . . .

Sec. 202. (a) There is hereby established the National Aeronautics and Space Administration (hereinafter called the "Administration"). The Administration shall be headed by an Administrator, who shall be appointed from civilian life by the President by and with the advice and consent of the Senate. Under the supervision and direction of the

President, the Administrator shall be responsible for the exercise of all powers and the discharge of all duties of the Administration ...

Man's Farthest Step
Into Space

MARSHALL MELIN

(*Sky and Telescope*, November 1958)

The last milestone before flight into interplanetary space was passed successfully on Oct. 12, 1958, when the lunar probe Pioneer 1 traveled to nearly 80,000 mi. from Earth, one-third of the distance to the Moon. When the instrument package separated from the third-stage rocket, it was traveling some 34,400 ft/sec, only 850 ft/sec too slowly to attain the orbit planned.

This velocity is nearly that of escape from the Earth. The narrowness of the margin is the true significance of the Pioneer rocket flight, for here is proof that spaceflight is possible with existing multistage rockets, if all their complex components can be made to function together.

FIG. 4. Preparations for the launch of Pioneer 1 (vehicle No. 127) at the Atlantic Missile Test Range, Cape Canaveral, Florida. The first-stage rocket (bottom) was an Air Force Thor; second stage, a Navy Vanguard; and third stage (below "USAF"), a solid-propellant Navy rocket. The Pioneer-1 probe (top) had a retrorocket intended to put an 84-lb payload into an orbit around the Moon. (U. S. Air Force photo.)

In this particular flight, the operations of the parts of the orientation system were not perfectly combined to achieve the delicate balance of speed and direction necessary for the precise orbit desired for the lunar probe. The direction of flight of the first-stage rocket was controlled by gyros, but these had a slightly higher drift rate than expected, with the result that the trajectory was some 3°.5 too high. This, in turn, meant that the drag by gravity was greater and the vehicle was slowed down by 850 ft/sec. This small difference so reduced the size of the missile's orbit that it attained its apogee point less than a day after launch, and plunged back into the Earth's atmosphere, to be destroyed like a meteor.

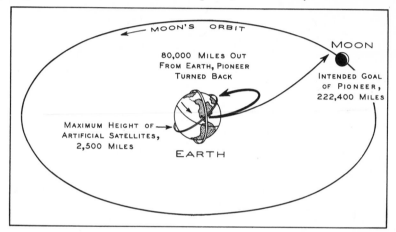

FIG. 5. Diagram (not to scale) of the Earth, satellite orbits, the intended orbit of Pioneer 1 to the Moon, and its turn-back at 80,000 mi.

This now-historic example demonstrates how critically the size of the orbit followed by a missile cast into space depends on its speed compared with the velocity of escape from the Earth. The latter is about 36,700 ft/sec at the Earth's surface and decreases as the square root of the distance from the Earth's center. At the Moon's average distance (60 Earth radii), the escape velocity is about 4700 ft/sec, whereas at the distance of 80,000 mi. attained by Pioneer it is some 8200 ft/sec.

A missile moving more slowly than escape velocity will travel in a closed path—a circle, or an ellipse from perigee to apogee—around the Earth's center. Should the perigee distance happen to be less than the Earth's radius, the orbit would come within the Earth itself, and the body would collide with the surface.

If we consider higher and higher missile speeds—still always less than the velocity of escape—the corresponding orbits are ellipses of greater and greater size. If the initial velocity were exactly that of escape, the body would travel outward on an open, parabolic orbit of infinite size, never to return. At still higher launch velocities, the trajectory would be a hyperbola, also of infinite size. The significant point, in relation to Pioneer, is how very sensitive the apogee distance is to initial velocity.

Two specifically lunar observations were intended for Pioneer:

The measurement of the Moon's magnetic field, whose intensity is entirely a matter of conjecture at present, and the scanning of the Moon's surface with an infrared-sensitive television device, from whose telemetered signals a crude map of the invisible far side of the Moon could be constructed.

For the latter, the far side of the Moon should be facing the sun, and this was the case when Pioneer was launched, as the lunar phase in mid-October was just past new.

The U.S. Air Force has recently been granted authority for two more lunar probes, in addition to the three originally allotted to it last spring. (Two of these three have already been fired, on August 17 and October 11.) The additional launches will be attempted after the first of the year, according to press reports, which also suggest that one of the vehicles may be sent to the vicinity of Venus.

THE LAUNCH OF PIONEER 1

Design considerations impose severe limitations upon rockets used in experiments like the lunar probe. Since the frame must be as light as possible, it cannot be constructed to withstand much lateral force. This is the reason for the nearly vertical flight path until much of the denser atmosphere is left behind. Thereafter, the rocket is turned nearly parallel to the Earth's surface.

The rocket thrust then works only against the inertia of the vehicle (p. 39), and not against the Earth's gravitational field.

In 140 sec of powered flight, the 50-ton Thor first-stage rocket, with a 150,000-lb-thrust Rocketdyne engine, rose to a height of 50½ mi. During this ascent, the direction of climb was established by swiveling the main booster combustion chamber on its gimbals, and differential control was achieved by small rockets attached to the main frame.

After cutoff of the first stage, with the space probe's speed about 10,000 mi/hr, the explosive separation bolts were detonated. The burned-out first stage and the cowling that had protected the later stages from air friction were thus cast off, and the second-stage liquid-fuel Aerojet-General 1040 propulsion system took over. Eight small rockets imparted spin to this 4000-lb assembly, which ascended to a height of 163.1 mi. The second-stage cutoff was controlled by an accelerometer, which had no way of detecting the improper climb angle.

The third stage used a solid-propellant rocket motor newly developed by the U.S. Navy's Allegheny Ballistics Laboratory. Weighing 400 lbs, this rocket provided about 2500 lbs of thrust, and carried the missile to a height of 221 mi. at third-stage burnout. Explosive bolts separated the Pioneer package, and a few minutes afterward its eight small vernier rockets were operated by command signal from the ground to provide small adjustments in attitude and velocity. Thereafter, the package was on its way, but its peak velocity of 23,300 mi/hr was some 570 short of that intended.

By command signal the final verniers were jettisoned. The reverse-acting fourth stage that failed to fire (because of lack of battery power) had a Thiokol engine developed by Space Technology Laboratory. This organization was in over-all charge of the entire missile assembly and of ground tracking.

TRACKING THE PROBE

The five primary radio-tracking stations were in continuous two-way teletype contact with one another and with the data-reduction center of Space Technology Laboratory at Inglewood, California. The nine Minitrack stations operated by the Naval Research Laboratory reported to their operations center at Cape Canaveral, which relayed the data to the California nerve center.

Because of the unprecedented distance to which Pioneer 1 was to travel, exceptional measures were needed to assure reception of its telemetered signals. Pioneer's transmissions on 108.06 megacycles—the radio frequency internationally agreed upon—were recorded by five of the largest steerable antennas in the world. These primary stations, which monitored the probe continuously up to 3:46 Universal time (UT) on October 13, minutes before its destruction in the atmosphere, are at Cape Canaveral, Florida; Millstone Hill, in Massachusetts; Jodrell Bank, near Manchester, England; Singapore, British Malaya; and Kanae, Hawaii.

The nine Minitrack stations that helped track Pioneer are located at Ancon, Peru; Antigua, Leeward Islands; Antofogasta and Santiago in Chile; Cotopaxi, Ecuador; Grand Turk, Bahamas; Havana, Cuba; Johannesburg, South Africa; and Woomera, Australia.

The stream of observations of the direction and line-of-sight velocity of Pioneer from these stations flowed into the data reduction center at Inglewood, where the figures were combined and analyzed with an

FIG. 6. One of the five primary tracking (Minitrack) stations around the world tuned to radio signals from Pioneer 1, the Kanae, Hawaii, installation employs a 60-ft paraboloid dish (left) and four sets of helical antennas (two seen at the right), all pointable on alt-azimuth mounts. These units monitored the down-link signals during Pioneer-1's descent over the Pacific on Oct. 23, 1958. (U. S. Air Force photo.)

IBM-704 electronic digital computer. From this came the distance- and speed-values that were so promptly relayed to the public.

The tracking stations made tape recordings of the five channels on which the observations obtained by Pioneer were being telemetered to Earth. Preliminary reports of some of the experiments were quickly made available, in announcements from the Pentagon.

TABLE 1. DISTANCES OF PIONEER 1 FROM THE EARTH'S CENTER

Date	UT	Miles	Date	UT	Miles
Oct. 11	8:42	4,000	Oct. 11	20:47	65,000
	9:48	12,300	Oct. 12	2:47	74,700
	10:47	20,300		4:47	77,700
	11:47	27,500		8:42	79,100
	14:47	44,400		20:47	54,400
Oct. 11	16:47	52,100	Oct. 13	3:46†	?

† Last radio observation before final fall.

Later Pioneer missions were far more successful. By 1976, the 140-lb Pioneer 6 had orbited the sun twelve times since launch in 1964, and had gathered a wealth of data on the solar wind, corona, magnetic field, and cosmic rays.

NOTE ON ROCKET STAGES: *Figure 4 shows that the "THOR Booster" which launched Pioneer 1 was a huge tank filled with propellant. The rapid burning of this propellant shoots gas out the jets that provide thrust (force) large enough to lift the whole assembly off the ground. As the propellant load decreases, the remaining structure is accelerated faster and faster, to an upward speed of several miles per second. When the booster tank becomes empty, there is no further need for it, and it is jettisoned before the second-stage rocket fires. With its smaller mass, the remainder can be accelerated more rapidly per pound of propellant burned, and there is a similar gain in thrust efficiency (p. 274) when the second-stage tanks are jettisoned. (This is an application of Newton's Laws of Motion: the acceleration, a, is equal to thrust or force, F, divided by mass, m; or F = ma.)*

Another reason for several rocket stages on a spacecraft is to provide changes in the direction of thrust. Unlike Pioneer 1, a satellite must be given a sideways impulse (thrust × time) after it is above the atmosphere, in order to put it into orbit around the Earth. When it is in orbit, it needs no further thrust—or rocket power—to stay up, just as the Moon needs no rocket power to stay in orbit. Apollo and other interplanetary spacecraft are usually put into a "parking orbit," then another stage is used to fire them at the right time into a trajectory taking them to the Moon, Mars, Venus, or other goal. Yet another stage—a retrorocket—is needed to put an Apollo spacecraft into lunar orbit, and another to start it on its return trip to Earth.

If the timing is inaccurate, the space probe fails, as often happened in the early space shots.—TLP

First Two Years of the Space Age

MARSHALL MELIN

(*Sky and Telescope*, November 1959)

Oct. 4, 1959, marked the second anniversary of the first successful launch of an Earth satellite. Table 2 lists all known attempts to send scientific payloads into space. This listing is incomplete, for the Soviet Union has not revealed its unsuccessful launch attempts.

In four firings, a total of seven major objects launched by the Russians have entered orbits around the Earth, and two long-range space probes have been sent aloft. The United States has made 33 attempts,

TABLE 2. SATELLITE AND PROBE LAUNCHES DURING THE FIRST TWO YEARS OF THE SPACE AGE

Launched		Nation	Name and Designation	Status	Notes
1957	Oct. 4	R	Sputnik I — 1957α2	Down about Jan. 4, 1958	Last radio signal Oct. 27, 1957. Rocket, 1957α1, down Dec. 1, 1957.
	Nov. 3	R	Sputnik II — 1957β	Down April 14, 1958	Last radio signal Nov. 10, 1957.
	Dec. 6	A	Vanguard	No orbit	First-stage malfunction.
1958	Feb. 1	A	Explorer I — 1958α	In orbit	Radio silent May 23, 1958.
	Feb. 5	A	Vanguard	No orbit	First-stage control malfunction.
	March 5	A	Explorer II	No orbit	Fourth-stage ignition failed.
	March 17	A	Vanguard I — 1958β2	In orbit	Solar-powered radio still operating. Rocket, 1958β1, in orbit.
	March 26	A	Explorer III — 1958γ	Down late June 1958	Radio silent June 16, 1958.
	April 28	A	Vanguard	No orbit	Relays for third-stage ignition failed.
	May 15	R	Sputnik III — 1958δ2	In orbit	Solar-powered radio still operating. Rocket, 1958δ1, down Dec. 3, 1958.
	May 27	A	Vanguard	No orbit	Second-stage cutoff at wrong orientation.
	June 26	A	Vanguard	No orbit	Second-stage cutoff premature.
	July 26	A	Explorer IV — 1958ε	In orbit	Radios silent Sept. 9 and Oct. 6, 1958.
	Aug. 17	A	Lunar probe	Failed	First-stage malfunction.
	Aug. 24	A	Explorer V	No orbit	Midflight collision of separated stages.
	Sept. 26	A	Vanguard	No stable orbit	Second-stage thrust low. Probably completed one revolution.
	Oct. 11	A	Pioneer I	High trajectory	Reached 70,700 mi. from Earth.
	Oct. 23	A	Beacon	No orbit	Payload separated from booster before burn-out.
	Nov. 8	A	Pioneer II	Failed	Third stage failed to ignite.

Date	Nation	Name	Status	Remarks
Dec. 6	A	Pioneer III	High trajectory	Reached 63,580 mi. from Earth.
Dec. 18	A	Atlas-Score — 1958ζ	Down Jan. 21, 1959	Radio silent Jan. 13, 1959.
1959 Jan. 2	R	Mechta — Artificial Planet 1	Solar orbit	443-day period. Radio contact to 373,125 mi. Rocket also orbiting.
Feb. 17	A	Vanguard II — 1959α1	In orbit	Radios silent March 15, 1959. Rocket, 1959α2, in orbit.
Feb. 28	A	Discoverer I — 1959β	Down early March 1959	Radio malfunction.
March 3	A	Pioneer IV — Artificial Planet 2	Solar orbit	407-day period. Radio contact to 407,000 mi. Rocket also orbiting.
April 13	A	Discoverer II — 1959γ	Down April 26, 1959	Last radio signal April 21, 1959.
April 13	A	Vanguard	No orbit	Second stage caused tumbling.
June 3	A	Discoverer III	Probably no orbit	No telemetry after second-stage firing.
June 22	A	Vanguard	No orbit	Pressure valve in second stage caused failure.
June 25	A	Discoverer IV	No orbit	Insufficient second-stage velocity.
July 16	A	Explorer	No orbit	Power supply for guidance failed. Destroyed by safety officer.
Aug. 7	A	Explorer VI — 1959δ2	In orbit	Solar-powered radio still operating. Rocket, 1959δ1, in orbit.
Aug. 13	A	Discoverer V — 1959ε	Down Sept. 28, 1959	Radio frequencies undisclosed.
Aug. 15	A	Beacon	No orbit	Booster fueling failed; orientation wrong.
Aug. 19	A	Discoverer VI — 1959ζ	In orbit	Radio frequencies undisclosed.
Sept. 12	R	Lunik II	On Moon	Radios ceased operation upon lunar impact Sept. 13, 1959. Final stage also on Moon.
Sept. 17	A	Transit I	No orbit	Third stage failed to fire.
Sept. 18	A	Vanguard III — 1959η	In orbit	Radios operating.
Oct. 4	R	Lunik III	In orbit	Radios operating. Rocket also in orbit.

This table is compiled largely from information supplied by the National Aeronautics and Space Administration. Under *Nation*, A designates American launchings; R, Russian. The U.S.S.R. has not released information on unsuccessful attempts. Dates are in UT.

12 of which have put 15 major objects in orbit; of five deep-probe launchings, three have returned useful information.

American space programs have recently been reorganized. The military work, in the hands of the Advanced Research Projects Agency since Feb. 7, 1958, has now been largely assigned to the Air Force; Project Midas, for infrared detection of ballistic missiles, and Project Samos, a satellite-reconnaissance system, are also under Air Force control. The Navy will supervise development of an orbiting navigation system under Project Transit, and the Army will handle Project Notus, a satellite communication system.

NASA has been responsible for the civilian space program since Oct. 1, 1958. Important in this agency's plans is a series of spaceflight vehicles: Delta, Scout, Vega, Centaur, and Nova, progressively scaled to larger and heavier space missions. When Scout is ready, perhaps in mid-1961, three British scientific satellites, now being developed by 10 teams in the United Kingdom, will be launched in a joint operation with the Americans.

The 1960 fiscal budget of the United States' space effort amounts to about $800 million, of which more than $500 million is allocated to NASA.

Mariners to Test
General Relativity

(Sky and Telescope, June 1970)

The Mariner 6 and 7 spacecraft that flew within 2130 mi. of Mars in the summer of 1969 are now traveling around the sun as tiny artificial planets, still being tracked by radio. In the spring of 1970 each of them passed behind the solar disk, allowing the application of a quantitative test of Einstein's General Theory of Relativity.

The idea behind the test was suggested six years ago by Irwin Shapiro, Massachusetts Institute of Technology (MIT), who proposed sending a radar pulse past the sun to Mercury or Venus when they are on the other side of the sun. According to relativity the velocity of light is less in a strong gravitational field, hence the round-trip travel time of a pulse should be increased by about 0.0002 sec.

The Mariner experiment is being conducted by John D. Anderson of Jet Propulsion Laboratory (JPL), and Duane O. Muhleman of California Institute of Technology (CalTech). On April 30, just before Mariner 6 passed behind the sun, JPL's 210-ft tracking antenna at Goldstone, California, beamed signals toward it with up to 400,000 watts of power. Striking the spacecraft antenna, the signals triggered a transponder which amplified and returned the signals to Goldstone. Similar operations were planned for Mariner 7 on May 10 just before it passed behind the sun.

When a Mariner is on the far side of the sun, the two-way radio contact requires about 45 min., yet this elapsed time can be measured to within a millionth of a second. Muhleman mentions that the final analysis of the data may take several months.

The Mariner spacecraft, designed for visits to nearby planets, are discussed more fully in Chapters 2 and 3.

This test of Einstein's theory, and several other such tests (see Vol. 1, Wanderers in the Sky), were positive, showing that massive objects like the sun distort the coordinate system of space and time. (Other accurate timing measurements show the relativistic effects of motion on clock rates, also.) As noted in Volume 8 of this series, Beyond the Milky Way, there are still some discrepancies when the theory is applied to very large regions of space such as clusters of galaxies, or to the whole universe.

During the 1960's, the engineering problems of the space age were solved, one by one; spacecraft stability was provided by multiple gyros, accurate guidance and control were perfected, space suits were developed for astronauts and cosmonauts, solar cells for electric power, meteor shields to prevent meteor damage, surface coatings to control spacecraft temperature, etc., etc. NASA turned to the scientific goals of space exploration. By 1967, both the Soviet Union and the U.S. had launched several manned spaceflights, using cosmonauts and astronauts who qualified by their experience as airplane test pilots. As shown in Chapters 2, 3, and 7, NASA had two programs of manned missions—Mercury spacecraft, carrying one astronaut per flight, and Gemini, carrying two— in preparation for the Apollo mission to the Moon. During these programs the primary goal was to perfect space technology and procedures.
—TLP

Three Astronomers Picked
to Become Astronauts

(*Sky and Telescope*, September 1967)

Of the 11 civilian scientists that NASA recently selected for training as space travelers from nearly 500 applicants, three have doctoral degrees in astronomy. The others are specialists in physics (two), geophysics, medicine (three), electrical engineering, and chemistry.

Among the trio is Karl G. Henize, 40, a veteran of the early satellite-tracking days at the Smithsonian Astrophysical Observatory (SAO) and currently active in analyzing celestial observations by Gemini astronauts. He is professor of astronomy at Northwestern University. Henize took his degree at the University of Michigan, and has worked at Mount Wilson and Palomar Observatories. Born in Cincinnati, Ohio, he is married and has three children.

Robert A. Parker is assistant professor of astronomy at the University of Wisconsin and received his doctorate from CalTech. He was born in New York City in 1936, is married and has two children.

Just this year 27-yr-old Brian T. O'Leary, a native of Boston, Massachusetts, was granted a graduate degree in astronomy by the University of California at Berkeley. Already he is enrolled in the NASA trainee program at the Space Sciences Laboratory of the university. He is married.

The new scientist-astronauts report for duty at the Manned Spacecraft Center, Houston, Texas, on Sept. 18, 1967. The 11 scientists increase the number of astronauts to 56. The final selection from a 65-man group was screened by the National Academy of Sciences.

The American Space Program for the 1970's

RAYMOND N. WATTS, JR.

(*Sky and Telescope*, May 1970)

Congress is now deliberating the authorization bill that will become the cornerstone upon which NASA will build its space program for the coming decade.

Last September, a special Space Task Group presented to President Nixon its proposals for this country's space activities. The President's March 7, 1970, recommendation to Congress, based heavily upon the Task Group's report, has been hailed as the most comprehensive space plan ever proposed. Calling for a NASA budget of $3.3 billion during the fiscal year 1971 (FY 71, which begins July 1970), the President's announcement emphasizes six broad areas:

1. Continued exploration of the Moon.

2. Exploration of the planets and the universe.

3. Development of reusable space shuttles to reduce costs of major launches and to service a manned space laboratory.

4. Extension of man's capability to function in space for increasingly longer periods of time.

5. Expansion of the practical applications of space technology to the benefit of mankind.

6. Encouragement of greater international cooperation in space.

Concerning the last item, NASA administrator Thomas O. Paine, who calls the 1970's program "very bold and forward looking," is already hard at work. He has made visits to Canada, Japan, and Australia, countries that already are cooperating with the United States in space research. His major problem is formalizing arrangements between our country and the Soviet Union, with which there is already a wealth of informal exchange and cooperation. Scientific journals circulate freely and scientists exchange visits. Unfortunately, the step to official joint efforts may be extremely difficult.

Manned Flight Schedules

The current Apollo program calls for voyages to the Moon every six months through the fall of 1971. By late 1972 it is hoped to put up an orbiting workshop under the name Skylab (formerly Apollo Applications Program—see Chapter 7). During its active lifetime of nearly a year, the workshop will be visited by several teams of astronauts for stays of up to 56 days. Thereafter, Apollo will be resumed with lunar voyages in 1974.

A major NASA effort during the 1970's is to develop a Space Shuttle, a reusable launch vehicle that may salvage at least part of the hundreds of thousands of dollars worth of hardware that is dumped into the ocean on each launching under present procedures. The Shuttle is envisaged as a two-stage vehicle with a cargo capacity of 20,000-50,000 lbs (see Chapter 10).

Interplanetary Exploration

There may be seven major missions to the planets during the decade, beginning next year with Mars orbiters that are expected to map photographically at least 70 percent of the red planet's surface.

1972 will be marked by our first attempt to explore Jupiter—two spacecraft flying past the giant planet at close range—and the launch of a space probe to pass close to both Venus and Mercury (see Chapter 4).

Two Viking missions to Mars have been postponed from 1973 to 1975. Each mission will consist of a pair of spacecraft, one to orbit the planet while the other descends for a soft landing (see Chapter 5).

The President's public announcement in March enthusiastically endorsed the Grand Tour concept in which a spacecraft visits Jupiter and then two or more of the outer planets. In each instance, Jupiter with its strong gravitational pull alters the spacecraft orbit, enlarging it so the spacecraft can continue far out into space to other favorably situated planets (see Chapter 6).

RESEARCH SATELLITES AND PROBES

Astronomers fare quite well under the reduced NASA budget. Four orbiting solar observatories (OSO) are scheduled between now and 1976; two orbiting astronomical observatories (OAO—this year and next); three interplanetary monitoring probes (IMP—in 1970, 1972, 1973); two international orbiters for ionospheric studies (1971, 1973); and two solar probes (1974, 1975).

During the first half of the 1970's, a host of practical working spacecraft will be launched. They include three Nimbus weather monitors, two meteorological satellites placed in synchronous orbits (24-hr period over the equator), and a geodetic satellite. In addition, four Earth-resources (ERTS) and applications-technology satellites (ATS) will provide data on instruments for studying the Earth and information about the Earth itself.

FINANCING THE SPACE PROGRAM

The funding level over the years for all of the foregoing projects has not been revealed—a point emphasized by critics of the President's space plans. It is clear that developing a Space Shuttle and mission to the outer planets will be costly. So will continuation of the Apollo program. Yet much of the required technology has already been developed and paid for, appropriate facilities exist, and trained, experienced crews are standing by. Proponents of the President's program argue that further slowdowns would cause uneconomical degradation of these valuable assets.

NASA's actual expenditures in FY 69 were $4,251,707,000, and in FY 70 (now closing) are set at $3,889,400,000. In FY 71 an even greater reduction occurs, down to $3,403,000,000. These changes have caused a continued drop in employment on NASA projects (including both the agency and its contractors). In June 1969, there were 218,300 jobs, and by June 1971, it is expected the number will be below 150,000.

To meet the cutback, NASA has made substantial changes in plans formulated as recently as a year ago. Production of Saturn-V launch vehicles has been suspended and the Electronics Research Center in Cambridge, Massachusetts, closed down.

However, if all goes reasonably well, the 1970's should see major advances in the exploration of space. It seems certain that by the close of this decade we will have vastly increased our knowledge of the Earth, Moon, and other planets.

European Satellite Plans

RAYMOND N. WATTS, JR.

(*Sky and Telescope*, October 1969)

The European Space Research Organization (ESRO) came into being in March 1964, "to provide for, and to promote, collaboration among European States in space research and technology exclusively for peace-

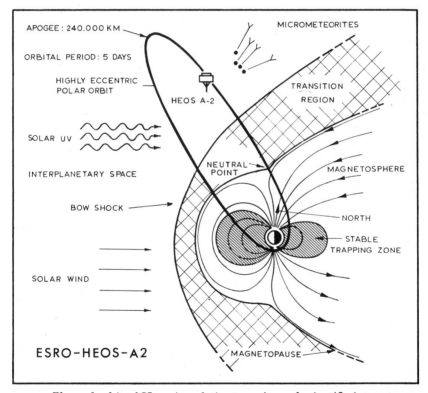

FIG. 7. Planned orbit of Heos A2 relative to regions of scientific interest near the Earth. The "stable trapping zone" near Earth is the Van Allen belt. Note the highly elliptical, inclined orbit (see Glossary) of Heos A2, the sun's ultraviolet (UV) light and solar wind coming from the left, the hatched transition region between bow shock (p. 172) and magnetosphere to the right, and micrometeorites in interplanetary space. (ESRO/ELDO *Bulletin* diagram.)

ful purposes." Since then, 22 different experiments have been carried in three satellites, ESRO 2, ESRO 1, and Heos A1. In addition ESRO has launched 76 rockets carrying 177 experiments.

With headquarters in Paris, France, the organization has centers in Noordwijk, Netherlands, at Darmstadt, West Germany, and at Frascati, Italy. By the end of 1968, ESRO employed 1119 people and had an annual budget of about 263 million French francs. This year's council chairman is the Dutch astronomer H. C. van de Hulst.

Last May ESRO concluded a year-long study to decide what kinds of spacecraft it will build in the near future. Two projects are already under way. One is called TD 1, scheduled for launch about February 1972. The other is Heos A2, set for December 1971.

Weighing about 450 kg, TD 1 is basically an astrophysical satellite with seven experiments: two large telescopes to detect stellar ultraviolet and infrared radiation, a gamma-ray spark chamber, and an x-ray spectrometer. These instruments will scan the entire sky once in six months. The others are a cosmic-ray telescope and detectors for solar x-rays and gamma rays (see Chapter 9).

Heos A2 will study cosmic-ray propagation in the solar system by measuring interplanetary fields and particles, and it will also monitor the boundary of the Earth's magnetosphere. This 102-kg satellite will be launched from California into a highly elliptical polar orbit (inclination near 90°, passing over the Earth's poles). The apogee will be situated not more than 30° from the north pole, and at a distance of 35 Earth radii (Fig. 7).

Although TD 1 is the most complex satellite yet developed in Europe, it cannot handle as many experiments as desired. Hence ESRO has decided to launch two additional spacecraft. ESRO 1b is scheduled for October 1969, ESRO 4 for September 1972.

Chinese Satellite

RAYMOND N. WATTS, JR.

(*Sky and Telescope*, June 1970)

On April 24, the People's Republic of China became the fifth country to place an artificial satellite in orbit using its own launch vehicle. (The others are France, Japan, the Soviet Union, and the United States.)

The official Peking news agency Hsinhua gave few details, and did not identify the launch site. However, the New York *Times* mentions reports that China's missile launch facility is at Shuang-cheng-tze in Inner Mongolia, about 400 mi. northwest of Lanchow.

The 380-lb satellite circles the Earth every 114 min. in an orbit inclined 68°.5 to the equator. Its apogee and perigee heights were said by the Hsinhua announcement to be 1481 and 273 mi., respectively. The spacecraft is too faint to be seen with the naked eye, but could be heard on 20.009 megahertz broadcasting the song "Tung Fang Hung" (The East is Red), the semiofficial anthem of Red China.

By 1970 the space effort had developed many earmarks of an international political party with a jargon of its own (NASA, ESRO, JPL = Jet Propulsion Laboratory, MSC = Manned Spacecraft Center, KSC = Kennedy Space Center, Heos = High-Energy Orbiting Satellite, Apollo missions to the Moon, Mariner missions to planets, Saturn and Atlas launch rockets, data acquisition, tracking network, etc., etc.) and what seemed to be large expenses. (But note that NASA's annual budget was only 1.9 percent of the U. S. federal budget—$3.75 billion out of $196.59 billion—very much smaller than defense- and welfare-program expenditures.)

During the 1960's the NASA program was dominated by competition with the Soviet Union—the "Moon Race"—which was followed with high interest by people all over the world, probably by the largest audience that any competition, including a war, ever had. But, after the first few televised Moon landings, Americans seem to have lost interest. They consider that the race is won, and ignore the space-engineering "spin-offs" (such as miniature-computer development, many electronic gadgets, medical findings) and the opportunities in space science shown in the remainder of this book.

Young people are familiar with interplanetary jargon, some of it illustrated in Figure 7: the solar wind of protons compressing the sunward side of Earth's magnetosphere, the Van Allen belt ("stable trapping zone") where electrons and protons oscillate from pole to pole in the Earth's magnetic field, and the hazardous micrometeorites and ultraviolet light incident on any spacecraft outside the Earth's protective atmosphere.

Another aspect of the space age is the growing range of international cooperation shown in the next three articles.—TLP

American Prospects
in Space

RAYMOND N. WATTS, JR.

(*Sky and Telescope*, May 1974)

Where do we go from here? Man has walked upon the Moon, spent months in space, and landed robot probes on other planets. What can we expect from the second 15 years of the space age?

The American Astronautical Society has just published *The Second Fifteen Years in Space*, edited by Saul Ferdman, and *Space Shuttle Payloads*, edited by George W. Morgenthaler and William J. Bursnall.

NASA's plans for the 1970's and '80's are based on a September 1969 report to the President (see p. 22), which recommended a "balanced manned and unmanned space program conducted for the benefit of all mankind." Among more specific objectives, it urged that the United States "develop new systems and technology for space operations with emphasis upon the critical factors of commonality, reusability, and economy, through a program directed initially toward development of a new space transportation capability."

NASA translated this: *Space Shuttle*. After some struggle, the space agency has secured congressional approval, and now massive resources are being devoted to developing this concept (Figs. 8, 153).

No one knows how much will really be spent on the Shuttle program, but current budget estimates indicate the scale. For the fiscal year beginning July 1, 1974, NASA has requested a total budget of $3.25 billion, which is $100 million more than the current year. Space Shuttle spending will jump from $475 million to $800 million, while the space program shows an over-all decrease of 7 percent. In other words, NASA plans to spend a larger fraction of its budget on aeronautics, and a much larger fraction of its space budget on the Shuttle.

To cut costs over the last few years, fewer trips to the Moon were scheduled. This made some vehicles available for the Skylab and Apollo-Soyuz (US-Soviet) programs, but spelled the end of American manned spaceflight until the Shuttle is ready for testing, perhaps in 1978. NASA continues to launch its unmanned spacecraft and planetary probes, but at a rate much reduced from the frantic pace of the mid-60's.

For example, NASA plans to continue developing a sophisticated Orbiting Solar Observatory (OSO) and the costly High Energy Astronomy Observatory (HEAO) series for more detailed measurement of celestial x-rays and gamma rays (see Chapter 9). The Viking program is still scheduled to land two spacecraft on Mars in 1976 (see Chapter 5), and fly-by probes will be sent to Jupiter and Saturn in 1977 (see Chapter 6).

Great strides have been made in rocket-engine design for the Space Shuttle. The main engine is to develop a thrust of 470,000 lbs and be able to restart 100 times. It should run for a total of 7½ hrs before being overhauled (at less than a third of the initial cost). Each Shuttle is to have three such engines.

A second engine called the aerospike is being developed for use in the Space Tug, a rocket-powered vehicle, carried by Shuttle (p. 401). It has an annular combustion chamber that surrounds a truncated spike "nozzle." Most of the thrust is produced by propellant flow from the combustion chamber expanding against the outside surface of the nozzle.

The Space Shuttle is about the size of a DC-9 jetliner. Weighing 170,000 lbs, it is 122 ft long and has a 78-ft wing span. Its cargo bay can accommodate a payload 60 ft long and 15 ft in diameter. Depending upon the orbit chosen, up to 65,000 lbs can be carried aloft (see Chapter 10).

FIG. 8. Space Shuttle looks like a delta-winged airplane on top of a 154-ft liquid-propellant booster tank with a solid-propellant rocket on either side.

Shuttle will be capable of maneuvering once it is in space. After staying aloft 7-30 days, it will plunge back into the atmosphere, at an angle of about 34° with the horizontal. In the Earth's lower atmosphere at subsonic speed, it will have a lift-to-drag ratio of about six, permitting it to maneuver like an airplane and finally to land at 185 mi/hr.

The normal astronaut crew on Shuttle will be four. From the flight deck the crew will control the spaceship and run various remote controlled manipulators to lift equipment out of the storage compartment or to retrieve objects from space. Below the flight deck are living quarters and room for up to six extra passengers.

The cabin atmosphere of oxygen and nitrogen will be maintained at about 14.7 lbs/in^2 and temperature between 65° and 80°F, with the cabin humidity, carbon dioxide, and odors controlled by lithium hydroxide and activated-charcoal filters.

Europe in Space:
The Emergence of ESA

J. KELLY BEATTY

(*Sky and Telescope*, May 1975)

In 1964, 10 countries formed the European Space Research Organization (ESRO), with Belgium, France, Italy, the Netherlands, Spain, Sweden, Switzerland, the United Kingdom, West Germany, and Denmark as members. ESRO was set up "to design and construct sounding-rocket payloads, satellites, and space probes," to procure launch vehicles, and to establish the necessary research, tracking, and data-processing facilities.

Also in 1964, six of these countries formed the European Space Vehicle Launcher Development Organization (ELDO). Until 1971, each member had to participate in all ESRO and ELDO programs, supplying funds in proportion to its national income. ESRO was able to place four satellites in orbit in 1968-69, launch three more in 1972, and send aloft 125 sounding rockets with scientific payloads over a nine-year period.

But there were several failures, and the members built 21 satellites independently of ESRO.

For these reasons, ESRO and ELDO have been restructured into the European Space Agency (ESA), approved at a meeting on April 15,

1975, of the representatives of 10 European nations. The new organization will be empowered to award contracts directly to industry.

Instead of being mandatory, most of ESA's current programs are *à la carte*, allowing member nations more freedom in financing specific projects. Thus, three programs begun in 1973 are being funded primarily by the nations undertaking them: Spacelab by West Germany, the Ariane launcher by France, and the Marots communications satellite by the United Kingdom.

From the start, ESRO concentrated its efforts on scientific satellites for the study of near-Earth space.

The first satellite placed in orbit was Iris, launched by NASA on May 17, 1968, from Vandenberg Air Force Base in California. Five devices in its payload counted energetic protons, electrons, and alpha particles in the magnetosphere and Van Allen belts; two other instruments monitored solar x-rays at wavelengths of 1-20A and 44-60A. ESRO 4, launched in 1972, travels in an orbit ranging from 1175 to 245 km (719-152 mi.) above the Earth's surface and established that in the upper atmosphere there is a seasonal variation in the abundance of argon, which is greatest over the sunlit pole.

The Heos-1 and Heos-A2 satellites were ESRO ventures to and beyond the Earth's magnetosphere, making measurements of magnetic fields, the solar wind, and cosmic radiation in interplanetary space (see Fig. 7).

Largest of the spacecraft built by ESRO is the TD 1A, launched on March 12, 1972, and now in a circular polar orbit 540 km high; its orbital plane precesses at the rate of 360°/yr, so that the spacecraft always travels nearly over the Earth's terminator (sunset-sunrise line).

TD 1A's prime scientific mission is to make a systematic sky survey at ultraviolet, x-ray, and gamma-ray wavelengths (Chapter 9). Two instruments measure solar x-rays and gamma rays and another studies charged particles. The data are being analyzed by groups in England, France, Holland, and Italy.

ESA plans to orbit four additional scientific satellites by the end of 1979:

Cos B, to be launched in June 1975, will carry a large gamma-ray telescope with a spark chamber to detect gamma-ray photons having energies of 20 million to 5 billion electron volts (see p. 358).

Geos, to be launched in mid-1976, will continue the exploration of the magnetosphere, monitoring changes in plasma, particle population, and electrical and magnetic fields.

IME involves two satellites, the "mother" to be developed by NASA, the "daughter" by ESA. Launched in 1977 aboard the same booster, these International Magnetospheric Explorers will travel 100-5000 km apart along the same highly elliptical orbit, rising to 140,000 km and descending to 300 km above the Earth's surface.

Exosat, designed as a successor to the successful TD 1A, will study celestial x-ray sources, making use of lunar occultations to determine positions.

In addition to these projects of its own, ESA is cooperating with NASA and the United Kingdom to develop the International Ultraviolet Explorer (IUE), controlled by scientists on the ground. It will observe the ultraviolet spectra of celestial objects with a 45-cm telescope in the 1150-3200A region, then relay the data in real time to stations at Greenbelt, Maryland, and Madrid, Spain.

ESA is also at work on four "applications" satellites: *Meteosat*, to be launched into 24-hr orbit in 1976, will be its first weather satellite; OTS and Marots are communications spacecraft, with 1977 launch dates; *Aerosat* in cooperation with the United States and Canada involves two air-traffic control satellites to be put into 24-hr orbits above the Atlantic Ocean in 1977 and 1979.

The United States and ESRO signed an agreement in 1973 giving the latter sole responsibility for the development and construction of a manned orbiting laboratory to be carried aloft in 1980 by NASA's Space Shuttle. Called *Spacelab*, this project marks the beginning of European involvement in manned space studies (see p. 402).

The Apollo-Soyuz Test Project

RAYMOND N. WATTS, JR.

(*Sky and Telescope*, October 1974)

For a decade or more, the Soviet Union and the United States developed and flew spacecraft designed to study the Earth, near-Earth space, the solar system, and deep space. However, throughout these parallel efforts, there was relatively little exchange of scientific and technical information.

In October 1970, NASA and the Soviet Academy of Sciences took steps that led to six meetings of space scientists alternately in Moscow and Houston. Finally, on May 24, 1972, President Nixon and Chairman Kosygin established the guidelines for cooperation between the two countries on studies of meteorology, the natural environment, near-

Earth space, the Moon, planets, space biology, and medicine. As a specific starting point, they agreed on an American-Soviet space rendezvous in 1975. After two more years of discussion it was decided that a Soviet Soyuz should rendezvous with an American Apollo some 150 mi. above the ground.

The larger and more powerful Apollo will carry aloft a special docking module, one end to be connected to the Apollo by standard NASA techniques, the other to have a special adapter for the Soyuz. As the spacecraft move toward each other, large guide plates will insure proper alignment and three capture latches will complete the initial contact. Eight structural latches will then make a firm connection, with rubber seals for a vacuum-tight fit between the two craft.

As plans now stand, the Soyuz crew, Aleksei A. Leonov and Valeri N. Kubasov, will leave the Baikonur launch complex in Kazakhstan at 12:20 UT on July 15, 1975. Their 6680-kg spacecraft will be launched northeasterly into an orbit inclined 51°.8 to the equator. The low and high points will be 188 km and 228 km, but the orbit will later be circularized at 225-km altitude.

The Soyuz has three major sections. At the forward end, the 1224-kg orbital module is 2.35 m. in diameter and 2.65 m. long; the crew use it for work and rest while in orbit. Behind this, the 2802-kg descent module houses the main controls and crew couches occupied during launch, descent, and landing. The instrument module is aft, carrying subsystems for power, communications, propulsion, and life support. Its weight is 2654 kg, and its 2.3-m. extent brings the spacecraft's over-all length to 7.15 m. (23.5 ft).

At 19:50 UT a Saturn-1B launch vehicle will carry the Apollo crew into orbit from Kennedy Space Center (KSC) in Florida with astronauts Thomas P. Stafford, Vance D. Brand, and Donald K. Slayton. The Apollo will have an orbital inclination of 51°.8 and initial perigee and apogee heights of 150 km and 167 km.

The command and service module will separate from the last-stage booster, make a U-turn, connect with the docking module (housed where the lunar-excursion module or LM would ordinarily be carried). By 9 hrs 14 min. GET (ground elapsed time measured from the Soyuz launch), command-module rocket-thrusters will be fired to avoid its bumping into the jettisoned booster.

Next, by a series of maneuvers, the Apollo crew must match their orbit to that of the Russians. If all goes well, at 49:55 GET the two craft will be ready for final approach and docking, scheduled for completion at 51:55 GET. The newly designed docking module is a cylinder 1.5 m. in diameter and 3 m. long, with an air-lock hatch at each end. It

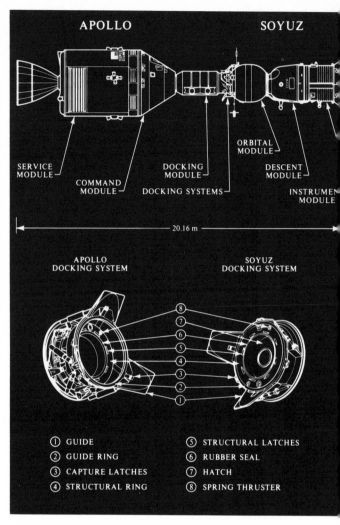

FIG. 9. The top diagram shows to scale the parts ("modules") of both Apollo and Soyuz, when docked. The lower diagram shows, at larger scale, the two docking systems, with eight important, matching parts. When Apollo gently touched Soyuz, the "capture latches" locked them together; during "hard docking" the structural latches pulled the two rubber seals into vacuum-tight contact. At separation, the "spring thrusters" pushed the two spacecraft apart. (NASA diagram.)

contains radio and TV communications equipment, antennas, heaters, and the displays and controls needed for transferring the crew. The two ships will spend about two days locked together. Before final separation, they will practice redocking, but at about 99:15 GET the Apollo will change its velocity by 1 m/sec to avoid further contact, after which each crew will conduct its own experiments. After about 43 hrs more, the Soyuz will land in Kazakhstan at about 142:00 GET, while the Apollo will stay aloft for approximately six days, then splash down in the Pacific Ocean.

The docking module had to be designed for the two ships' very different atmospheres—the result of a traditional difference between American and Soviet space technology. In orbit, Apollo astronauts breathe pure oxygen at a reduced pressure of 5 lbs/in², while Soyuz cosmonauts have a normal atmosphere (which NASA will adopt for future manned craft, such as the Space Shuttle).

The docking module can hold two crewmen simultaneously. It has hatches with controls on both sides at each end. While the spacecraft are docked, the Soyuz pressure will be reduced to 10 lbs/in², making it possible for cosmonauts to transfer from Soyuz to Apollo atmosphere without taking time in the air lock to breathe pure oxygen and force nitrogen from their blood. The Apollo pressure will remain at 5 lbs/in².

The most important benefits from the Apollo-Soyuz Test Project are long range. It is a first step toward other international missions on a larger scale. Also, with a compatible technique of rendezvous and docking, either NASA or the Soviet Union will be able, in the future, to rescue the other's spacecraft in distress.

The project also provides an opportunity to conduct scientific experiments. Soviet plans are not yet available, but NASA has announced a number of experiments:

• An ultraviolet-absorption experiment to measure the concentrations of atmospheric constituents at the 250-km orbital altitude has been designed by T. M. Donahue at the University of Pittsburgh, using a retroreflector on the Soyuz.

• A search for sources of extreme-ultraviolet radiation in the night sky. Such sources may include stellar coronas, defunct pulsars, and stars that are accreting matter. The instrument is a grazing-incidence telescope, mounted outside the Apollo spacecraft, with an ultraviolet detector at its focal point (see Fig. 140, Chapter 8).

• The helium glow of the night sky, giving the distribution of helium in interplanetary space and the penetration of interstellar helium into the solar system. Both of these were provided by C. S. Bowyer, University of California at Berkeley.

• Soft x-rays from both the Earth and the sky. Herbert Friedman of the Naval Research Laboratory built a detector to be mounted in a bay of the Apollo service module. No systematic sky survey has been made in the energy range of 100-1000 electron-volts.

Apollo is to carry a small electric furnace with which a number of metallurgical tests will be made, one to see whether cast materials with improved properties can be made under o-g conditions.

A number of biomedical experiments are scheduled, including one on white blood cells to see how resistance to infection may change during a prolonged space mission (see Chapter 7).

More on Apollo-Soyuz

RAYMOND N. WATTS, JR.

(*Sky and Telescope*, February 1975)

In July, when Deke Slayton crawls from an orbiting Apollo spacecraft into a Soviet-built Soyuz, it will be a gala day indeed. He and the two other American astronauts will be offered Ukranian borscht, a piquant Georgian mutton soup called kharcho, and a green Russian sorrel soup. The second course of spiced veal, chicken, paté, ham, and sausage will be followed by prunes with nuts, fruitcake, and fruit juices. This banquet in space will culminate several years of intensive effort by Soviet and American space engineers.

A major step in the program was taken on Dec. 2, 1974, at 12:40 a.m. Moscow time, when Soyuz 16 carried Anatoli V. Filipchenko and Nikolai N. Rukavishnikov aloft for a 6-day flight to test modifications of their spaceship made especially for the linkup with an Apollo craft.

Tracking by NASA radar showed that Soyuz 16 was not in the desired initial orbit, but ranged in height from 165 km to 250 km. (Some time ago Soviet engineers explained that an initial Soyuz orbit was apt to be imprecise.) After maneuvers, through 18 revolutions, Soyuz 16 settled into a circular orbit 220 km high, as planned.

The newly designed docking adapter, carried for the first time on a Russian spacecraft, was tried, using a special latch ring to represent the Apollo side of the docking mechanism. A key test was the reduction of cabin pressure from 14.7 lbs/in^2 to 10, which will be necessary next July to compensate partially for the Apollo's pure-oxygen pressure of 5 lbs/in.^{2} In addition, the Soviet life-support system's capacity has been increased to take care of four men while one of the American astronauts is visiting in July. Soyuz 16 landed on December 8 at 11:04 a.m. Moscow time.

As this volume goes to press, the Apollo-Soyuz Test Project has achieved its cooperative goal, and produced several significant scientific results. For the first time, the Soviet Union broadcast TV coverage of the Soyuz launch from the huge Tyuratam Cosmodrome near the city of Leninsk, and the landing in Kazahkstan, with a brief and effective firing of retrorockets. TV audiences in both East and West also saw the first handshake in space between Cosmonaut Leonov and Astronaut Stafford. Aside from some difficulties with the TV camera on Soyuz, the mission went smoothly and started a new era of Soviet-U.S. cooperation in space science. There was an uncomfortable ten minutes during the Apollo reentry, when thruster fuel was sucked into the CM and forced the three astronauts to don gas masks. The scientific results (extreme-ultraviolet and soft x-ray measures) are covered on pages 353 and 375.

Two other cases of international cooperation should be mentioned. In 1974 NASA launched the ANS satellite for the Netherlands (see p. 375) and in April 1975 the Soviet Union launched the 730-lb Aryabhatta satellite for India. Both of these, built by national institutions, were designed to do x-ray astronomy, and the Indian Scientific Satellite Project plans further launches in 1977.—TLP

2

Early Flights and the Hazards of Space

G Forces and Weight in Space Travel

FRITZ HABER

(*Sky and Telescope*, February 1953)

In 1492, mankind was still struggling with the idea that the Earth was not a flat disk, but a globe. Columbus was not the one to discover this fact, but was one of its most ingenious and successful promoters, as his results show. The story goes that in his efforts to obtain support in his plans, he was occasionally confronted with the argument that nobody could live on the other side of the globe because he would just "fall off." A few centuries later Isaac Newton could theoretically disprove such embarrassing accidents; however, Columbus tried it long before and proved that the same firm stand existed on the soil of America as it did on the soil of Spain.

Columbus rightly believed that his weight kept him from falling off the globe; he had no reason to doubt the constancy of weight. It took a few more centuries to produce noticeable variation of weight.

It is common belief that weight is constant and that the National Bureau of Standards sees to it that "a pound is always a pound" (or a kilogram always a kilogram, equal to 2.21 lb). Since aviation and rocketry have made the variation of weight a daily experience, it is apparent that a revision of the concept of weight is necessary.

In daily life if you want to know how heavy a certain piece of meat is, you weigh it on a scale. That seems very simple, but it can become quite complicated if that particular piece of meat happens to be full of life and is your energetic youngster. Such small fry like to bounce up and down, making it almost impossible to read accurate weight on the scale.

This shows the condition necessary in measuring the weight of a body. Scale and body must be kept quiet or, in more scientific terms, they must not be accelerated, a rule clarified and explained by Newton, the founder of classical mechanics. Newton stated that any acceleration (speeding up) of a body requires a force acting on the body equal to the product of the acceleration times the mass of the body. Acceleration is the rate of change of speed in amount or direction. Another Newtonian law establishes that each active force has its reactive counterpart of the same amount but opposite direction. According to this law, a force acting on a body and accelerating it has a counterpart ("reaction") called the force of inertia. If a force of inertia is produced by the acceleration of gravity at the surface of the Earth, then one speaks of a force of inertia of "one g." This is the reason the forces of inertia are sometimes called "g forces."

These forces can be measured in the same manner as weight, that is, on scales—and that was exactly what you did when you put your bouncing youngster on the scale. He was accelerating himself up and down, thus producing forces of inertia of variable amount, which showed on the dial. Think of a spring scale with a piece of metal on it. The dial reads a weight of, say, 100 lbs. If we transfer scale and metal to a point way out in space where no gravitational field exists, and then accelerate the scale at 1 g, we would read a force of inertia of 100 lbs. If the entire setup were enclosed in a box with no means of detecting motion or acceleration, how could you differentiate between weight and force of inertia? The indicator on your dial shows 100 lbs and "that's it." The 100 pounds could be called weight, or it could just as well be called force of inertia.

Physicists say that a body has two different kinds of mass: the gravitational mass, which displays itself as the attraction between two bodies— and inertial mass, a quantity which becomes apparent as a force of inertia during an acceleration. This is the reason for differentiating between weight and force of inertia. Normally, weight is identified with the force of attraction between gravitational masses, while force of inertia is considered to be different from weight.

We call the weight obtained on the surface of the Earth under unaccelerated conditions the "normal weight" of a body. Under any other conditions, forces of inertia are added to or subtracted from the normal weight, resulting in an amount of weight that can be anything from zero to many times the normal weight.

Everyone has experienced this either when going around a corner in a car, riding an elevator, or sitting in a train that is brought to a sudden stop. In all these instances, your body weight was different from your normal weight.

What really is the weight of a body and how can it be determined? There are many possible and correct answers, but one—in my opinion— is the clearest: The weight of a body is equal to the force of its support, and is independent of the force of gravity. An airplane in the air is supported by aerodynamical lift. The pilot can vary this lift and make it three or four times as great as the normal weight of the airplane. The craft then weighs three or four times its normal weight.

In all cases of weightlessness the identification of weight with support becomes exceptionally clear if a body is left without any support whatsoever; then we can state immediately that the body is without weight.

Now visualize a spaceship at the moment of takeoff. The rocket engine has been started and is developing a certain amount of thrust. It is imperative that this thrust be greater than the normal weight of the ship, otherwise it would not be lifted off the ground. A thrust somewhat greater than the normal weight lifts the ship in an accelerated motion. The thrust of a rocket engine can be considered constant, whereas the total mass of the rocket decreases because fuel is burned and discharged during the operation of the rocket engine. A constant propulsive force acting on a rocket of decreasing mass produces an ever-increasing acceleration of this rocket. This results in ever-increasing forces of inertia; that is, increasing weight for everything within the rocket during the period of takeoff.

A rocket engineer is anxious to see his ship go fast because the economy of rocket propulsion is improved by a high speed. In order to obtain high speed as quickly as possible, it is necessary to have high acceleration; that is, with great weight of the crew. However, there is a limit to this; the human body in a prone position can tolerate an 11-fold increase of its weight for 2-3 min., and in supine position a 14-fold increase for the same time. Breathing becomes very difficult due to the tremendous weight of parts of the chest. Engineer and doctor must cooperate on this point in order to achieve the highest performance of man and machine.

After the rocket has attained the desired speed, its engine is cut off. At this time the thrust and the support become zero; therefore the rocket and everything in it are without weight. The speed, the shape of the trajectory, or any gravitational field plays no role.

How does it feel to be weightless? Everyone at some time has been weightless; after jumping off a chair or a diving board, you become weightless during the jump because there is no force supporting you. During a quick drop in an elevator you experience, momentarily, the peculiar "stomach-lifting" sensation which is the relaxation of the tissue which carries the normal weight of the inner organs. The only difference between the elevator experience and that of the space traveler is that your weightless state lasted only for a very short time—1-2 sec at the most—while the future astronaut will encounter it for much longer periods of time.

It is unknown what a prolonged state of weightlessness does to the human body. We can rule out any severe disturbances of circulation, but should expect some trouble in orientation and muscular co-ordination (see Chapter 7). Man has a wide range of adaptive abilities, and there is a high probability that he will overcome these difficulties. It is necessary to gather further experimental information on this subject.

It has been proposed that artificial gravity be provided by rotating the spaceship and thus producing weight at the circumference wall. Such a method, however, is not practical for a small spaceship; the rate of spinning must be high and, since the center of the ship is the center of rotation, there is no weight in the center. Weight increases rapidly toward the wall, making motion quite difficult. The rotating spaceship, however, has its merits for providing weight if the ship can be made large enough. Then the speed of rotation can be smaller and the weight gradient along the radius is no longer a source of trouble (p. 420).

The variation of weight will be the strangest among the many new experiences of the future space traveler. Man lives with the weight of his body from the first moment of his life, and weight is so much a part of him that he almost forgets about it. Scientists will work together to supply the spaceship with everything to which a man is accustomed in his daily life. He will have air to breathe, food to eat, the climate will be suitable, the lights should simulate day and night, but there is one thing man must leave behind and that is his weight.

We now know, 22 years later, that "g forces" are extreme during three phases of any space mission: high-g forces during launch, o-g in orbit or on unpowered flight to the Moon or to another planet, and high-g forces during re-entry into the Earth's atmosphere. More recent work with the giant centrifuge (in which a man or instruments are swung around a central axis on a 50-ft arm capable of 30 g) at the Johnson Space Center (JSC) in Houston confirms Haber's 14 g that the human body can stand for short intervals, although the astronauts were usually asked to take only 7 g in training. The "no-weight" o-g condition was enjoyed by many cosmonauts and astronauts, particularly by Alan Bean and Joe Kerwin, who did a sort of space ballet on Skylab 2, where they had room to tumble and do flying leaps.

The re-entry force, when a spacecraft is suddenly slowed down by the atmospheric drag, was uncertain until 1960, when two well-instrumented NASA satellites (Discoverers 13 and 14) were brought down in the Pacific and recovered. Records showed that slightly convex nose cones decelerated the spacecraft, starting at 75-mi. altitude with a force of 14 g. For several minutes each nose cone was heated to 3000°F or 4000°F, but its thick layers of "ablative material" burned away (ablated) as planned, leaving the insulated capsule behind it relatively cool.—TLP

Man in Space

MARSHALL MELIN

(*Sky and Telescope*, March 1959)

Although at least two years will intervene before the United States can place its first astronaut in orbit, NASA has let contracts to McDonnell Aircraft Corp. for designing and building a re-entry vehicle (to return from orbit to Earth), and plans the selection, by a committee of aeromedical specialists, of the first astronaut.

The candidates for the mission are engineers or physical scientists (p. 21) who are also trained test pilots.

Some of the stresses that this man must face can be simulated in advance in terrestrial laboratories. The most important physiological factor is the effect of high accelerations upon blood circulation, especially the supply to the brain and central nervous system.

Without doubt, a main function of the first astronaut will be to serve as the subject of intensive medical experiment. During the flight, he will be under constant surveillance.

The safety of this one man aloft is the present concern of hundreds of medical scientists who have summarized the limits of human tolerance to the air composition and pressure that may occur in a space capsule (Fig. 10).

Cosmic rays also present a serious hazard. The particles responsible for the Van Allen radiation belt (see Fig. 7), and its biological effects

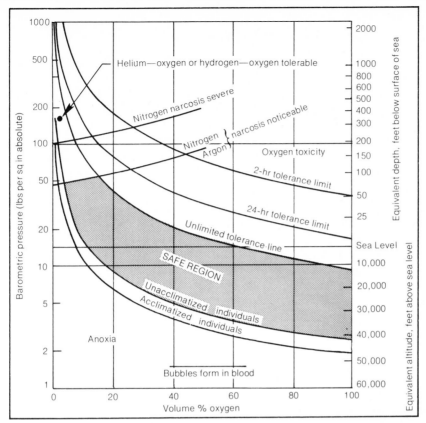

FIG. 10. Human tolerance to atmospheric environment in terms of percent oxygen in the cabin "air," and pressure. Normal air at sea level is 20 percent oxygen, 79 percent nitrogen, and 1 percent argon at 14.7 lbs/in² (with various amounts of water, CO_2, and contaminants). The lower part of the diagram refers to low-pressure environments in aircraft and spacecraft; the upper part to high-pressure in submarines. Various effects, such as "anoxia" (brain damage due to lack of oxygen) are marked at the pressure values and oxygen-content values where they take place. (Rand Corporation diagram.)

require study. Perhaps the first manned flight will be planned for so low an altitude that little shielding will be needed.

The space capsule for Project Mercury will be raised into orbit by an Atlas booster, to which later stages will be added. The capsule is to be provided with an escape rocket in case something goes wrong during launch. Another feature will be stabilizing jets for controlling its orientation in space.

One part of the capsule assembly requires a very precise adjustment— the retrorocket. The acceleration it imparts to start the re-entry phase must be closely controlled. If re-entry is too steep, atmospheric drag will be too abrupt, and the pilot may be crushed under his own weight during the descent.

In the proposed design, the astronaut will be supported by a couch contoured to fit his back during the high accelerations of launching and re-entry. During the final descent, the large heat shield is to be jettisoned, as a drag parachute opens at an altitude of perhaps 20 mi. Then a landing parachute will act to slow the vehicle to a low velocity, and a cushion inflated to absorb the shock of landing.

One space hazard generally overlooked by the public is "radiation," which affects living organisms, electronic instruments, and photographic materials. The term "radiation" covers cosmic rays, which are much stronger above the atmosphere, the Van Allen-belt particles (protons and electrons) illustrated in Figure 7, and high-energy solar-wind particles shot out of the sun in increased numbers when there is a flare on the solar surface (see Chap. 8). Recent studies (Science, Jan. 24, 1975, p. 263) show that astronauts on Apollo flights to the Moon were subjected to heavy cosmic-ray exposure, particularly when they were outside the shielding spacecraft on the lunar surface.

Spacecraft structure provides better shielding for Van Allen-belt radiation and solar wind, although a "dent" in the magnetosphere off the eastern coast of Brazil—the "South Atlantic Anomaly"—is a place to be avoided with photographic and electronic gear, shielded or not. (The high-intensity Van Allen-belt radiation there penetrates several inches of metal shielding.)

NASA also made an effort to avoid having astronauts outside spacecraft on EVA (Extra-Vehicular Activity) just after a solar flare (p. 318). High-energy particles arrive at Earth or Moon some 30 min. after the flare is seen in visible light, so NASA established in 1967 a flare patrol with seven TV cameras mounted on 65-ft towers in Australia, Hawaii, Mexico, South Africa, Spain, and Texas. After being set each morning, the cameras followed the sun throughout the day and provided warning when a bright flare appeared on the solar image.

Meteoroids—the small particles and chunks of rock and iron that produce meteor trails when they hit the Earth's atmosphere—also pose a hazard to spacecraft (see Vol. 1, Wanderers in the Sky). Luckily, most meteoroids are small, and Harvard astronomer Fred L. Whipple had collected data on the relative numbers, from dust-size micrometeoroids to meteoroids of several grams' mass. He also showed that a thin protective shield or "bumper" a few centimeters outside the spacecraft walls would protect the walls by exploding small meteoroids before they could puncture.—TLP

Meteoroid Damage
to Spacecraft

(*Sky and Telescope*, March 1964)

How great is the risk that a space vehicle's skin will be punctured by meteoritic impact? The danger now appears to be about 3000 times less than indicated by calculations made seven years ago, writes Fred L. Whipple, director of the Smithsonian Observatory, in the *Journal of Geophysical Research*. The downward revision comes partly from improved data about the numbers and masses of meteoroids, and partly from a better understanding of hypervelocity-impact effects.

Whipple gives the following formula as his present best estimate for the number N of meteorite punctures that will occur per second in 1 m.^{2} of thin aluminum that is P cm thick:

$$\log N = -4.02 \log P - 13.33.$$

For a thin aluminum skin 0.01 cm (0.004 in.) thick, one puncture per square meter can be expected in 2¼ days. The corresponding times for 0.1 and 1 cm are 65 and about 700,000 yrs, respectively. A steel skin should be much safer still, extending these intervals about 10 times. Whipple points out that the uncertainty in these predictions is also large—approximately a factor of 10.

Data on "hyperballistic impacts" were obtained by several vacuum guns built by NASA. One at Goddard Space Flight Center (GSFC) shot microscopic grains of an iron compound at 150 km/sec.

Uncertainty about the meteoroid hazard was reduced by studies of actual meteoroid impacts on spacecraft.—TLP

Explorer-23
Micrometeoroid Results

RAYMOND N. WATTS, JR.

(*Sky and Telescope*, September 1968)

"How many meteoroids are there in a given volume of space near the Earth?" In search of answers, NASA has launched a number of satellites carrying special micrometeoroid detectors, notably the Pegasus and the S-55 series of Explorers.

Three satellites of the Pegasus series went into orbit during February, May, and July 1965. Explorer 23 went into orbit during 1964 and the results have been compiled by Robert L. O'Neal of Langley Research Center.

The S-55 satellite was 24 in. in diameter and 92 in. long and weighed 213 lbs. Apogee and perigee heights were about 980 km and 464 km and the satellite was spinning about 150 times/min. It carried around its circumference 210 stainless-steel cells containing helium under pressure. Whenever the craft encountered a meteoroid that punctured a cell, the resulting gas leak caused a pressure-sensitive microswitch to open, indicating the puncture. Each cell was a semicylinder about 1 in. in radius and 7½ in. long made of half-hard 302 stainless steel. The skin thickness was 25.4 microns (0.001 in.) for 70 cells, twice that for 140 others, to provide separate counts in two ranges of impact energies. Each cell had an effective area of about 0.01 m.^{2}, and was no longer sensitive once a puncture had occurred.

At frequent intervals during the year, a ground station interrogated the satellite's telemetry to find how many switches had opened. Thus the exposed area was known at any time and the average puncture rates for the two thicknesses. In all, there were 1298 interrogations.

During the year that ended Nov. 7, 1965, the 25-micron cells recorded 50 punctures. This indicated that an area of 1 m.^{2} on the satellite had been hit once every 2.8 days, on the average, by an object with enough energy to pierce stainless steel of that thickness. The 140 thicker cells reported 74 punctures during the same year, corresponding to 1 micrometeroid penetration each m.^{2} every 4.8 days.

These puncture rates from Explorer-23 data are close to those from the beryllium-copper cells aboard Explorer 16.

Although his meteoroid "bumper" fully protects a spacecraft from micrometeoroid penetration, Whipple's estimates of the probable number of meteoroid impacts on a spacecraft near the Earth and Moon have been revised drastically downward, and the hazard seems to be much less than originally feared.

*The following articles describe a variety of early space missions, most of them in preparation for Apollo landings on the Moon. A remarkably large number went according to plan, but the ones that failed show the hazard of unreliable equipment.—*TLP

America's Space Effort Accelerates

(Sky and Telescope, March 1962)

The Moon remains the primary target in the United States' space program for the immediate future, although several meteorological and communications satellites, an orbiting astronomical observatory, and probes to Venus and Mars are now planned by NASA.

A broad look into NASA's current and future activities was given in January by Robert C. Seamans, Jr., its associate administrator. He told about five areas of research, summarized here.

Lunar Exploration. Three series of unmanned lunar flights are expected: Ranger, Surveyor, and Prospector. The first is specifically to study the Moon's internal structure by landing a seismometer-transmitter combination. Ranger firings have been planned for this year and next.

Surveyor flights are a follow-up to Ranger for 1963 through 1965. They entail soft landings of packages with 350 lbs of instruments in several areas on the Moon's visible side for study of the surface conditions. From orbital flights, some Surveyors will map the lunar surface in preparation for future manned landings. The parent Surveyor missile will weigh about 2500 lbs for landing missions and somewhat less for the orbital flights.

Still further studies will be undertaken in the unmanned Prospector program.

Human Spaceflight. Beginning with John Glenn's much-postponed flight, five trips of three laps each around the Earth are scheduled in the Project Mercury program this year. Technical improvements should permit 1-day orbital flights in late 1962 or early 1963.

NASA has a new project called Gemini, to follow Mercury in 1963-64. It will develop the technique of manned spaceflight rendezvous (joining two spacecraft in orbit). Two passengers are to remain aloft a week or more.

Project Apollo goes far beyond Mercury, as it aims at a 2-week round trip to the Moon by three spacemen, who will be landed there for on-the-spot lunar exploration. This achievement can come only after many preliminary steps: Earth-orbiting flights, circumlunar passes, and perhaps in-space rendezvous.

Planetary Probes. This year a modified Ranger is expected to explore the vicinity of Venus, passing by the planet. Similar fly-bys are planned to Venus and Mars in March and November 1964.

Meteorological Satellites. The Tiros series of satellites has exceeded expectations as monitors of world weather events. Tiros III discovered and tracked 50 tropical storms during the hurricane-typhoon season of 1961.

Three more Tiros craft will be launched, in addition to the Nimbus series that begins later this year. More versatile than Tiros, Nimbus' cameras and other atmospheric sensors will always face the Earth, and its polar orbit will enable it to view every part of our planet twice a day, thereby assisting both long- and short-range weather predicting.

Communications Satellites. The first Echo satellite, a huge aluminized plastic sphere, has already proved that it is possible to transmit messages across the oceans by reflecting radio signals from 1000 mi. up.

Financed by the American Telephone and Telegraph Co., two Telstar launches are proposed this year. This endeavor of private industry seeks to establish message-relay stations in orbits 5000 mi. high, to cover large areas of the Earth at one time.

During this year and next, under Project Relay, NASA will orbit a 125-lb satellite to receive and transmit a band of radio frequencies wide enough for one TV channel. Great Britain, Brazil, France, and West Germany will participate with the United States in this project, which will have worldwide benefits.

The Syncom satellite, a narrow-band (no TV) active repeater system stationed in a 24-hr orbit, may also be launched late in 1962.

RANGER 3 FLIES PAST MOON

Space-age difficulties beset America's most recent attempt to reach the Moon. After a seemingly perfect launch from Cape Canaveral on January 26, the two-stage Atlas-Agena-B rocket combination gave its payload, Ranger 3, excessive speed; the probe passed about 23,000 mi. ahead of the Moon and went on into a circumsolar orbit somewhat larger than the Earth's.

Until a Ranger mission or a further flight from the Soviet Union succeeds, the only invasions of the Moon's "privacy" are those achieved by Lunik 2, which landed on the Moon in September 1959, and Lunik 3, which in October that year took photographs of about two thirds of the Moon's far side (see p. 84 and Vol. 1).

Later successful Ranger flights to the Moon are discussed in Chapter 3 and the planetary flights in Chapters 4, 5, and 6. Seamans' 1962 report

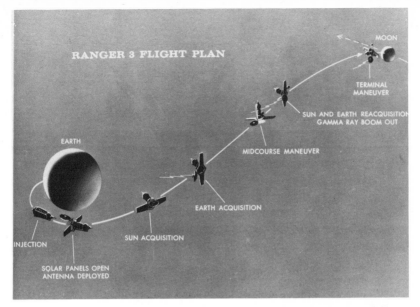

FIG. 11. Planned sequence of events along Ranger-3's flight path to the Moon. After the solar panels were folded out, the spacecraft was oriented (by gas jets) so that they faced the sun ("Sun acquisition"). Then the radio antenna was pointed toward Earth ("Earth acquisition"). The midcourse maneuver at 100,000 mi. from Earth was intended to correct the trajectory, and the terminal maneuver to aim at the selected impact point on the Moon and to eject (at 10 mi. above the lunar surface) a 98-lb capsule with retrorocket to "soft land" at 100 mi/hr. Other Rangers followed similar flight plans. (NASA diagram.)

FIG. 12. Ranger-3 spacecraft, showing its parts and instruments. (NASA diagram.)

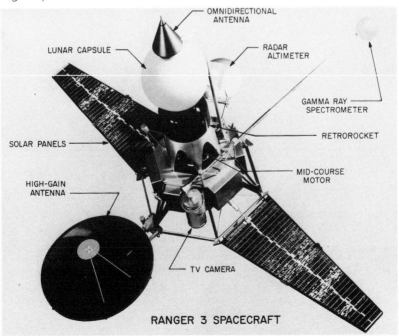

shows the wide variety of NASA space missions, including the "applications" flights directly beneficial to terrestrial residents—for weather forecasting, navigational aids, and communications.

The "active" communications satellites were a great improvement over the "passive" reflectors (Echo). They receive, amplify, and rebroadcast radio and TV across large distances. Syncom 1 failed, but Syncom 2 was put into "geosynchronous orbit" in July 1963 at 22,500 mi. above the Atlantic, where it has a 24-hr period and remains over the equator at the same longitude permanently. NASA launched more Syncoms and a series of Early Bird communications satellites for the Comsat Corporation, established by the U.S. Congress to set up a worldwide communications network. In 1966, the thirteenth active satellite, Lani Bird 1, failed to reach its intended position over the Pacific, but was soon replaced by Lani Bird 2. Later, NASA started the Applications Technology Satellite (ATS) series, which served several interests, including communication. In 1975, ATS 6 was moved into geosynchronous position over Kenya to provide public TV programs for the largely illiterate rural population over a wide area of South Africa and India. In early 1976, Sat Com 1 was put in geosynchronous orbit for U.S.-Latin American communications. This RCA transponder has twice the capacity of Syncom 2.

In the meantime, preparations for a Moon landing went on with a few failures.—TLP

The Flights of
Gemini 6 and 7

RAYMOND N. WATTS, JR.

(*Sky and Telescope*, February 1966)

Ill-starred Gemini 6 had twice failed to leave its launch pad, but finally got into orbit and back to Earth again with near-perfect precision. Two

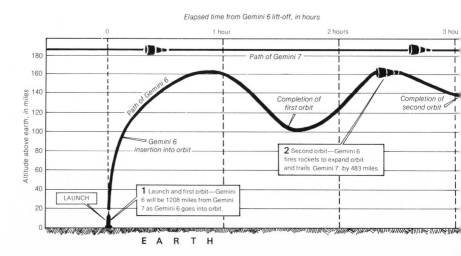

Elapsed time from Gemini 6 lift-off, in hours

days later Gemini 7 splashed down in the Atlantic recovery area, after setting a new endurance record.

The story began on Oct. 25, 1965, when a special Agena was launched as a rendezvous vehicle for Gemini 6, but the rocket disintegrated, forcing cancellation of Gemini's launch 42 min. before zero hour.

Gemini 6 was rescheduled for rendezvous with Gemini 7 during the latter's 2-week endurance flight, which started with a smooth firing of its Titan-2 booster on December 4 at 2:30:03 p.m. EST.

Gemini 6 was ready to go on the 12th. With astronauts Walter M. Schirra, Jr., and Thomas P. Stafford aboard, the Titan booster roared to life on schedule at 9:54 a.m. EST, then abruptly shut down only 2.2 sec before lift-off was due. The astronauts waited calmly for 1½ hours before they could be freed from the capsule. Had Schirra decided to use the escape system, there would have been no chance within months for another rendezvous attempt.

It was learned later that an electrical plug had pulled out prematurely. A computer in the blockhouse monitoring the launch "knew" that something was wrong, turned off the engines, and held the booster on the pad.

Moreover, while refueling and checking the Titan, technicians found that a tiny cap left in a fuel line would have restricted flow enough to have spoiled either of the first two attempts to send up Gemini 6.

In the early hours of December 15, Schirra and Stafford climbed aboard for a third attempt, and this time left the ground, at 8:32 a.m. EST.

A little more than 6 hrs later, as the two craft passed high over the Pacific Ocean, Gemini 6 and 7 cautiously approached within a foot of

FIG. 13. Plot of Gemini altitudes above the Earth for 6 hrs after Gemini-6 launch, showing Gemini 7 at constant (185-mi.) altitude in circular orbit. The insets show views of this orbit from the northern side, and how Gemini 6 spiraled out for the successful rendezvous. (*New York Times* diagram.)

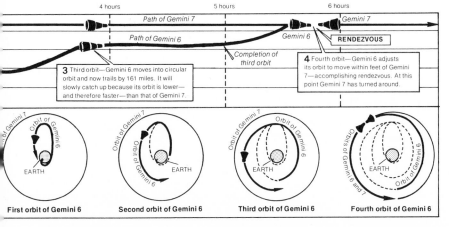

one another. To accomplish this, Gemini 6 had precisely performed several critical maneuvers to change its orbital plane slightly and to adjust the size of its orbit. The rendezvous was a notable demonstration of the precision control that is possible with Gemini spacecraft, and the medical results derived from the two-week voyage of astronauts Frank Borman and James A. Lovell, Jr., on Gemini 7 will have a significant bearing on plans for the first flight to the Moon.

Gemini Docking in Space

RAYMOND N. WATTS, JR.

(Sky and Telescope, May 1966)

A major event in manned space exploration—successful rendezvous and docking—has been overshadowed by the near disaster that followed.

The launches of Gemini 8 and its Atlas-Agena target vehicle were all but perfect; the approach and rendezvous in space of the two craft went as planned; and 6½ hrs after the Gemini launch the manned craft was firmly docked with its Agena target vehicle. Only 28 min. later, both craft were in serious trouble.

Following a pattern like that flown by Gemini 6 (Fig. 13), Gemini 8 maneuvered to within 2 ft of the Agena, then radioed for permission to dock. With a slight forward push from its thrusters, astronaut Neil Armstrong nudged the Gemini's nose into the Agena docking cone, and a mooring latch hooked the two craft tightly together.

After a check of equipment, the two astronauts were scheduled to perform a powered maneuver, turning the joined craft around. Suddenly, the two ships began to roll. Since Armstrong and his companion, David R. Scott, were unable to correct the spin, they decided to separate their craft from the Agena. As they did so, their rolling became even more violent, while the Agena began to tumble end over end.

Armstrong radioed to Earth that one of the attitude-control thrusters was stuck, and that he would have to activate the re-entry control system to stop the spinning.

The astronauts finally brought their craft under control, and reported to the ground that they were nearly out of fuel for all control systems. Therefore, a quick decision was made at the Mission Control Center in Houston to land Gemini 8 in the Pacific Ocean about 400 mi. southeast of Okinawa. Aided by a new on-board computer designed for the Apollo lunar landing program, the Gemini splashed safely into the sea.

NASA officials established that the probable cause of the difficulty was a short circuit, which caused a 25-lb thruster to turn on unexpectedly, resulting in both roll and yaw (end-over-end spin) at nearly one turn/sec.

NASA and the Gemini crews were well advised to use caution in these early rendezvous trials (which led to routine docking of spacecraft a few years later); a space collision would surely have been fatal.

The Gemini program ended with Gemini 12 in November 1966, bringing to 16 the number of NASA manned spaceflights (Table 3). It established the feasibility of docking, of leaving a spacecraft on EVA, and of astronauts taking astronomical photos (see Chaps. 7 and 8). Buzz Aldrin's EVA on Gemini 12 showed that gadgets designed to help the astronaut in o-g really work—a tether to hook the weightless man at the place where he needed to work, hand holds, Velcro pads, tethers for tools, and improved space suits (see p. 72).

The NASA Apollo program, using three-man spacecraft suitable for trips to the Moon, was under way in 1966, and many other successful space shots were launched.—TLP

TABLE 3. PRE-APOLLO NASA MANNED SPACEFLIGHTS

Mission	Astronauts	Time in orbit		
MERCURY		h	m	s
3[1]	Shepard	0	15	22
4[1]	Grissom	0	15	37
6	Glenn	4	55	23
7	Carpenter	4	56	05
8	Schirra	9	13	11
9	Cooper	34	19	49
GEMINI				
3	Grissom, Young	4	53	00
4	McDivitt, White	97	56	11
5	Cooper, Conrad	190	56	01
7	Borman, Lovell	330	35	13
6	Schirra, Stafford	25	51	24
8	Armstrong, Scott	10	42	06
9	Stafford, Cernan	72	20	56
10	Young, Collins	70	46	45
11	Conrad, Gordon	71	17	08
12	Lovell, Aldrin	94	33	—

[1] Suborbital flights.

Zond 5 Returns from the Moon

RAYMOND N. WATTS, JR.

(*Sky and Telescope,* November 1968)

Although Soviet authorities have released few details as yet, the difficult mission performed by the unmanned spacecraft Zond 5 is another major step toward taking men and instruments to the Moon and back.

Zond 5 was launched on Sept. 15, 1968, from the Baikonur Cosmodrome in Central Asia. After leaving its parking orbit around the Earth, Zond 5 received a radioed mid-course trajectory correction on the 17th and arched around the Moon early the next day, passing about 1210 miles from the lunar surface.

Even before Soviet spokesmen announced that the spaceship was returning to Earth, this was made known by Sir Bernard Lovell, from tracking data obtained with the 250-ft radio telescope at Jodrell Bank in England.

Zond 5 re-entered the Earth's atmosphere on September 22. No retro-rockets were used, but the craft was slowed by air resistance and by a parachute which gently lowered it into the southern Indian Ocean about 2500 mi. east of the Cape of Good Hope, where a Soviet recovery ship picked up the floating spacecraft that same day.

This was the first Russian water landing.

In addition to telemetering its instrument readings, Zond 5 recorded them on tape, which has far better signal-to-noise ratio than radio signals.

Another Successful OSO

RAYMOND N. WATTS, JR.

(*Sky and Telescope,* October 1969)

The sixth Orbiting Solar Observatory was successfully launched from Cape Kennedy, Florida, on August 9, 1969. The 640-lb satellite was carried into a nearly circular orbit 350 mi. above the Earth by a Delta-N rocket.

Aboard OSO 6 are seven experiments weighing 227 lbs. They came from institutions in the United States, Italy, and England. After 11 days of testing, all were found to be operational; they are now returning data on the sun from far above our atmosphere.

Like its predecessors, OSO 6 has two sections. One spins about 30 times a minute to stabilize the spacecraft; the other points continuously at the sun. The spinning nine-sided aluminum "wheel" carries five of the experiments and the basic spacecraft-support equipment. The stabilized "sail" section, 23 in. by 44 in., has the other two experiments and solar-power cells.

With a pointing accuracy of better than 1 arc-min. and the ability to scan selected areas of the sun's disk, OSO 6 is more versatile than previous craft in the series.

Roster of Space Activity

RAYMOND N. WATTS, JR.

(*Sky and Telescope*, January 1969 and February 1970)

Table 4 summarizes launches of satellites and space probes during 1968 and 1969. It contains a total of 146 entries, 17 referring to events of 1967 too late to be included in that year's table. Previous summaries of this kind have appeared in each January issue of *Sky and Telescope* since 1964; the first five years of the space age were summarized in March 1963.

The first column gives the official designation, assigned as nearly chronologically as possible by the World Warning Agency on behalf of COSPAR. An intended spacecraft is not included unless its lifetime was at least 90 min. Gaps in the numbering indicate routine test flights and classified military launchings by the Department of Defense.

The Soviet Cosmos series, omitted for 1969, includes both military reconnaissance satellites and scientific spacecraft.

In the second and third columns are given the popular name and the country of origin. Australia and the European Space Research Organization (ESRO) make their first appearance in 1969. The dates of launch and of when the satellite re-entered the atmosphere (or landed on the Moon) are found in the columns headed "Launch" and "Decay."

The next four columns describe the orbit of each spacecraft. The elements are based upon NASA's *Satellite Situation Report*, which is compiled by Goddard Space Flight Center from data supplied by NORAD and the Smithsonian Astrophysical Observatory. Many satellites maneuvered in space, but the table does not include these orbital changes.

The last column tells the primary purpose of the craft.

TABLE 4. SPACE ROSTER FOR NOVEMBER 1967 THROUGH DECEMBER 1969

Designation	Popular name	Country	Lifetime Launch	Lifetime Decay	Perigee (km)	Apogee (km)	Period (min.)	Inclination (°)	Experiments
1967-110A	Cosmos 190	USSR	11/3/67	11/11/67	195	323	89.6	65.6	
1967-111A	ATS 3	US	11/5/67		35,772	35,815	1436.4	0.4	Earth photos in color
1967-112A	Surveyor 6	US	11/7/67	11/10/67	Landed on Moon				Photos and tests on lunar surface
1967-113A	Apollo 4	US	11/9/67	11/9/67	370	371	88.3	32.7	Unmanned orbital flight
1967-114A	ESSA 6	US	11/10/67		1410	1488	114.8	102.1	Meteorology
1967-115A	Cosmos 191	USSR	11/21/67	3/2/68	269	489	92.1	70.9	
1967-116A	Cosmos 192	USSR	11/23/67		745	756	99.8	74.0	
1967-117A	Cosmos 193	USSR	11/25/67	12/3/67	199	335	89.8	65.6	
1967-118A	WRESAT	Australia	11/29/67	1/10/68	170	1250	98.9	83.3	Atmospheric physics
1967-119A	Cosmos 194	USSR	12/3/67	12/11/67	201	312	89.6	65.6	
1967-120A	OV3 6	US	12/5/67		408	439	93.0	90.6	Radiation studies
1967-123A	Pioneer 8	US	12/13/67		Heliocentric orbit				Interplanetary particles and fields
1967-123B	TTS 1	US	12/13/67	4/28/68	293	482	92.3	32.9	Tracking
1967-124A	Cosmos 195	USSR	12/16/67	12/23/67	204	352	90.0	65.6	
1967-125A	Cosmos 196	USSR	12/19/67	7/7/68	221	856	95.4	48.7	
1967-126A	Cosmos 197	USSR	12/26/67	1/30/68	227	475	91.5	48.4	
1967-127A	Cosmos 198	USSR	12/27/67		895	950	103.4	65.1	

Designation	Name	Country							
1968-01A	Surveyor 7	US	1/7/68	1/10/68	Landed on Moon				Photos and tests on lunar surface
1968-02A	Geos 2	US	1/11/68		1080	1574	112.2	105.8	Geodetic beacon
1968-03A	Cosmos 199	USSR	1/16/68	2/1/68	159	221	88.2	65.5	
1968-06A	Cosmos 200	USSR	1/19/68		518	538	95.1	74.0	
1968-07A	LM-1/Ascent	US	1/22/68	1/24/68	Elements not available				Apollo lunar-module test
1968-07B	LM-1/Descent	US	1/22/68	2/12/68	167	336	89.5	31.6	Apollo lunar-module test
1968-09A	Cosmos 201	USSR	2/6/68	2/14/68	202	327	89.7	64.9	
1968-10A	Cosmos 202	USSR	2/20/68	2/24/68	211	456	91.2	48.4	
1968-11A	Cosmos 203	USSR	2/20/68		1186	1203	109.2	74.0	
1968-13A	Zond 4	USSR	3/2/68		Elements not available				
1968-14A	OGO 5	US	3/4/68		271	148,186	3795.9	31.1	Earth's space environment
1968-15A	Cosmos 204	USSR	3/5/68		271	843	95.7	70.9	
1968-16A	Cosmos 205	USSR	3/5/68	3/13/68	197	292	89.3	65.6	
1968-17A	Explorer 37	US	3/5/68		510	883	98.7	59.4	Solar radiation
1968-19A	Cosmos 206	USSR	3/14/68		598	640	97.1	81.2	
1968-21A	Cosmos 207	USSR	3/12/68	3/24/68	201	320	89.7	65.6	
1968-22A	Cosmos 208	USSR	3/21/68	4/22/68	202	278	89.2	64.9	
1968-23A	Cosmos 209	USSR	3/22/68		871	945	103.1	65.3	
1968-24A	Cosmos 210	USSR	4/3/68	4/11/68	198	374	90.2	81.3	
1968-25A	Apollo 6	US	4/4/68	4/4/68	183	184	88.2	32.5	Unmanned test flight
1968-26A	OV1 13	US	4/6/68		556	9317	199.5	100.0	Radiation studies

Table 4 (Continued)

Designation	Popular name	Country	Lifetime Launch	Lifetime Decay	Perigee (km)	Apogee (km)	Period (min.)	Inclination (°)	Experiments
1968-26B	OV1 14	US	4/6/68	11/10/68	560	9935	207.8	100.0	Radiation studies
1968-27A	Luna 14	USSR	4/7/68		Orbit around Moon				
1968-28A	Cosmos 211	USSR	4/9/68	11/10/68	199	1545	102.1	81.0	
1968-29A	Cosmos 212	USSR	4/14/68	4/19/68	180	200	88.3	51.6	Linked with Cosmos 213
1968-30A	Cosmos 213	USSR	4/15/68	4/20/68	186	254	89.1	51.6	Linked with Cosmos 212
1968-32A	Cosmos 214	USSR	4/18/68	4/26/68	199	370	90.1	81.3	
1968-33A	Cosmos 215	USSR	4/18/68	6/30/68	255	403	91.1	48.4	
1968-34A	Cosmos 216	USSR	4/20/68	4/28/68	195	265	89.1	51.8	
1968-35A	8th Molniya 1	USSR	4/21/68		414	39,719	713.2	64.0	Communications repeater
1968-36A	Cosmos 217	USSR	4/24/68	4/26/68	150	182	87.6	62.2	
1968-37A	Cosmos 218	USSR	4/25/68	4/25/68	Elements not available				
1968-38A	Cosmos 219	USSR	4/26/68		225	1747	104.7	48.4	
1968-40A	Cosmos 220	USSR	5/7/68		677	755	99.0	74.0	
1968-41A	Iris	ESRO	5/17/68		330	1090	98.9	97.2	Solar, cosmic radiations
1968-43A	Cosmos 221	USSR	5/24/68		214	2082	108.3	48.4	
1968-44A	Cosmos 222	USSR	5/30/68	10/11/68	281	520	91.3	70.9	
1968-45A	Cosmos 223	USSR	6/1/68	6/9/68	221	979	89.9	72.9	
1968-46A	Cosmos 224	USSR	6/4/68	6/12/68	167	311	89.1	51.8	
1968-48A	Cosmos 225	USSR	6/11/68	11/2/68	248	519	92.2	48.4	

1968-49A	Cosmos 226	USSR	6/12/68		597	642	96.8	81.2	
1968-51A	Cosmos 227	USSR	6/18/68	6/26/68	202	271	89.2	51.8	
1968-53A	Cosmos 228	USSR	6/21/68	7/3/68	203	241	88.9	51.6	
1968-54A	Cosmos 229	USSR	6/26/68	7/4/68	222	328	89.8	72.9	
1968-55A	Explorer 38	US	7/4/68		5851	5859	224.4	120.6	Radio astronomy
1968-56A	Cosmos 230	USSR	7/5/68	11/2/68	283	544	92.8	48.4	
1968-57A	9th Molniya 1	USSR	7/6/68		396	39,806	714.6	65.0	Communications repeater
1968-58A	Cosmos 231	USSR	7/10/68	7/18/68	206	391	89.6	64.9	
1968-59A	OV1 15	US	7/11/68	11/6/68	153	1800	104.6	89.8	Radiation studies
1968-59B	OV1 16	US	7/11/68	8/19/68	145	556	91.5	89.8	Radiation studies
1968-60A	Cosmos 232	USSR	7/16/68	7/24/68	200	355	89.4	65.3	
1968-61A	Cosmos 233	USSR	7/18/68		199	1505	101.9	81.9	
1968-62A	Cosmos 234	USSR	7/30/68	8/5/68	208	295	89.5	51.8	
1968-66A	Explorer 39	US	8/8/68		673	2533	118.2	80.6	Atmospheric density
1968-66B	Explorer 40	US	8/8/68		680	2533	118.3	80.6	Atmospheric physics
1968-67A	Cosmos 235	USSR	8/9/68	8/17/68	203	283	89.3	51.8	
1968-68A	ATS 4	US	8/10/68	10/17/68	219	767	94.5	29.0	Meteorology
1968-69A	ESSA 7	US	8/16/68		1432	1475	114.9	101.7	Meteorology
1968-70A	Cosmos 236	USSR	8/27/68		590	627	96.8	56.0	
1968-71A	Cosmos 237	USSR	8/27/68	9/4/68	199	323	89.7	65.4	
1968-72A	Cosmos 238	USSR	8/28/68	9/1/68	203	210	88.6	51.7	
1968-73A	Cosmos 239	USSR	9/5/68	9/13/68	198	262	89.1	51.8	

TABLE 4 (Continued)

Designation	Popular name	Country	Lifetime Launch	Decay	Perigee (km)	Apogee (km)	Period (min.)	Inclination (°)	Experiments
1968-75A	Cosmos 240	USSR	9/14/68	9/21/68	203	283	89.3	51.8	
1968-76A	Zond 5	USSR	9/14/68	9/21/68	Orbit around Moon and Earth				Returned experiment package to Earth
1968-77A	Cosmos 241	USSR	9/16/68	9/24/68	198	326	89.7	65.4	
1968-79A	Cosmos 242	USSR	9/20/68	11/13/68	260	404	91.2	70.9	
1968-80A	Cosmos 243	USSR	9/23/68	10/4/68	206	297	89.5	71.3	
1968-81A	OV2 5	US	9/26/68		Elements not available				Radiation studies
1968-81B	ERS 28	US	9/26/68		175	35,732	629.5	26.3	Radiation and metallurgy
1968-81C	ERS 21	US	9/26/68		35,776	35,785	1435.8	3.0	Radiation and metallurgy
1968-81D	Les 6	US	9/26/68		Elements not available				UHF propagation
1968-82A	Cosmos 244	USSR	10/2/68	10/2/68	140	158	87.4	49.6	
1968-83A	Cosmos 245	USSR	10/3/68		272	841	92.0	70.9	
1968-84A	ESRO I	ESRO	10/3/68		259	1528	102.8	93.7	Atmospheric physics
1968-85A	10th Molniya 1	USSR	10/5/68		429	39,639	711.9	64.8	Communications repeater
1968-87A	Cosmos 246	USSR	10/7/68	10/12/68	145	317	89.1	65.3	
1968-88A	Cosmos 247	USSR	10/11/68	10/19/68	215	343	89.9	65.4	
1968-89A	Apollo 7	US	10/11/68	10/22/68	229	306	89.9	31.6	Manned orbital flight
1968-90A	Cosmos 248	USSR	10/19/68		473	543	94.7	62.2	
1968-91A	Cosmos 249	USSR	10/20/68		491	2158	112.1	62.3	

Designation	Name	Country	Date	Perigee	Apogee	Period	Inclination	Purpose
1968-93A	Soyuz 2	USSR	10/25/68	191	229	88.6	51.7	Manned orbital flight
1968-94A	Soyuz 3	USSR	10/26/68	183	205	88.3	51.7	Manned orbital flight
1968-95A	Cosmos 250	USSR	10/30/68	753	845	100.6	74.0	
1968-96A	Cosmos 251	USSR	10/31/68	170	226	88.3	64.7	
1968-100A	Pioneer 9	USA	11/8/68	Heliocentric orbit				Interplanetary fields, rays
1968-101A	Zond 6	USSR	11/10/68	Circled Moon; returned to Earth				Lunar photography
1968-103A	Proton 4	USSR	11/16/68	248	465	91.6	51.5	Interplanetary particles
1968-109A	Heos A	ESRO	12/5/68	4339	224428	6792.7	28.2	Field and radiation studies
1968-110A	OAO 2	USA	12/7/68	765	777	100.3	35.0	UV observations
1968-114A	ESSA 8	USA	12/15/68	1417	1464	114.6	101.8	Meteorology
1968-116A	Intelsat 3 F-2	USA	12/19/68	35,770	35,790	1435.9	0.7	Communications repeater
1968-118A	Apollo 8	USA	12/21/68	166	540,355	24,399.0	30.7	Manned translunar flight
1969-1A	Venera 5	USSR	1/5/69	Entered Venus' atmosphere				Soft landing on Venus
1969-2A	Venera 6	USSR	1/10/69	Entered Venus' atmosphere				Soft landing on Venus
1969-4A	Soyuz 4	USSR	1/14/69	213	224	88.8	51.7	Manned rendezvous, docking
1969-5A	Soyuz 5	USSR	1/15/69	196	212	88.6	51.7	Manned rendezvous, docking
1969-6A	OSO 5	USA	1/22/69	536	561	95.6	32.9	Solar studies
1969-9A	Isis A	Canada	1/30/69	599	3525	128.4	88.4	Atmospheric research
1969-11A	Intelsat 3 F-3	USA	2/6/69	35,786	35,809	1435.4	1.3	Communications repeater
1969-14A	Mariner 6	USA	2/25/69	Heliocentric orbit				Mars fly-by
1969-16A	ESSA 9	USA	2/26/69	1430	1505	115.2	101.8	Meteorology
1969-18A	Apollo 9	USA	3/3/69	176	462	90.0	33.5	Manned orbital flight

TABLE 4 (Continued)

Designation	Popular name	Country	Lifetime Launch	Decay	Perigee (km)	Apogee (km)	Period (min.)	Inclination (°)	Experiments
1969-29A	Meteor	USSR	3/26/69		632	686	97.9	81.1	Meteorology
1969-30A	Mariner 7	USA	3/27/69		Heliocentric orbit				Mars fly-by
1969-35A	11th Molniya 1	USSR	4/11/69		483	39,595	712.1	64.9	Communications repeater
1969-37A	Nimbus 3	USA	4/14/69		1080	1138	107.4	99.9	Experimental meteorology
1969-37B	EGRS 13	USA	4/14/69		1072	1133	107.3	99.9	Geodesy
1969-43A	Apollo 10	USA	5/18/69	5/26/69	Barycentric orbit				Manned circumlunar flight
1969-45A	Intelsat 3 F-4	USA	5/22/69		35,226	35,671	1418.9	0.5	Communications repeater
1969-51A	OGO 6	USA	6/5/69		400	1087	99.6	81.9	Geophysical studies
1969-53A	Explorer 41	USA	6/21/69		378	213,849	4840.0	83.8	Interplanetary fields, rays
1969-56A	Biosat 3	USA	6/29/69	7/7/69	356	387	92.0	33.5	Primate in space
1969-58A	Luna 15	USSR	7/13/69	7/21/69	Orbit around Moon; hit Moon				Unannounced
1969-59A	Apollo 11	USA	7/16/69	7/24/69	Lunar landing; returned to Earth				First manned lunar landing
1969-61A	12th Molniya 1	USSR	7/22/69		496	39,526	711.0	64.9	Communications repeater
1969-64A	Intelsat 3 F-5	USA	7/26/69		269	5399	146.7	30.3	Communications repeater
1969-67A	Zond 7	USSR	8/7/69	8/14/69	Circled Moon; returned to Earth				Lunar photography
1969-68A	OSO 6	USA	8/9/69		491	553	95.1	32.9	Solar studies
1969-69A	ATS 5	USA	8/12/69		35,762	36,898	1463.9	2.7	Experiment package technology
1969-83A	ESRO 1B	ESRO	10/1/69	11/23/69	294	378	91.2	85.1	Atmospheric physics
1969-84A	Meteor 2	USSR	10/6/69		619	676	97.6	81.2	Meteorology

Designation	Name	Country	Launch date		Perigee	Apogee	Period	Inclination	Remarks
1969-85A	Soyuz 6	USSR	10/11/69	10/16/69	194	229	88.8	51.7	Manned rendezvous, docking
1969-86A	Soyuz 7	USSR	10/12/69	10/17/69	200	217	88.4	51.6	Manned rendezvous, docking
1969-87A	Soyuz 8	USSR	10/13/69	10/18/69	215	278	89.4	51.6	Manned rendezvous, docking
1969-88A	Intercosmos 1	USSR	10/14/69		253	626	93.3	48.3	Atmospheric research
1969-97A	GRS-Azur	Germany	11/8/69		386	3149	121.9	102.9	Magnetosphere
1969-99A	Apollo 12	USA	11/14/69	11/24/69	Lunar landing; returned to Earth				Second manned lunar landing
1969-101A	Skynet A	UK	11/22/69		34,695	36,678	1431.0	2.4	Communications repeater
1969-110A	Intercosmos 2	USSR	12/25/69		206	1200	98.5	48.4	Ionospheric studies

Table 4 is but a sample of the hundreds of space launches in the 1960's. Both sides in the space race suffered failures, as recounted in the following articles, showing that hazards due to defective equipment were only slowly eliminated. These troubles were minor compared with the fire that killed three astronauts, Gus Grissom, Ed White, and Roger Chaffee, while rehearsing procedures for the first manned Apollo mission on Jan. 27, 1967. The three men were locked in the spacecraft with a pure-oxygen atmosphere when an electrical spark ignited several flammable materials in the cabin. NASA thus learned by catastrophe that it must prohibit flammable materials and enforce better reliability tests on electrical connections. NASA also decided to use a 60-40 mix of oxygen-nitrogen during Apollo launches, later in the mission to be gradually replaced by pure oxygen. (Shuttle will have a more normal cabin atmosphere—see p. 405).

Five others of the 73 astronauts have lost their lives in non-space-related accidents—airplane crashes and auto accidents—two of them in 1967. It was a bad year for cosmonauts also.—TLP

The Soyuz-1 Disaster

RAYMOND N. WATTS, JR.

(Sky and Telescope, June 1967)

From Moscow on April 23, 1967, the Tass international news service released a telegram:

"The spaceship Soyuz 1 was orbited today at 0335 hours Moscow time. This is its pilot's, Colonel Komarov's, second spaceflight. In October 1964, he piloted the ship Voskhod."

Thirty-six hours later, Tass reported: "The CPSU Central Committee, the Presidium of the USSR Supreme Soviet, and the Council of Ministers have announced with great sorrow that Colonel-Engineer Vladimir Komarov, Hero of the Soviet Union, perished tragically while completing the test flight of spaceship Soyuz 1 today."

The launching was from the Baikonur Cosmodrome in Kazakhstan, and appears to have developed difficulty after a few orbital revolutions. After circling the globe 18 times, the craft successfully re-entered the atmosphere, but crashed to Earth. Tass explained that "the spaceship safely passed the most difficult braking stretch in the dense layers of the atmosphere. However, when the main cupola of the parachute opened at an altitude of 7 km, the straps of the parachute got twisted and the spaceship descended at a great speed which resulted in Komarov's death."

A less tragic failure occurred after the launch of Apollo 6, an un-manned test flight, when the second-stage-rocket motors shut down 2 min. prematurely and failed, 3 hrs later, to re-ignite, leaving Apollo 6 in an elliptical orbit 112-216 mi. high. However, ground controllers were able to bring the empty Command Module (CM) to a successful splash-down in the Pacific after a 10-hr flight.—TLP

Successful Apollo Mission

RAYMOND N. WATTS, JR.

(Sky and Telescope, December 1968)

In the first manned test flight of the Apollo series, Walter M. Schirra, Jr., Donn F. Eisele, and R. Walter Cunningham safely completed a 260-hr, 4.5-million-mi. journey that carried them 163 times around the world.

Before lift-off, the Apollo spacecraft atop its Saturn 1-B booster and Saturn IV-B second stage towered 224 ft above its concrete pad at Cape Kennedy. The launch was accomplished smoothly on Oct. 11, 1968, at 11:03 a.m. EDT.

When the first stage dropped away 2.4 min. later, the spacecraft and its attached second stage together were 113 ft long and weighed 69,000 lbs. Apollo 7 circled the Earth almost twice in an orbit between 176 mi. and 142 mi. high, before jettisoning the second stage.

One of the many tasks of the crew was to perform a rendezvous with the second stage on October 12. As a safety measure, the spacecraft had earlier dumped its excess propellants. Then, by firing two bursts of its large engine, the astronauts brought the Apollo 7 to within 70 ft of the Saturn IV-B at 5 p.m. EDT, over the Pacific Ocean.

Other scheduled activities included navigational exercises sighting landmarks and stars, with the space sextant. Attached to this sextant is a wide-field telescope for general viewing and for identification of navi-gational marks. However, when the sextant was moved, "white sandy particles" clouded the view, and only a blur of light could be seen when the instrument was pointed near the sun, as if there were drops of moisture inside the telescope and a light leak that admitted solar glare. This malfunction was not dangerous, because Apollo 7 was guided from the ground.

The 11-day period that Schirra, Eisele, and Cunningham spent in orbit was the second longest to date. (Gemini 7 had 14 days in 1965.) On October 22 at 7:11 a.m. EDT, the Apollo spacecraft dropped into the water 325 mi. south of Bermuda.

Detailed medical studies of the three astronauts are being made for clues to possible hazards in much longer flights. Protracted lack of exercise can cause deleterious cardiovascular effects. Also, experiments conducted by the Air Force have shown that a spacecraft environment lowers resistance to disease. The Apollo-7 astronauts were x-rayed before and after their mission to test the degree of bone demineralization.

The highly successful Apollo flights to the Moon are covered in Chapter 3, but Apollo 13, a near disaster, failed in its mission.—TLP

Apollo 13 Makes It
Back to Earth

RAYMOND N. WATTS, JR.

(*Sky and Telescope*, June and July 1970)

The ill-fated Apollo-13 spaceship was the focus of worldwide attention in mid-April 1970. After a last-minute crew change caused by a German-measles scare, the launch and the first part of the trip to the Moon went smoothly. From blast-off at 2:15 p.m. EST on April 11 until 10:08 p.m. on April 13, the astronauts concerned themselves with routine chores.

Astronauts James A. Lovell, Jr., Fred W. Haise, Jr., and John L. Swigert, Jr., had just finished a telecast to Earth when they felt a shudder run through their spacecraft. Their instruments soon told them that they were in mortal danger.

No. 2 oxygen tank in the service module had ruptured, depriving the electricity-generating fuel cells of their crucial oxygen supply. As a result, the Command Module was soon without power. Swift action by both ground controllers and crew resulted in the conservation of as much battery power as possible, and the astronauts themselves transferred to the Lunar Module.

Later, they fired the Lunar Module's engine to change their trajectory so that they would loop around the Moon and head back toward Earth.

But their troubles were not over. Supplies of water and oxygen were critically short. Moreover, the temperature fell to near freezing, adding physical discomfort to the astronauts' danger.

FIG. 14. The Apollo-13 Service Module (SM) photographed from the Command Module (CM) just after it was jettisoned before splashdown. The explosion had blown away an entire panel, exposing the interior, including two of the fuel cells just forward (left) of the damaged parts. At right is the SM's propulsion system and nozzle. (NASA photo.)

Despite their many hardships and handicaps, the Apollo-13 astronauts guided their Command Module (CM) to a precise Pacific splashdown at 1:08 p.m. on April 17. Shortly before re-entry, they had jettisoned the Service Module, which disintegrated in the Earth's atmosphere.

Also jettisoned was the Lunar Module's Snap-27 nuclear generator, containing 8.36 lbs of radioactive plutonium in a graphite canister (Fig. 106). This is believed to have fallen into the Pacific Ocean northeast of New Zealand, somewhere near latitude 25° S, longitude 174° W, in about 2700 fathoms of water. It was not expected to break on impact, hence is not an immediate danger.

One finding of considerable interest to selenologists was produced by the 15-ton Saturn IVB jettisoned by Apollo 13 to smash into the Moon at 8:09 p.m. EST on April 14, while traveling almost vertically downward at 9300 mi/hr. The impact point, southwest of the crater Lansberg in Oceanus Procellarum, was 87 mi. WNW of the seismometer planted by the Apollo-12 astronauts (see p. 129).

Only 30 sec later this seismometer recorded a sharp onset of moonquakes that continued for 3 hrs, 30 min.

Furthermore, on April 14 the neutral-particle detector deployed next to the seismometer recorded an influx of particles only 22 sec after the Saturn IVB crashed. This may mean that the impact scattered material at least 87 mi. with velocities of more than 14,000 mi/hr.

NASA's Apollo-13 review board, headed by Edgar Cortright, has determined the probable cause of the explosion in the Service Module on

April 13. Even though the SM burned up during re-entry, the board believes that two switches for a small heater in the destroyed oxygen tank were damaged before launch.

During an Apollo mission the oxygen tank contains both liquid and gaseous O_2. As the oxygen is used in the spacecraft, pressure in the tank is maintained by having the heater vaporize more of the liquid.

In a prelaunch test of Apollo 13, too high a voltage seems to have been applied to the switches, melting the contacts so the switches were closed permanently.

During the flight, the heater remained on and may have reached a temperature as high as 1000°F, crumbling insulation off the wires. Finally, when the craft was 205,000 mi. from Earth, the bare wires caused arcing and an explosion; the oxygen tank was destroyed, blowing out the side panel of the SM.

OAO B Fails to Orbit

RAYMOND N. WATTS, JR.

(*Sky and Telescope*, January 1971)

The Orbiting Astronomical Observatory program suffered a severe set-back when, on Nov. 30, 1970, OAO B failed to go into orbit around the Earth. Early indications from telemetry and tracking data suggest that a part of the protective shroud failed to separate properly. Its weight was too great for the Centaur second-stage engines to sustain, and the spacecraft plunged back into the atmosphere.

The OAO-B project, said to cost $98 million, is described in Chapter 8, p. 344.—TLP

The First Laboratory in Space—The Ill-Fated Soyuz 11

RAYMOND N. WATTS, JR.

(*Sky and Telescope*, August and September 1971)

The Soviet Union's second attempt to establish a manned orbiting space station had been an immense success until disaster struck on June 30, 1971, when the three cosmonauts were found dead in the just-landed

Soyuz 11. The tragedy occurred at the end of their return to Earth from the Salyut laboratory that was their home for 22 days beginning June 7. They had spent a record-breaking 24 days in space.

The unmanned Salyut had been launched on April 19. Then Soyuz 10 had carried three astronauts aloft, but although they managed to dock with Salyut, they never entered it. The problem may have been dissimilar pressures, preventing the crewmen from opening the hatch. In any event, the Soyuz-10 crew had to return to Earth on April 24.

On June 6, a new crew and spacecraft were sent up under command of Lt. Col. Georgi Dobrovolsky, with Vladislav Volkov as flight engineer and Viktor Patsayev as test engineer. Docking with Salyut was completed on June 7.

The cosmonauts set about making the electrical and hydraulic connections, paying particular attention to equalizing the pressure on both sides of the entrance hatch. It was opened, and Patsayev crawled into the space laboratory, where he connected the communications systems and was soon joined by Volkov. Together they powered up the Salyut, which had been dormant for more than a month.

The first indication that this might be a lengthy mission came on June 8, when Salyut's thrusters were fired to enlarge its orbit and reduce the pronounced effect of atmospheric drag on so large a satellite. The perigee height was eventually raised from 210 km to 255 km, apogee from 243 km to 277 km. On this basis it was then predicted the cosmonauts could safely remain on board until early in July.

The combined Salyut-Soyuz orbiting space station weighed over 27 tons. It had a length of about 20 m. (66 ft) and a maximum diameter of 4 m. Inside, it had several separate areas, with some 3500 cu ft of usable work space (about the volume of a typical 40-ft house trailer).

Although there is no indication that this particular orbiting laboratory will be reused, Soviet space experts have said that in the near future scientific equipment will be exchanged—carried aboard and removed as the Soyuz crews are ferried to and from Salyut-type satellites (see Chapter 7).

The flight plan called for photographic and spectrometric observations of the Earth below.

Once settled aboard Salyut, the three crewmen staggered their work schedules so that round-the-clock experiments could be made. In one biological project, frog eggs were hatched. . . . In a hydroponic (soilless) garden, the cosmonauts grew flax, onions, and cabbages. . . .

Crew members' pulse, respiration, and metabolism were constantly monitored, and blood samples were taken at intervals.

On June 29 the three men aboard Salyut were directed to return to

Earth in Soyuz 11. The two craft were unlinked without difficulty at 21:28 Moscow time. But ground communication with the crew ceased with the completion of the firing of the retrorocket that braked Soyuz into its descent, to near Karaganda in Soviet Kazakhstan.

"According to the program," said Tass, "after aerodynamic braking in the atmosphere, the parachute system was put into action, and . . . the soft-landing engines were fired. The flight of the descending apparatus. ended in a smooth landing in the preset area."

A recovery helicopter came down alongside. When the hatch of Soyuz 11 was opened, the lifeless bodies of Dobrovolsky, Volkov, and Patsayev were found in their seats.

In mid-July the special commission of Soviet scientists appointed by the Communist Party Central Committee to investigate the tragedy reported that rapid decompression of the command module was the direct cause of death.

The flight was proceeding normally and crew members were performing all their duties properly until about 30 min. before landing. Although the spacecraft was structurally sound when inspected after its automatic landing, a rapid decompression occurred during this interval. The quick change from the 15-lb/in² cabin atmosphere to the near-zero pressure of outside space caused virtually instantaneous death of all three men.

The commission's preliminary announcement did not explain the loss of pressure, but *Aviation Week and Space Technology* notes that the Soyuz design has a direct connection between the Command Module and the so-called Orbital Module (Salyut). Because the pressure is virtually the same in both chambers, there appears to be no way to check for leaks in the hatch once it is closed prior to the separation of the two craft.

When the cosmonauts separated their Command Module from the Orbital Module in preparation for re-entry, a leak in the equalization valve at the center of the hatch could have caused the rapid drop in pressure.

It is reported that at least ten Soviet test pilots lost their lives in pre-mission testing and training, showing that the Russians are willing to take risks with their pilots' lives. NASA lost several astronauts in airplane accidents and one in an auto accident, but we know of no other "space-related" deaths due to fire or loss of cabin pressure. However, a major program (see Chapter 7) and huge investment were threatened by an accident during the launch of NASA's Skylab in May 1973. The NASA organization was innovative and flexible enough to repair the damage and to proceed more or less on schedule.—TLP

Skylab's Troubled Flight

RAYMOND N. WATTS, JR.

(Sky and Telescope, July 1973)

After a virtually perfect countdown, on May 14, 1973, at 1:30 p.m. EDT, the Saturn-5 booster carried the Skylab off the launch pad and up through a bank of low clouds. There was a slight vibration 63 sec later, as the vehicle passed through maximum aerodynamic pressure. Like any launch vehicle, the Saturn encountered increasing air drag as it accelerated through the atmosphere, but it was also climbing into thinner and thinner air. Consequently, the drag forces reached a brief maximum before declining again.

The entire Skylab mission was to be seriously threatened by that vibration, although it was so slight that it went undetected until a detailed analysis was made of telemetry data. Everything still seemed to be going according to plan as commands from an onboard computer released the protective metal shrouds covering the forward end of the space vehicle. The Apollo Telescope Mount (ATM) swung into position at right angles to the main body, and the ATM's four windmill-like solar panels deployed.

When Skylab attained a nearly perfect orbit ranging between 274 mi. and 271 mi. above the Earth's surface, ground controllers began to feel confident that the mission was opening successfully. But about 41 min. after liftoff, there was no indication that the two large panels of solar cells had properly folded out from the main body of the orbital workshop. It was soon apparent that something was wrong and that the mission was in jeopardy.

Working feverishly, ground technicians and engineers deduced that something was amiss with the heat shield intended to protect the orbital Workshop from the searing rays of the sun. During launch, this white-painted metal shield was held close against the spacecraft's skin. Once in orbit, it was to be released and pushed a few inches away by torsion bars. For some as yet unknown reason, the shield had deployed at the worst possible moment, and it was torn away in the slipstream as Skylab raced through the atmosphere only 40,000 ft above the launch pad.

Further detective work suggested that when the heat shield tore loose, it probably jammed or damaged the solar panels that later failed to deploy. To plan emergency repairs, the astronauts' launching was postponed from May 15 to May 20. During a round-the-clock effort at NASA's space centers in Huntsville and Houston, as well as at several aerospace companies that had built Skylab parts, teams struggled to invent, perfect, and build substitute heat shields that could be carried up to Skylab by the astronauts. Other groups replanned mission activities to conserve electricity, so that the space station could be operated with only the power generated by the ATM solar-cell panels, which could provide a peak power of 10,480 watts. (The orbital workshop panels that failed to deploy would have generated 12,400 watts more.)

The Skylab story was similar to a TV thriller; temperature rose to 190°F inside, food supplies were going bad, and noxious fumes were coming off plastics—the $2 billion space workshop seemed lost. Like TV heroes, NASA specialists came to the rescue. Controllers turned the spacecraft so that the part without heat shield was away from the sun, and the temperature dropped to 110°F. The cabin atmosphere was vented, getting rid of the noxious gases, and competing teams produced sunshades to cover the suspected gap in the heat shield. The ripped heat shield and jammed solar panel were reproduced on the full-scale Skylab models at Huntsville, Alabama, and Houston, Texas, so that Skylab-1 astronauts could practice the repairs necessary, and the materials and tools they would need were manufactured and loaded in the Skylab-1 Command-Service Module.

On May 25, Charles Conrad, Joseph Kerwin, and Paul Weitz were launched in that CSM from KSC, and 8 hrs later they were in orbit, surveying Skylab from a distance of a few hundred feet. They found one solar panel gone, the other jammed in its folded position by a small strip of metal that could not be pried loose by Weitz, using a 10-ft hook on EVA. After docking and a crew rest period, Weitz entered Skylab and found no poison gas, temperatures from 50° to 100°F. Then, all three astronauts pushed a large (24-ft) umbrella out through an "air-lock" on the sunward side to shade the surface where the heat shield was gone (Fig. 15). The Workshop temperature dropped to 98°F. On May 30 it was down to 82°F, and the astronauts started their scheduled work. The power problem remained until, on June 2, two astronauts clambered out on EVA and unjammed the 30-ft solar panel, using a shears to cut the metal strip. The remainder of Skylab's successful story is told in Chapter 7.

Seven years earlier, spaceflight became sophisticated enough that astronauts could leave the protective spacecraft and "space walk" on EVA. This preparation for lunar exploration required a space suit to protect a man from the vacuum of space. The "Life Support System" was a major bio-engineering development, starting in 1934 with Wiley Post's attempts to fly aircraft to higher and higher altitudes. A book, *Suiting Up for Space*, by Lloyd Mallan (John Day, N. Y., 1971) describes

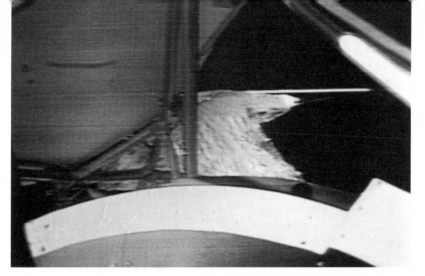

FIG. 15. The "umbrella" sunshade being pushed out (through an airlock) to protect the sunward side of Skylab, as viewed by TV through the window in the Apollo CM docked to Skylab. The sunshade was made at JSC with a top layer of aluminized mylar, a middle layer of laminated nylon, and a bottom layer of thin nylon. (NASA photo.)

the development undertaken primarily by the rubber companies and a brassiere manufacturer. The model used by Ed White in mid-1965 was much more bulky than those later worn by Apollo astronauts on the Moon, and by the Skylab crew. The cosmonauts had a less sophisticated model connected to the Voskhod spacecraft, but it allowed another first for the U.S.S.R.—TLP

Man in Space

RAYMOND N. WATTS, JR.

(*Sky and Telescope*, May 1965)

The first man to face the rigors of space without the protection of a rigid spacecraft climbed cautiously out of his airlock on March 18, 1965. Soviet cosmonaut Alexey Leonov then removed the protective cover from the lens of a television camera, revealing his space gymnastics to the whole world.

Inside Voskhod 2, Pavel Belyayev supervised the operation, communicating with Leonov via a connecting cable. Leonov spent 20 min. in space, 10 of them floating free from the capsule except for a slim 16-ft tether connecting him to the spacecraft's life-support system.

Voskhod 2 had been launched from the Baikonur Cosmodrome 90 min. earlier, and circled the Earth every 90.9 min. in an orbit inclined

65° to the equator. Apogee, the highest to date for manned flight, was 308 mi. above the Earth's surface; perigee was 106 mi.

Leonov's space suit did not burst—hence the space walk was a success—but it "bloated" at 6 lbs/in² inside air pressure, and he could not squeeze back through the airlock. After 6 min. of trying, Belyayev reduced the suit pressure to 3.5 lbs/in², the bloated suit diminished somewhat in size, and Leonov got back inside Voskhod 2.

A few months later, Ed White became the first astronaut to walk in space, on EVA from Gemini 4, piloted by James A. McDivitt. The mission had other troubles (failing to dock with a defective target rocket), but White's space suit worked, although dew formed inside the visor, interfering with vision. After another EVA, in which two men transferred from Soyuz 5 to Soyuz 4, the Russians ceased their efforts on EVA and have since made no announced space walks (as of 1976). On the 3-man Soyuz-Salyut missions, they omitted space suits altogether, in order to save space and weight. Western space experts say that this resulted from the Soviet's failure to build a larger booster rocket for launching 3-man missions (p. 281). About fifty years ago, the Russian space expert, Tsiolkovsky, had predicted that a space suit would be essential to a far-reaching space program.

NASA continued improving the space suit on Gemini-10, 11, and 12 missions. These suits were all connected to the spacecraft with an umbilical cord carrying oxygen and removing CO_2 and water. Finally, on Apollo 9, Rusty Schweickart tested the first independent space suit, the one used with minor improvements on all later NASA flights.

The purpose of NASA's space-suit development during the 1960's was the manned exploration of the Moon, covered in Chapter 3. As the $40-billion Apollo program progressed, some scientists criticized the manned space effort, saying, "Instruments would be cheaper and would get just as many data." In response, Professor Harry Hess of Princeton wrote in Science for June 14, 1963: "Instruments cannot replace the man. . . . The man can look around and at a glance pick the significant items which might be examined. . . . He can discriminate, find, and interpret the unexpected. . . . Great technological advances commonly have been a by-product of wars. An almost equal motivation can be provided by the space program without the disastrous effects of combat. A major breakthrough is just over the horizon in communications . . . miniaturization of electronics . . . improved meteorological services . . . advances in medicine. . . . The first Sputnik resulted in a vigorous rejuvenation of secondary education in the U.S. . . . The spirit of a nation is of supreme importance. The goal has been set. Let's get on with it."—TLP

3

Exploring the
Moon and
Its History

For twelve years, from 1961 to 1973, the space effort was dominated by NASA's Apollo Program. At first, the goal in politicians' minds was to win the race to the Moon. The Soviet Union kept interest alive by winning several subordinate contests[1]—the first man in orbit, the first space walk (see p. 73), the first space shot to hit the Moon (p. 19), and the first photos of the Moon's far side. But NASA put the first man (and 11 others) on the Moon, brought back almost half a ton of lunar material for study on Earth, left half a dozen instruments operating there, and mapped the lunar surface as accurately as the Earth's.

Space scientists in both countries shifted the emphasis to lunar exploration—explaining such known features as maria, craters, highlands, rilles (Fig. 44), and lava flows—and using evidence of lunar history in studies of the origin and history of the solar system. The first of these

[1] The man probably responsible for all this was Sergy Korolyov, the top engineer in the Soviet space organization when he died in 1966. He was a friend of Khruschev and "sold" the Soviet Premier on the many successful flights. He supervised the design and building of the medium-size booster rocket that has been used for all major launches since 1964, despite several unsuccessful efforts in the early '70's to launch a larger one.

two scientific goals was served by Ranger and Zond flights. Some of the earlier results are presented in Volume 1, Wanderers in the Sky, and in the first chapter of this volume.

NOTE ON VIEWING LUNAR PHOTOGRAPHS: If you have trouble seeing the vertical relief in the sharply shadowed photos reproduced here, try turning the picture upside down. The small round craters are, of course, depressions, and the sunlight is coming from the side where a dark shadow extends out onto the crater floor.

For positions of surface features on the Moon's front side, see Figure 16 (a map from Vol. 1, Wanderers in the Sky), with features listed on the next two pages.—TLP

LUNAR MAP LOCATING TABLE

For each crater in the alphabetical list that accompanies the lunar map are given here the selenographic longitude and latitude, respectively, of the co-ordinate grid intersection that is nearest to the crater. This permits quick location of features that are known by name only.

No.	Lon	Lat	No.	Lon	Lat	No.	Lon	Lat	No.	Lon	Lat
1	+10	-20	76	0	-70	151	+30	-10	226	-10	-40
2	+10	-10	77	+10	-50	152	+30	+10	227	-10	-30
3	-30	-20	78	+30	-10	153	+40	-40	228	+30	-50
4	+10	0	79	-60	0	154	+10	+10	229	+30	+40
5	0	-10	80	+30	+30	155	-40	+10	230	-10	+50
6	+10	+40	81	-10	-10	156	-20	-30	231	+10	-20
7	0	-30	82	+30	+20	157	-10	+40	232	+20	+20
8	+20	-20	83	-50	-30	158	-30	-70	233	+10	-30
9	0	-20	84	+20	0	159	0	-10	234	+70	-60
10	0	-10	85	+50	+60	160	-70	+20	235	+30	+30
11	+10	-30	86	0	-20	161	-30	0	236	-50	+30
12	+60	0	87	-30	+30	162	-70	-30	237	+50	+20
13	+20	+10	88	0	-50	163	-10	0	238	+10	+10
14	0	+30	89	+20	-10	164	-20	+30	239	0	-10
15	0	+60	90	-30	+30	165	-30	0	240	0	-30
16	-50	+20	91	+10	-10	166	+60	-10	241	-60	+60
17	0	+30	92	-40	-30	167	-10	-20	242	-20	+20
18	+20	+50	93	-70	-20	168	-40	-30	243	+20	-30
19	0	-20	94	-40	0	169	-60	-40	244	-30	-30
20	+20	-50	95	+60	+50	170	-40	-10	245	0	-30
21	+50	+50	96	0	+70	171	-20	+40	246	+50	-30
22	0	+30	97	-10	+10	172	0	-40	247	-60	+10
23	+10	-20	98	-30	-10	173	+10	-50	248	-20	0
24	+20	-50	99	+20	+40	174	+10	-50	249	-70	+50
25	-70	-70	100	-30	+20	175	+30	-30	250	0	0
26	+20	-40	101	+40	-40	176	+10	+30	251	+50	-40
27	-30	-50	102	+10	-40	177	+30	+20	252	-70	0
28	+30	-20	103	+20	-20	178	-70	0	253	+40	+30
29	+60	+30	104	0	-40	179	-20	-50	254	+20	+10
30	+50	+40	105	+60	+10	180	-20	-20	255	+30	-30
31	+20	+20	106	-40	0	181	+20	+10	256	+20	-20
32	-40	-60	107	-20	+60	182	+50	+20	257	+50	-20
33	-30	+50	108	+30	-20	183	+30	-10	258	-10	-40
34	+50	-50	109	-20	-10	184	+40	-10	259	0	-40
35	-50	-10	110	+50	+40	185	-10	-50	260	-30	-60
36	-10	+60	111	+60	-40	186	-40	+40	261	-50	-40
37	-10	-20	112	-10	0	187	+10	+10	262	-40	-50
38	-20	-60	113	-40	-20	188	+30	-70	263	-10	0
39	0	-30	114	-10	-30	189	+30	+20	264	-70	+20
40	+50	-70	115	+70	+40	190	+70	-40	265	-40	+50
41	+40	-20	116	-20	+10	191	+30	0	266	+20	-70
42	0	+60	117	+10	-20	192	-30	+50	267	+60	-30
43	-20	-10	118	+60	+30	193	+10	-30	268	+20	+10
44	+50	-20	119	+10	-30	194	-30	+20	269	-10	+10
45	+10	+10	120	+50	-10	195	+20	+20	270	+60	-30
46	-40	+50	121	+10	0	196	-30	-30	271	+10	-40
47	+70	-70	122	+10	-30	197	-50	-20	272	+50	+60
48	-20	-20	123	-70	-10	198	+60	+40	273	+60	+40
49	+60	+30	124	-40	+30	199	+50	0	274	-70	+20
50	+30	+40	125	-10	-10	200	+40	-40	275	+20	-20
51	+10	+40	126	+40	-10	201	+20	+70	276	+50	+10
52	-30	-30	127	+70	+30	202	-30	+10	277	+10	+40
53	+30	-10	128	-30	-40	203	0	-40	278	0	-20
54	-30	-30	129	+10	-10	204	+50	-20	279	+30	-10
55	-70	+10	130	-50	-10	205	0	-70	280	0	+60
56	-30	-70	131	-40	+50	206	-10	0	281	-10	+30
57	0	+40	132	+60	-30	207	+30	-60	282	+30	0
58	+20	-20	133	-20	-40	208	0	-40	283	0	0
59	-70	+10	134	-20	+40	209	+40	-30	284	-10	-40
60	-50	-20	135	-10	-30	210	+40	-60	285	0	+10
61	+20	-30	136	+10	-50	211	+30	-40	286	+60	-20
62	+50	+40	137	+40	+50	212	+70	-50	287	-60	-30
63	+30	+30	138	-30	-10	213	0	-40	288	-40	-30
64	-20	-30	139	-50	+20	214	-10	-10	289	+30	+20
65	+10	-50	140	0	-10	215	0	0	290	+40	-50
66	-40	-40	141	-40	+60	216	0	-10	291	0	-30
67	-10	-60	142	-20	-30	217	-20	-10	292	-20	-30
68	+60	+30	143	-70	0	218	+50	+20	293	0	-30
69	+50	-10	144	-30	-20	219	+60	-20	294	-20	-40
70	-30	+50	145	0	0	220	-30	+70	295	+20	-30
71	+70	+10	146	-40	+60	221	-60	-50	296	-20	-30
72	0	+20	147	0	0	222	-70	-40	297	0	-60
73	+50	-20	148	-30	+10	223	+60	+10	298	+20	-30
74	-20	+10	149	+70	-30	224	+30	-30	299	-50	-60
75	-70	-20	150	+20	0	225	+50	0	300	-50	-20

FIG. 16. Karel Andel's drawing of the Moon with selenographic latitude- and longitude-grid superimposed. Lists of mountains, valleys, and craters, and the craters' coordinates (lat. and long.) are given on the previous and following pages.

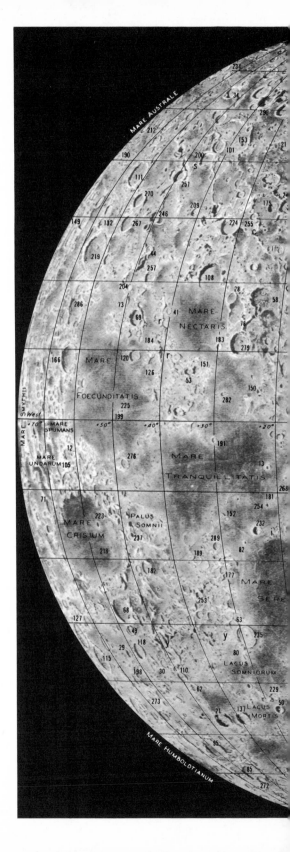

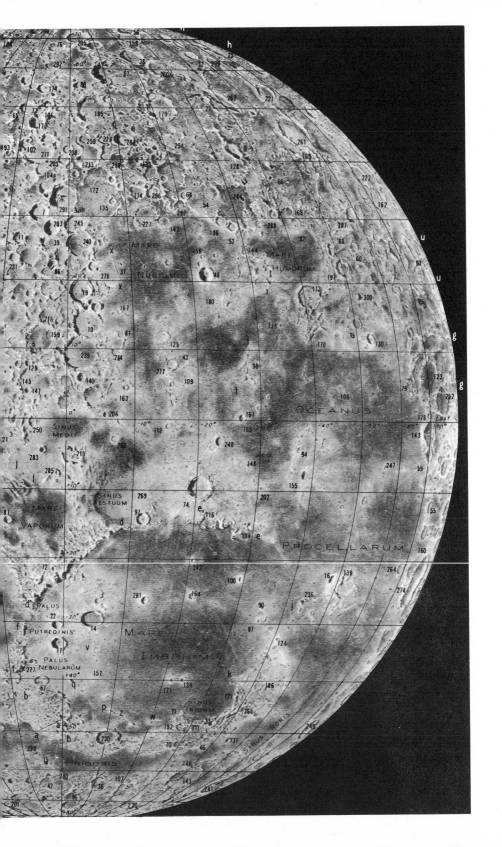

MOUNTAINS AND VALLEYS

a. Alpine Valley
b. Alps Mts.
c. Altai Mts.
d. Apennine Mts.
e. Carpathian Mts.
f. Caucasus Mts.
g. D'Alembert Mts.
h. Doerfel Mts.
i. Haemus Mts.
j. Harbinger Mts.
k. Heraclides Prom.
l. Hyginus Cleft
m. Jura Mts.
n. Laplace Prom.
o. Leibnitz Mts.
p. Pico
q. Piton
r. Pyrenees Mts.
s. Rheita Valley
t. Riphaeus Mts.
u. Rook Mts.
v. Spitzbergen
w. Straight Range
x. Straight Wall
y. Taurus Mts.
z. Teneriffe Mts.

LUNAR CRATERS

1. Abenezra
2. Abulfeda
3. Agatharchides
4. Agrippa
5. Albategnius
6. Alexander
7. Aliacensis
8. Almanon
9. Alpetragius
10. Alphonsus
11. Apianus
12. Apollonius
13. Arago
14. Archimedes
15. Archytas
16. Aristarchus
17. Aristillus
18. Aristoteles
19. Arzachel
20. Asclepi
21. Atlas
22. Autolycus
23. Azophi
24. Baco
25. Bailly
26. Barocius
27. Bayer
28. Beaumont
29. Bernouilli
30. Berzelius
31. Bessel
32. Bettinus
33. Bianchini
34. Biela
35. Billy
36. Birmingham
37. Birt
38. Blancanus
39. Blanchinus
40. Boguslawsky
41. Bohnenberger
42. Bond, W. C.
43. Bonpland
44. Borda
45. Boscovich
46. Bouguer
47. Boussingault
48. Bullialdus
49. Burckhardt
50. Bürg
51. Calippus
52. Campanus
53. Capella
54. Capuanus
55. Cardanus
56. Casatus
57. Cassini
58. Catharina
59. Cavalerius
60. Cavendish
61. Celsius
62. Cepheus
63. Chacornac
64. Cichus
65. Clairaut
66. Clausius
67. Clavius
68. Cleomedes
69. Colombo
70. Condamine
71. Condorcet
72. Conon
73. Cook
74. Copernicus
75. Crüger
76. Curtius
77. Cuvier
78. Cyrillus
79. Damoiseau
80. Daniell
81. Davy
82. Dawes
83. De Gasparis
84. Delambre
85. De la Rue
86. Delaunay
87. Delisle
88. Deluc
89. Descartes
90. Diophantus
91. Dollond
92. Doppelmayer
93. Eichstädt
94. Encke
95. Endymion
96. Epigenes
97. Eratosthenes
98. Euclides
99. Eudoxus
100. Euler
101. Fabricius
102. Faraday
103. Fermat
104. Fernelius
105. Firmicus
106. Flamsteed
107. Fontenelle
108. Fracastorius
109. Fra Mauro
110. Franklin
111. Furnerius
112. Gambart
113. Gassendi
114. Gauricus
115. Gauss
116. Gay-Lussac
117. Geber
118. Geminus
119. Gemma Frisius
120. Goclenius
121. Godin
122. Goodacre
123. Grimaldi
124. Gruithuisen
125. Guericke
126. Gutenberg
127. Hahn
128. Hainzel
129. Halley
130. Hansteen
131. Harpalus
132. Hase
133. Heinsius
134. Helicon
135. Hell
136. Heraclitus
137. Hercules
138. Herigonius
139. Herodotus
140. Herschel
141. Herschel, J.
142. Hesiodus
143. Hevelius
144. Hippalus
145. Hipparchus
146. Horrebow
147. Horrocks
148. Hortensius
149. Humboldt, W.
150. Hypatia
151. Isidorus
152. Jansen
153. Janssen
154. Julius Caesar
155. Kepler
156. Kies
157. Kirch
158. Klaproth
159. Klein
160. Krafft
161. Landsberg C
162. Lagrange
163. Lalande
164. Lambert
165. Landsberg
166. Langrenus
167. Lassell
168. Lee
169. Lehmann
170. Letronne
171. Leverrier
172. Lexell
173. Licetus
174. Lilius
175. Lindenau
176. Linné
177. Littrow
178. Lohrmann
179. Longomontanus
180. Lubiniezky
181. Maclear
182. Macrobius
183. Mädler
184. Magelhaens
185. Maginus
186. Mairan
187. Manilius
188. Marinus
189. Maraldi
190. Marinus
191. Maskelyne
192. Maupertuis
193. Maurolycus
194. Mayer, Tobias
195. Menelaus
196. Mercator
197. Mersenius
198. Messala
199. Messier
200. Metius
201. Meton
202. Milichius
203. Miller
204. Monge
205. Moretus
206. Mösting
207. Mutus
208. Nasireddin
209. Neander
210. Nearchus
211. Nicolai
212. Oken
213. Orontius
214. Palisa
215. Pallas
216. Parrot
217. Parry
218. Peirce
219. Petavius
220. Philolaus
221. Phocylides
222. Piazzi
223. Picard
224. Piccolomini
225. Pickering, W. H.
226. Pictet
227. Pitatus
228. Pitiscus
229. Plana
230. Plato
231. Playfair
232. Plinius
233. Pontanus
234. Pontécoulant
235. Posidonius
236. Prinz
237. Proclus
238. Protagoras
239. Ptolemaeus
240. Purbach
241. Pythagoras
242. Pytheas
243. Rabbi Levi
244. Ramsden
245. Regiomontanus
246. Reichenbach
247. Reiner
248. Reinhold
249. Repsold
250. Rhaeticus
251. Rheita
252. Riccioli
253. Römer
254. Ross
255. Rothmann
256. Sacrobosco
257. Santbech
258. Sasserides
259. Saussure
260. Scheiner
261. Schickard
262. Schiller
263. Schröter
264. Seleucus
265. Sharp
266. Simpelius
267. Snellius
268. Sosigenes
269. Stadius
270. Stevinus
271. Stöfler
272. Strabo
273. Struve
274. Struve, Otto
275. Tacitus
276. Taruntius
277. Theaetetus
278. Thebit
279. Theophilus
280. Timaeus
281. Timocharis
282. Torricelli
283. Triesnecker
284. Tycho
285. Ukert
286. Vendelinus
287. Vieta
288. Vitello
289. Vitruvius
290. Vlacq
291. Walter
292. Weiss
293. Werner
294. Wilhelm I
295. Wilkins
296. Wurzelbauer
297. Zach
298. Zagut
299. Zuchius
300. Zupus

Lunar Results from
Rangers 7 to 9

GERARD P. KUIPER

(Sky and Telescope, May 1965)

On March 24, 1965, Ranger 9 struck the Moon, only 2.8 mi. from the center of the target area 10 mi. NE[1] of the central peak of Alphonsus. In the 18 min. prior to impact, it had transmitted to Earth some 5814 pictures of the lunar surface of a quality that exceeded even the excellent records made by Rangers 7 and 8.

Altogether, the three Ranger missions had contributed over 17,700 photographs. But mere numbers do not tell the story. What matters is the quality of the records, the interest of the objects photographed, the broad coverage of extremely varied terrain, the stereo views provided for contour mapping, and above all, the high resolution of the last several dozen frames of each set.

To an astronomer who has attempted through telescopic observations to understand the over-all structure of the lunar surface and the processes that brought it about, this avalanche of new data means the beginning of a new era in lunar exploration. To him the Moon will never look the same again.

What has been learned about the Moon is indicated by the Ranger-9 photographs (Figs. 19 and 20) reproduced in this supplement, and some comments regarding them. To this may be added data drawn from the Experimenters'[2] report (Ref. 1) on the Ranger-7 mission, and data on recent Hawaiian lava flows and volcanoes. It will be at least two years until our analysis can be completed with any reasonable degree of thoroughness. Approximately half a year was needed to draw up the Experimenters' scientific report. As this work progressed and went through two or three drafts, the views of the members gradually converged, though complete unanimity would not be expected in so broad and so new an area. The views expressed here are therefore necessarily my own.

[1] In this article Kuiper uses the terms *east* and *west* in the astronautical sense, which is opposite to the selenographical usage. In his convention, Alphonsus is east of Guericke.

[2] The Experimenters' Team consists of Ewen A. Whitaker and the author (University of Arizona), Eugene Shoemaker (U.S. Geological Survey), Harold C. Urey (University of California at La Jolla), and Raymond L. Heacock (Jet Propulsion Laboratory).

My five main conclusions are:

I. The moon has no cover of *cosmic dust* even 1 mm thick, which would have obliterated the various sharply defined color provinces and other photometric detail.

II. The maria are *lava flows* or a succession of them. Seven flows have been mapped in a limited region of Mare Imbrium. Each has a characteristic color; different ones may have different colors; the oldest (lowest) tend to be the reddest. The Imbrium flows are 20-200 m. thick; one is 200 km long. They have terminal walls, as do terrestrial basaltic flows.

III. As is well known, the mare floors show both positive features (mountains, hills, ridges) and negative features (craters of various types, depressions, rilles). The *ridges* are found to be strips of uplifted mare floor, across which both albedo and color are continuous, caused by dikes that formed along structural planes. Major ridge *mountains* may also form, by extensive extrusions of lava or cinder-cone formation. These mountains have craters along their crests; they are brilliant white compared to the dark mare floor and therefore may be covered with a white substance such as the remarkable snowlike central peak of Alphonsus.

The negative features are, first, the *primary impact craters*, whose number on the maria increases roughly fourfold each time the crater diameter is halved. That is, if the craters are arranged in groups of 1-2, 2-4, 4-8, 8-16 m., . . . 1000-2000 m., and so forth, each group covers roughly the *same* area of the mare, and their total area contribution is about 1 percent. Next are the *secondary impact craters*, which swarm around the large primary impact craters, in size starting at about 1/10–1/20 the diameter of the primary, and increasing in number with decreasing diameter, somewhat faster than is the case for primary craters. Secondary craters drop off in numbers very rapidly beyond about three crater diameters from the primary, but some isolated secondaries may occur far out.

Besides primary and secondary impact craters, a large class of new objects was found on the maria (and on the floors of Alphonsus and Ptolemaeus): *collapse features*. These shallow depressions, as well as the abundant *dimple craters*, are not old, eroded craters; they are due to collapse; that is, to internal causes.

We appear to observe here a karst-type landscape, which dimple-crater regions closely resemble in appearance and scale (Ref. 2). Some of the 500,000 karst-type depressions in the United States are seen in Figure 18.

There are several processes that could have led to karst-type depressions on the lunar maria: *a.* Hot lavas have appreciable vapor pressure, causing extensive frothing when the lava is exposed to a vacuum, with

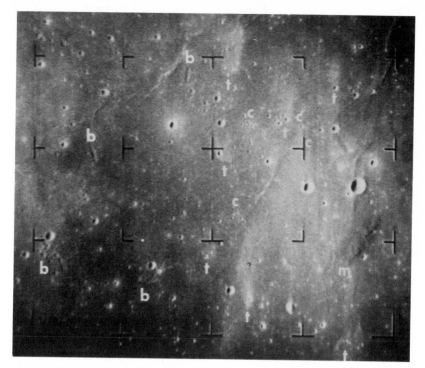

FIG. 17. Part of Mare Cognitum from Ranger-7 photo A183. The added letters *c* and *t* identify clusters of small secondary craters and sections of rays from Copernicus Crater (far off the top, north) and Tycho Crater (far off bottom). Ejecta from Bullialdus Crater are marked *b*; *m* is a dark mountain range, younger than the mare; *v* marks two volcanoes at the toe of a white mountain ridge. (JPL-NASA photo.)

a 10- to 100-fold increase in bulk volume. In laboratory experiments, many cavities so produced collapse after cooling. Cooling of subsurface magmas on the Moon must have caused widespread collapse. *b.* In terrestrial lava fields, collapse often follows relocation or drainage of magmas. *c.* The volume of the solid is less than the liquid; of cool rock less than rock near the melting point. *d.* Escape of gases. These four mechanisms are equivalent to the removal of limestone by solution in terrestrial karst formations. Cave-ins of the ceilings of cavities will lead to *dimple craters*. On the basis of this identification, the sinkholes and dimple craters are early post-mare in age.

IV. *Crater rays* (bright streaks extending outward from some craters) can be studied well on the Ranger-7 records. Figure 17 shows several short Tycho and Copernicus rays, designated by *t* and *c* respectively, the the first running up (north) from the letters, the second running down (south). Their locations in Mare Cognitum are readily found on Earth-based photographs of the entire mare. Each ray element fits into a con-

FIG. 18. This snow-covered landscape in southern Indiana, with its shallow depressions and undulations, is a good example of "karst topography," relatively barren land underlain by limestone, in which caves and sinkholes and underground river channels have been dissolved, and into which the surface drainage sinks. (Photo courtesy of Martin S. Burkhead, Indiana University.)

tinuing pattern, made somewhat difficult to trace because the two ray systems intersect and interlace in Mare Cognitum. Each ray element starts with a small cluster of whitish secondary craters, with the ray element pointing away from the main crater, Tycho or Copernicus.

In the Ranger-7 report it is shown by Ewen A. Whitaker and the author that these findings may be understood if the bright-ray craters were caused by the impact of *comets* (p. 137); that outlying fragments of the comet are probably responsible for the bright secondaries in the crater rays; and that the gases generated by the comet impact caused a blast that blew the debris from the secondary explosions downwind from the central crater.

V. The *fine structure of the mare floor* is revealed in the last four Ranger-7 P-camera photographs. The sculptor Ralph Turner has produced, for the region best covered, a precise model at scale 1:500. . . .

The near absence of *rocks* on the lunar surface has caused much comment, and seemed at first in remarkable contrast to the frequency of impact craters. It is readily seen, however, that this is to be expected from the limited strength of the lunar-surface rock. Laboratory tests suggest that the bearing strength of lunar-surface rock froth is about 1-10 kg/cm^2, or 1-10 $tons/ft^2$. Such material will bury a denser rock of 1 $m.^3$ or larger if tossed onto the lunar surface from a distance greater than 100-200 m. Deeper layers will presumably be more resistant, since they are likely to have been formed from very fluid magmas of the type that produced the extensive nearly horizontal flows.

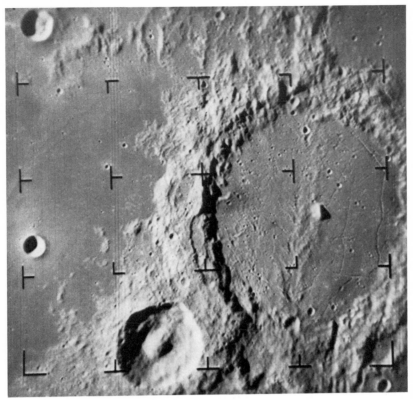

FIG. 19. The crater Alphonsus fills the right half of this picture taken 170 sec before impact, when Ranger 9 was 258 mi. above the lunar surface. The area is about 120 mi. by 110 mi., with north up. Part of Mare Nubium at left has much lower crater density than the floor of Alphonsus, which has also a dendritic ("fishbone") spine running north just left of the central peak, and several dark halos near the rim (at "2-o'clock, 4:30, and 9-o'clock" positions). (JPL-NASA photo.)

Therefore, rocks large enough to have been visible on the lunar surface, if originating anywhere except in the immediate vicinity (within 100 m.) will have caused small craters, and will not now be found resting on the surface.

The target area for Ranger 9 selected for the first day of the launch window was the floor of the crater Alphonsus . . . on the line connecting the central peak with the dark-halo craters inside the northeastern rim (right in Fig. 19), about one third of the distance from the peak. (The distance selected was one third, not one quarter, to minimize the chance of the spacecraft impacting within the shadow of the peak and causing the loss of the last and presumably most valuable frames.)

There were, of course, several additional scientific objectives: the structure of the crater floor, the rilles, the central dendritic spine on the floor, the Imbrium system of grabens (downfaulted strips) and ridges in the crater walls, and the relationship to its neighboring craters. The crater Arzachel is apparently older than Alphonsus. It has a low-level, complex floor and an enormous central peak. While the relationship of floor level to peak elevation is plausible enough (Ref. 3), it is not at all clear why the five large craters near the center of the Moon's disk (Hipparchus and Albategnius included) have such different floor levels. This major problem, plus that of the several lava lakes scattered among the "big five," at a variety of levels, calls for precision contour mapping, which Ranger 9 has now made possible.

The first of the A frames of Ranger 9 is reproduced as Figure 19. It shows the dendritic central spine through Alphonsus to be composed of ridges that apparently have extruded from a system of cracks that is still partly preserved as rilles. On these rilles are located eight prominent and some small dark-halo craters which show very dark at full moon. The dark halos surrounding these craters might have been caused either by lavas or by cinders and ash. The Ranger-9 photographs have shown that the second alternative applies.

Igneous activity is by no means limited to these dark craters and the central spine. Many linear ridges, some quite dark, especially on the northern crater wall and running nearly radially onto the crater floor, are clearly of internal origin also. Furthermore, the great majority of the negative features on the Alphonsus floor are found to be due to collapse. . . . The direction of the spine being toward Mare Imbrium, it is supposed that a major fracture system caused by the Imbrium impact is responsible for the location of this volcanism.

The most remarkable feature of Figure 19 is undoubtedly the almost featureless central peak. If the record is copied with more contrast, some small bright spots become visible and a slightly mottled pattern appears over the entire mound.

A closer view of the northeastern sector of Alphonsus' wall and of four dark-halo craters not illustrated here shows that the texture of the mountains is not unlike that of the central peak, though there is more structure. The soft contours of the hills are remarkable; they do not look like broken rock masses tossed upward and sideways by a gigantic impact. It is presumed that the walls of this pre-mare crater originally consisted of displaced unconsolidated crustal material that during the period of maximum lunar heating was metamorphosed by steam and other vapors rising from the deeper layers. The fact that a prominent lava lake exists at a higher level within the wall indicates that subsurface mobility of the magmas was quite limited—a conclusion that is compatible with the different lava levels (p. 82). A dark-halo crater placed squarely on the rille has clearly filled part of the rille and has deposited its dark material on the crater floor.

A further remarkable observation on Figure 19 is that the rilles have rounded edges. This would indicate erosion that probably occurred in the earliest history after the crater floor solidified; or else, it could mean incomplete melting of surface materials, which subsequently slumped into subsurface fissures.

Figure 20 shows the last three frames with the impact point marked. It is expected that Ranger 9 produced a crater approximately one-half the diameter of the largest one shown on the third frame.

The last picture was only about 13-percent transmitted back to Earth when the spacecraft was destroyed by impacting the Moon. The partial frame obtained is of extraordinary interest, however, because of its high resolution, about 25 cm. It shows nine small rock masses on the surface, two of which have been identified in Figure 20.

All these originated in the largest crater, which is 150 ft in diameter. The shadow lengths show that the rocks are partly buried in the Moon. If the density of these rock fragments is assumed, the bearing strength of the floor of Alphonsus can be derived. For a bulk density of two, the average strength of the floor is about 1 ton/sq ft. This is probably adequate for spacecraft soft landings.

Much hard work lies ahead before the treasure of pictorial records produced by the Ranger program may be regarded as having been

  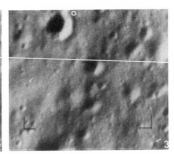

FIG. 20. The last three photos transmitted by Ranger 9 before impact, each marked by a white circle where the impact took place beside a 25-ft crater. The last (right) is from ¾ mi. above the surface, 0.453 sec before impact, covers about 155 ft x 125 ft, and shows craterlets as small as 2.5 ft. White arrows mark two rocks casting shadows. (JPL-NASA photos.)

reasonably utilized. The examinations so far made leave no doubt that Ranger has opened up a new era in lunar science.

REFERENCES

1. "Experimenters' Analysis and Interpretation," JPL *Technical Report* No. 32-700, Ranger VII, Vol. II, Feb. 10, 1965.
2. W. D. Thornbury, *Principles of Geomorphology*, pages 321 and following (John Wiley, New York, 1954).
3. G. P. Kuiper, *Proceedings* of the National Academy of Sciences, 40, 1108, 1954.

Zond-3 Photographs of the Moon's Far Side

YURI N. LIPSKY

(*Sky and Telescope*, December 1965)
Translated from the Russian by Martin P. Lopez-Morillas

Investigation of Earth's only natural satellite remains an important aspect of planetology. A powerful new tool became available in the fall of 1959, when Luna 3 became the first space probe to photograph another heavenly body from nearby and to televise the pictures back to Earth. (See Vol. 1, *Wanderers in the Sky*.) At that time more than 10 million km² of the other side of the Moon were photographed.

However, about 20 percent of the lunar surface—the eastern sector of the normally invisible side—remained unknown. It was the task of the Soviet probe Zond 3 to fill this gap.

Zond 3 was placed in a parking orbit around the Earth on July 18, 1965, and went on to reach the Moon's vicinity in 33 hrs. Photography began on July 20, when the craft's distance from the Moon was 11,570 km.

Because a smaller part of the far side was to be photographed, it was possible to select more favorable lighting conditions. While the photographs were being taken, the sun stood directly over the north rim of the crater Riccioli, surface features in the unknown segment were obliquely illuminated, and the long shadows revealed the topography extremely well.

The probe approached the Moon from the western edge as seen from Earth, and moved toward the morning terminator, which at that time was near longitude −166°. The terminator virtually coincided with the boundary of the region previously photographed by Lunik 3.

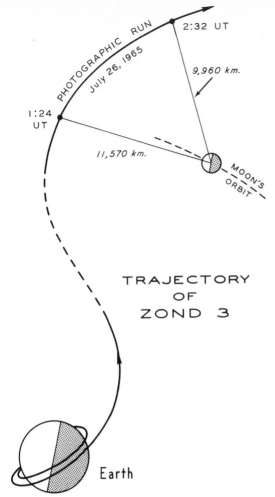

2:32 UT

PHOTOGRAPHIC RUN
July 26, 1965

9,960 km.

1:24
UT

11,570 km.

MOON'S
ORBIT

FIG. 21. The path of Zond 3 (not to scale) from near-Earth "parking orbit" to fly-by of the Moon on July 20, 1965. (Sternberg Astronomical Institute, Moscow, diagram.)

TRAJECTORY
OF
ZOND 3

Earth

Hence the new pictures cover practically all of the part previously un-recorded. They also show a substantial part of the visible hemisphere, making it possible to locate reliably the newly discovered formations on the standard selenographic coordinate system.

The small-size telephoto unit aboard Zond 3 included an $f/8$ camera with a focal length of 106.4 mm. The exposures on 25-mm film were 1/100 and 1/300 sec, at intervals of roughly 2¼ min. About 25 shots were obtained in a 68-min. camera run.

During the picture taking, the distance of Zond 3 from the Moon decreased to 9220 km, but increased to 9960 by the end of the series. The direction of the probe from the center of the Moon changed by an angle of 60° during the picture taking. (See Fig. 21.)

The film was photographically processed immediately, and then fed directly to the transmission system. This began to function on July 29 when Zond 3 was more than 2,200,000 km from Earth.

Scans of the photographs were transmitted earthward via a highly directional parabolic antenna. The transmission system could operate in various modes. In particular, all frames were first sent with low-resolution rapid scan, at a rate of one every 2¼ min. This made it possible for us on Earth to select the most important and interesting frames.

Subsequent transmission required 34 min. per frame, with a resolution of 1100 lines and wide photographic latitude. These pictures clearly show transitions of tone, permitting recognition of detail. To make possible accurate photometric measurement of lunar formations, a calibration scale was impressed on each picture.

This year's pictures fully confirm the earlier conclusion that there is a major dissimilarity between the visible and invisible hemispheres: the far side has fewer maria and is in general lighter and more mountainous. Whereas the northern part of the Moon's earthward side abounds in maria, the northern part of the far side consists of one gigantic conti-

FIG. 22. Heavily cratered surface of the previously unmapped region to the left of Mare Orientale on the Moon's far side, photographed by Zond 3 at 2:25 UT, July 20, 1965. The bright disk at lower left is for photometric calibration. The black line and other streaks are defects. (Sternberg Astronomical Institute, Moscow, photo.)

nent, substantially larger than the antipodal southern continent on the near side.

From the 1959 photographs it was concluded that the far side of the Moon is more densely spread with craters than is the visible face. This concentration has been fully confirmed. . . . The smallest craters distinguishable in the Zond-3 pictures are 3 km across.

Other interesting features on the Moon's far side are the numerous ring-shaped concavities which we call *thalassoids*. These are large depressions with diameters up to more than 500 km. In size and shape they are comparable to maria, but differ in having conspicuously crater-scattered floors and in lacking the characteristic dark color of maria.

What Luna 9 Told Us
about the Moon

YURI N. LIPSKY

(*Sky and Telescope*, November 1966)
Translated from the Russian by Anatole Boyko

The successful soft landing on the Moon by the Soviet Union's Luna-9 spacecraft on Feb. 3, 1966, enabled earthmen for the first time to see millimeter-sized details of the lunar surface. Previously, Ranger space probes had transmitted pictures showing formations a hundred to a thousand times smaller than terrestrial observatories could photograph. In turn, the Luna-9 photographs pushed the limit to features a hundred to a thousand times smaller still.

The flight of Luna 9 consisted of four steps: orbit around the Earth (perigee height 173 km, apogee 224 km, inclination 57°), acceleration toward the Moon (total weight entering lunar trajectory was 1583 kg), lunar orbit, and soft landing in Oceanus Procellarum. This soft landing was the first in the history of cosmonautics.

About an hour before landing, at an altitude of about 8000 km, the spacecraft and its engines were reoriented and the booster jettisoned. The braking engines were started on command of the radio altimeter, at an altitude of about 70 km; in 48 sec they reduced the station's velocity from 2600 m/sec to a few meters per second. It was of prime importance to jettison the braking mechanism, so that the lander could put down on virgin ground. This was accomplished, and the retrosystem landed nearby.

Luna 9 came to rest near the western shore of Oceanus Procellarum, at longitude 64°22′ W, latitude 7°08′ N. The landing area is in the im-

FIG. 23. A Russian artist's preflight concept of Luna 9 opened up for operation on the lunar surface. The panoramic camera on top is about 2 ft above the surface. (Novosti Press Agency drawing.)

mediate neighborhood of the broken ridge that ends southeast of the crater Cavalerius F. Thus, the photographs taken by Luna 9 do not refer to either a pure mare or pure mountain surface.

As it sat on the Moon, Luna 9 transmitted three panoramic pictures of its landing area, with the sun's altitude about $7°$, $14°$, and $27°$.

Just before the third transmission at 16:00 UT on February 4, Luna 9 shifted its orientation. The cause of this displacement is difficult to determine. It involved a small change of the station's inclination (about $6°$) and a several-degree turn in azimuth. Perhaps Luna 9 had some mechanical effect upon the soil. If its position after landing was not very stable, even a small force could make it slip.

For stereo viewing of lunar microrelief, Luna 9 had three precisely spaced mirrors in the field of view of its optical scanning system. These mirrors produced stereoscopic images of six strips of lunar surface. In addition, the above-mentioned displacement of Luna 9 provided a sufficient base line for a further stereoscopic study of the panoramic photographs.

The center of the television camera lens was about 60 cm above the station base. The vertical extent of the camera field was about 29°. Since the lunar surface was the primary subject of interest, the cameras were tilted slightly downward, so that the field extended 11° above and 18° below the plane perpendicular to the axis of camera rotation. The resolution (see Glossary) was about 0°.06 per picture element. At a distance of 1.5-2 m, features measuring 1.5-2 mm were resolvable. Complete transmission of each panorama took about 100 min.

The pictures from Luna 9 revealed for the first time the small-scale structural peculiarities of the lunar surface. While they contain images of some craters a few meters across, resembling those photographed by Rangers, the details of greatest interest range from a few millimeters to tens of centimeters in size.

It was previously known that the brightness of lunar features increases to a sharp maximum near full moon, and this had been interpreted as small-scale shadowing well simulated by a structure of randomly oriented fibers, separated by several times their thickness, and also by "fairy-castle" structures—fluffy configurations of very low density.

It had been conjectured that very fine dust from meteoritic impacts on the Moon could produce these structures—in the lunar vacuum the particles could stick together at their points of contact. In the transmitted pictures, there are no fine-fibered structures, no fairy castles, no heaps of angular fragments, sand, or loose dust or like forms, which many Earth-based observers had supposed to form the Moon's microrelief. Moreover, the photographs show that almost all the objects have regular forms, not branched or random ones. Many depressions of various sizes are clearly visible, and smaller craterlets overlap larger ones. There are considerable numbers of small furrows, some of them intersecting, and many rocklike objects.

These rocks (as we shall call them) could not have formed from the fusion of small particles in the lunar vacuum; instead, their forms and sharp outlines indicate monolithic structure. They could not be meteorites since any impact at 15-20 km/sec would cause an explosion and leave a depression. The rocks rest on the ground almost without depressing it, indicating a comparatively low velocity of fall. Hence these rocks could only have been ejected from nearby craters produced by meteoritic impact or volcanic eruption.

G. P. Kuiper (p. 80) assumed that the bearing strength of lunar surface soil is about 1-10 kg/cm², on the basis of laboratory studies, and estimated that a denser rock of 1 m³ or larger would be buried, if tossed onto the lunar surface from a distance greater than 100-200 m. The

FIG. 24. First picture transmitted by Luna 9 from Oceanus Procellarum on Feb. 4, 1966, with the sun 7° above the lunar horizon. Note that the camera was tilted—the left side of the picture is high. The rock in the foreground is about 6 in. across, and 6 ft from the camera. (Novosti Press Agency photo.)

rock in the foreground of Figure 24 is about 20 cm long. It has scarcely sunk into the soil at all. We assume that the sinking amounts to 2 cm, the angle of ejection of this rock as 45° above the horizontal, the distance it has been thrown as 200 m., and the rock's weight in lunar gravity about 0.2 kg. Then the impact force was 1 ton. From the dimensions of the rock, we find that the average pressure at impact was 2-3 kg/cm². Thus, the soil readily stands a pressure of a few kilograms per square centimeter, and the rock has borne this pressure without cracking.

This agrees well with Kuiper's estimate of 1-10 kg/cm² for the strength of the lunar soil. In reality, it may be 5-10 times greater. Indeed, ejecta from craters scatter over many tens of kilometers. It is not unlikely that the rocks seen in the Luna-9 pictures traveled several kilometers. Probably the flight trajectories were rather steep, for there are no indications of skid marks. (Steep trajectories favor the idea that the rocks are fragments of dense subsurface strata, which were not exposed to lunar erosion before ejection—explaining why the surfaces of many rocks differ in microstructure from the soil on which they rest.)

Other important evidence of the strength of the lunar soil is provided by the circumstances of the Luna-9 soft landing. Before its lobes unfolded, the station was spherical, and its weight under lunar gravity was about 16 kg (96 kg on Earth). The impact load was borne by a small area of ground, yet so little sinking occurred that hours later even a weak mechanical effect could displace the station.

Another important inference from the Luna-9 photographs is the absence of a fine dust cover at the landing site. Luna 9 refutes the dust hypothesis once and for all, since, in the area covered by its panoramic photographs, no dust can be detected. Moreover, the following argument, advanced by Kuiper, is very convincing to us. Luna 9 ought to have attracted surface dust electrostatically, had any been available. Yet the parts of the station visible in the photographs have no noticeable dust cover. The television head and optical components were absolutely free of dust, both immediately after landing and at the end of transmission.

News has arrived of Surveyor 1's soft landing in Oceanus Procellarum, and of Lunar Orbiter's photographs of the Moon from close range. The author takes this opportunity to congratulate the participants in these magnificent experiments. Study of the information transmitted by Luna 9, Surveyor 1, and Lunar Orbiter will extend scientific knowledge of our eternal satellite, which nowadays already seems within hand's reach.

The lunar close-up photos were exciting, and Luna 9 put to rest the fear that lunar landers might sink into 10-20 ft of powdery dust. Geologists were fascinated with the nature of rock and soil never exposed to water and wind erosion, as on Earth, and astronomers were interested in the crater records of many past impacts on the lunar surface. NASA produced many more photos of high quality, and details of a bouncy landing.—TLP

Results from Lunar Orbiter 2

RAYMOND N. WATTS, JR., AND JOSEPH ASHBROOK

(*Sky and Telescope,* January 1967)

Five space vehicles have been put into orbits around the Moon during the past few months: the Soviet Union's Lunas 10, 11, and 12, and the United States' Lunar Orbiters 1 and 2. The Moon's surface is being given detailed scrutiny, in final preparation for the first attempts to place men there.

The primary purpose of the Lunar Orbiters is photographic reconnaissance of possible landing sites for Apollo astronauts. These are on the near side of the Moon, within a strip centered on the lunar

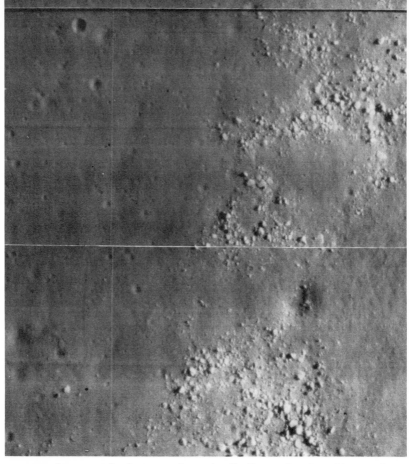

FIG. 25. An area 780 ft x 950 ft at Apollo Prime Landing Site 2, photographed with Orbiter-2's telephoto lens on Nov. 19, 1966, and transmitted to Goldstone receiver 8 hrs later. The boulders make this site unsuitable for LM landing. (NASA photo.)

equator and extending 45° on each side of the center of the visible disk. A relatively smooth and level tract is needed for an Apollo Landing Module to touch down safely and later to return its passengers to their mother ship.

The primary mission of Orbiter 2 was the photography of 13 potential landing sites.

Figure 25 shows a very small part of one of these, Site 2 in Mare Tranquillitatis, unsuitable for LM landing because of boulders.

Prior to this second picture-taking mission, Orbiter 1 was crashed on Oct. 29, 1966, onto the far side of the Moon near 162° E. Before launching Orbiter 2, scientists at the Jet Propulsion Laboratory commanded Lunar Orbiter 1 to fire a small rocket so that the recoiling Orbiter would strike the Moon's surface, terminating its radio signals, which would otherwise interfere with those from the second satellite.

Orbiter 2 was lifted from its Cape Kennedy launch pad by an Atlas-Agena rocket on Nov. 6, 1966. The Agena placed the 850-lb mooncraft into a 115-mi-high parking orbit and fired again, 11 min. later, to send the vehicle toward the Moon.

On the 10th, the on-board retrorockets fired to put the craft into a circumlunar orbit ranging between 1150 mi. and 130 mi. above the Moon's surface. During the next 5 days, Orbiter 2 did not take pictures, but served as a passive probe of the lunar gravitational field, just as Orbiter 1 had done during its initial revolutions.

With its egg-shaped pressure shell, the photographic equipment weighs 150 lbs. Its two cameras view the Moon through a protective window of quartz, which is covered, except during actual picture taking, by a thermally insulating flap.

The medium-resolution lens is an 80-mm Xenotar working at $f/5.6$, with shutter speeds of 1/25, 1/50, and 1/100 sec. For high-resolution photography, an $f/5.6$ Paxoramic lens of 24-in. focus, weighing less than 16 lbs, is employed. The 200-ft roll of 70-mm film is enough for 211 pictures with each lens.

The photographic schedule was arranged to yield the maximum topographic information concerning the 13 primary sites. Overlapping pictures from the medium-resolution camera will be used as stereo pairs. The high-resolution coverage is not stereoscopic, and hence the pictures will be analyzed individually by photometric methods, yielding detailed relief maps.

For the primary sites, 184 frames were budgeted. The remaining 27 frames were used on secondary sites, some on the far side of the Moon.

Picture taking was carried on from November 18 to 25, and sample frames were transmitted intermittently to check on the cameras' performance.

After photography had ended, a complete readout of the film began that was to last until December 13. The process was lengthy, since pictures could be sent only when the spacecraft was in sunlight, with its high-gain antenna pointed earthward. Sending each full frame, consisting of one medium-resolution and one high-resolution picture, required about 43 min.

The transmissions ended unexpectedly on December 6, when the spacecraft failed to respond to signals from Earth. However, more than 97 percent of the exposed frames had already been relayed.

The excellent performance by Orbiter 2 augurs well for the remaining three spacecraft of this series that NASA intends to place into circumlunar orbits. Quite apart from the primary goal of picking a safe landing point on the Moon, these missions should yield astronomers rich cartographic and selenological material. Improved measurements of the Moon's gravitational field and shape should tell us much about the internal structure of our natural satellite, perhaps enabling us to make a decision between rival theories of its evolution.

Surveyor 3 on the Moon

RAYMOND N. WATTS, JR.

(Sky and Telescope, June 1967)

With a jolt, Surveyor 3 hit the Moon on April 19, 1967. After a couple of hops, the three-legged spacecraft settled on Oceanus Procellarum in a crater some yards across. Having successfully completed its 65-hr flight, Surveyor 3 made ready to start a series of experiments that included photographing its surroundings and digging into the Moon's soil. Weighing only 620 lbs, Surveyor 3 carried a pointable TV camera and flat mirrors that enable its camera to "see" underneath the spacecraft. It also has a small scoop on a movable mechanical arm (Fig. 26).

FIG. 26. The Surveyor-3 surface sampler, a steel-tipped digger on an arm extendable to 5 ft from the spacecraft. It is being examined by R. F. Scott and F. Roberson of the Surveyor Surface Sampling Experiment Team. (JPL-NASA photo.)

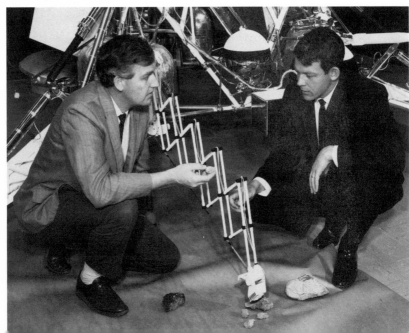

FIG. 27. Surveyor-3 TV pictures of its sampler carrying 2 in³ of lunar soil (left), later dumped on the lander's footpad for examination (right). Note the footpad imprint top left, showing that Surveyor 3 bounced to its final position. (JPL-NASA photos.)

This steel-tipped digger is called a surface sampler. Driven by four small motors, it can reach out over an area of about 20 sq ft. By monitoring the current flow to the motors, technicians on Earth can determine the weight of a lifted object, or calculate how hard the scoop is pressing down against the soil. On April 21, the scoop controls were tested and the arm was flexed.

As soon as it was certain that the arm was functioning properly, the scoop was pressed into the ground and its imprint photographed. Later, the scoop, which is about 5 in. long and 2 in. wide, was plunged into the lunar soil and several short trenches were dug. The scientists wanted to see whether the edges would cave in and also to look at material that the scoop deposited on one of Surveyor's footpads, as shown in Figure 27.

FIG. 28. The black box is Surveyor-7's radiation sampler, hanging above the lunar surface (left photo) by a jammed nylon cord. The digger arm is moved over (center photo) by JPL controllers, and pushes the box to the surface (right photo), where it was able to analyze the soil. (JPL-NASA photos.)

After studying the first results from the digging experiment, Jet Propulsion Laboratory scientists suggested that the lunar soil at the landing site behaves somewhat like coarse, damp beach sand. It is firm enough to support the weight of a structure far larger than Surveyor 3, yet can be dug easily.

Initial tracking data indicated that Surveyor landed within 2½ mi. of its intended target, which was at 23° 10' W, 3° 20' S.

Data from strain gauges mounted on the spacecraft's legs have indicated that it jumped high off the ground twice after its initial contact. JPL engineers guessed that the vernier rockets used to control Surveyor's final descent did not shut off when they should have and therefore "launched" the craft from the Moon's surface.

Surveyor 7, last of the series, landed without a jump in the hilly region north of Tycho on Jan. 9, 1968. It carried a small (6 in. × 6 in.) box containing a "radiation sampler" to be lowered onto the lunar surface where it could detect major constituents of the soil. The nylon cord carrying the box stuck, and JPL controllers successfully used the scoop to push the box to the ground, as shown in Figure 28. From data telemetered to JPL, Anthony Turkevich (University of Chicago) found that highland soil differs from the mare material, analyzed in the same way on Surveyor 5, by having smaller proportions of iron, nickel, and cobalt.

An important result of the Surveyor landings was Turkevich's measures of the chemical elements in the lunar soil, which showed marked differences from the composition of the Earth and solar system as a whole. The analyzer box contained a radioactive Curium source, which irradiated the soil with alpha particles that were scattered by elements in the soil. By counting the alphas coming off at various scattering angles, Turkevich was able to show that lunar material is richer in aluminum, calcium, and titanium, and poorer in sodium, magnesium, and iron than the Earth is (see p. 148).

Another significant discovery made at this time concerned the structure of the Moon.—TLP

Lumps inside the Moon

(*Sky and Telescope*, October 1968)

Mass concentrations of dense material lie beneath the surface of the Moon, report Paul M. Muller and William L. Sjogren of the Jet Propulsion Laboratory. Their evidence for the existence of these "mascons" comes from small changes noted in the motion of Lunar Orbiter 5, when very precise radio Doppler tracking (p. 98) was continued for 80 consecutive circuits of the Moon.

At the time, the spacecraft was traveling in a nearly polar orbit around the Moon once each 3 hrs 11 min., closest approach to the surface being about 100 km in latitude 2° N. The tracking data during the 1½ hrs of lowest altitude of each circuit were processed by computer, in order to detect any abnormal acceleration of the spacecraft. In this way, a gravity-anomaly map of virtually all the nearer hemisphere could be constructed.

Five conspicuous mascons, each about 50-200 km in extent and perhaps 50 km below the surface, were detected under the five large circular "seas" on the near side—Imbrium, Serenitatis, Crisium, Nectaris, and Humorum.

Mascons are absent from the lunar "continents" (bright highlands) and from the irregularly shaped "seas," such as Oceanus Procellarum, Mare Tranquillitatis, and Mare Fecunditatis.

The presence of these features under every ringed mare, except Sinus Iridum, and their relative absence elsewhere, suggest a physical relationship. JPL investigators suggest that each mascon may be the remains of an asteroid-size body which caused its associated mare by impact.

The mascons were unexpected, and provoked many scientific discussions about the rigidity of the lunar crust necessary to hold them up, and about the way they were formed. The "Doppler radar" used in their discovery deserves some mention; it is one of the significant contributions by radio engineers to the space program.

The Doppler effect, described more fully in Volume 7, Stars and Clouds of the Milky Way, is best illustrated by the change in pitch of an automobile horn or a fire-engine siren as it passes you—from higher pitch as it approaches to a lower pitch as it recedes. Christian Doppler explained this in 1843 by noting that a sound source or a light source gives out a definite number of pulses or waves per second. As it approaches you, you receive (hear) more waves per second (higher frequency or pitch) because the source is a bit closer for each succeeding wave— the later waves reach you sooner than they would if the source remained at a fixed distance. If the waves travel at speed c and the source is approaching you at speed v, the period P (interval between waves) is decreased by Pv/c. The frequency f (waves per second) is just 1/P, so the increase in frequency $\triangle$F is fv/c.

This Doppler formula, $\triangle$f/f = v/c, holds for light waves and radio waves, where f and c are very large numbers (millions of waves per second, and 186,000 mi/sec = 30 billion cm/sec) for S-band radio waves. The Lunar Orbiters, and later spacecraft in orbit around the Moon, were equipped with radio "transponders" that replied to radio signals from Earth by broadcasting back at a precisely known frequency f. The ground receivers can measure the received frequency f + $\triangle$f with high accuracy. Sjogren and Muller thus knew f, $\triangle$f, and c, from which they could calculate v, the orbiter's velocity toward or away from the Earth, with an accuracy of a few millimeters per second. After correcting for the known rotation of the Earth and the known motion of the Moon relative to the Earth, they could plot an orbiter's motion around the front side of the Moon with very high accuracy. It was here that they noticed small speed-ups and slow-downs as the orbiters passed large circular maria, and reasoned that these extra accelerations were caused by mass concentrations in the maria. Later Doppler measures of lunar orbiters produced accurate gravity maps showing further mascons.

The Apollo Program progressed according to schedule, with three astronauts on each mission. For instance, in December 1968, Apollo 8 circled the Moon for 20 hrs with astronauts Frank Borman, James Lovell, and William Anders. From 70 mi. above the surface they obtained hundreds of photographs, some of the Moon's far side, but the main purpose of the flight was to check equipment and procedure. The success of these flights evidently surprised leaders in the Soviet space organization. They had planned a manned lunar flight, but decided that it was likely to fail, cancelled it, and concluded that NASA's Apollo program would fail also.

After five preparatory flights, testing every conceivable misstep in the 8-day mission, Apollo 11 carried three men, two of whom would make the "great leap for mankind." The background preparation and "feel" of the mission is beautifully expressed by Michael Collins in his book Carrying the Fire (Farrar, Strauss, Giroux, N.Y., 1974).—TLP

The First Men
on the Moon

ROBERT HILLENBRAND

(*Sky and Telescope*, September 1969)

For the first time in history, men have set foot upon another world. Whether Apollo 11 was a triumph of the human spirit, science, engineering, or management can be debated, but July 20, 1969, is a date to be remembered.

The safe return of Thomas Stafford, John Young, and Eugene Cernan from their successful Apollo-10 mission around the Moon gave the green light for Apollo 11. The flight in May was a dress rehearsal for the one in July in every respect except that no landing was attempted. The complete Apollo system of spacecraft equipment had been flown three times without a single failure of any major component.

Only an eyewitness can appreciate how enormous the Moon rocket is. Including the Apollo perched on top, the three-stage Saturn 5 towered 363 ft above the launch pad. Nearby loomed the huge "VAB," the 525-ft-high Vehicle Assembly Building covering 8 acres, inside which the Saturn-Apollo configuration had been assembled.

Three hours before blast-off, Neil Armstrong, Edwin Aldrin, Jr., and Michael Collins (in their space suits) rode in a van from the Operations building to board their ship—the tiny Command Module (CM) near the top of Figure 29. Launch came at 9:32 a.m. EDT. At —9 sec the automatic sequencer started the Saturn's five giant F-1 engines, and fuel rushed into the chambers at the rate of 15 tons/sec. The access arm of the control tower swung back, the tie-down bolts exploded, and the great rocket gradually began to rise.

Aboard the spacecraft, all went well as it traveled for about 2½ hrs in a parking orbit some 120 mi. high. Then the Saturn third-stage engine was restarted over the mid-Pacific to insert the craft in its moonward trajectory.

Apollo entered its circumlunar orbit on July 19 at 1:22 p.m. EDT when behind the Moon. The engine of the service propulsion system was fired to slow the craft by about 2900 ft/sec. Two revolutions later, a second, shorter burn rendered the orbit almost circular and about 69 mi. high.

FIG. 29. Apollo-11 CSM (lower center), 350 ft above the KSC launch pad, with the access arm from the mobile launcher on the left, and the mobile service structure behind it. The white, 23-ft launch-escape rocket on top of Apollo 11 would be fired to save the astronaut crew if there were an accident or dangerous development during launch. (It is later jettisoned.) (NASA photo.)

FIG. 30. The Apollo-11 crew: Neil Armstrong, 38, commander; Michael Collins, 38, CSM pilot; and Edwin Aldrin, Jr., 39, LM pilot. (Dow Chemical Company photo.)

Early on July 20, Armstrong and Aldrin made a systematic check of the LM (dubbed "Eagle") in which they were about to ride down to the lunar surface. Late in the 12th revolution, just before the spacecraft went behind the Moon, the astronauts were given a go-ahead for undocking.

When the craft reappeared from behind the Moon at 1:50 p.m., the CM (containing Collins) had already separated from the LM, and the two ships were only feet apart. At 2:12, Collins fired the maneuvering rockets to move about 2 mi. away.

At 3:08 p.m., when the LM was again behind the Moon, its on-board guidance- and navigation-computer triggered a 29.8-sec engine firing that slowed the lander, dropping it into a lower orbit 50,000 ft above the lunar surface. At that altitude the main retrorocket came on for the powered descent toward the target area 250 mi. away in Mare Tranquillitatis.

During this approach, after the radar altimeter had been switched on, a problem arose when the Eagle's on-board computer repeatedly warned that it was being overloaded. (It was later realized that the radar-selector switch was set on the wrong mode, but the computer and radar managed to hold out.)

When the Lunar Module was down to 7200 ft, its attitude jets fired to tilt it upright, so that Armstrong and Aldrin got their first view of the landing area when it was only 3½ min. away. When down to 300 ft, Armstrong took over manual control, guiding the craft over a small rubble-littered crater to a smooth patch beyond and touching down gently.

The spot was approximately in latitude 0°.8 N, longitude 23°.5 E, only about 4 mi. from the center of the intended landing area. The time

FIG. 31. View from the LM toward the Apollo-11 landing site in Mare Tranquillitatis, just north (right) of the crater Moltke at top center. At lower left is the (out-of-focus) thruster on the LM. The crater at lower right is Maskelyne. (NASA photo.)

was 4:17:40 p.m. EDT when Armstrong radioed back: "Houston. Tranquillity Base here. The Eagle has landed."

After rest period and EVA preparations, Armstrong climbed down from the spacecraft to set foot on the Moon at 10:56 p.m. with the remark: "A small step for a man; one giant leap for mankind." Aldrin followed at 11:14. The fine-grained lunar soil gave crisp boot prints, yet was so firm that the landing vehicle's legs had sunk only 2-3 in. into it. On the desolate featureless plain, the local time was early morning, so the astronauts and the Eagle cast long westward shadows. The astronauts estimated that the boulder-strewn crater that Armstrong had maneuvered to avoid lay about ½ mi. east of them.

In all, Armstrong and Aldrin spent 2 hrs, 21 min. outside the spacecraft. They walked around, collected about 50 lbs of rock and soil samples, set up a small seismograph to record moonquakes and meteorite impacts, erected a laser retro-reflector, and hung a foil sheet on a frame to collect solar particles (see p. 103).

Walking outside, Aldrin noted that the descent engine's blast had turned up the surface dirt only beneath the nozzle, leaving a midget crater. He remarked also that the surface soil was quite shallow, for in obtaining core samples by ramming a hollow tube into the ground, he met with great resistance after 2-3 in.

Their work done, the two men began to discard the equipment they no longer needed, which was now excess weight. Then they rested in the capsule.

Lift-off from the Moon was the most critical stage of the mission, for a malfunction could maroon the explorers beyond any hope of rescue. But the ascent stage lifted off properly at 1:54 p.m. EDT on July 21, after a lunar visit of 21 hrs and 37 min. The capsule started off vertically, but thruster jets tipped it to fly on a gradually ascending trajectory.

Meanwhile, Collins had been circling the Moon in the "Columbia"— the Command and Service Module (CSM)—69 mi. over the surface.

The LM ascent stage was successfully inserted into Columbia's orbit, some minutes behind it. After 1½ revolutions the Eagle caught up, and docked with the command ship at 5:35 p.m. EDT.

The astronauts carried their gear and the lunar samples into the Columbia from the Eagle, before releasing the latter craft to drift in orbit around the Moon. Originally it had been intended to send the discarded Eagle off into a solar orbit, but the 60 lbs of fuel remaining in its tanks were insufficient.

Finally, with a 2½-min. burst of its main engine, the Columbia left its circumlunar orbit on July 22 at 12:56 a.m. and headed for home.

The remainder of the trip, re-entry, and recovery followed time-tested procedures, and went without a major hitch. The Apollo-11 CM splashed down near local dawn in the central Pacific on July 24 at 12:50 p.m. EDT, about 250 mi. south of Johnston Island.

There the familiar routine ended. For the first time an interplanetary quarantine was in effect, to guard against the remote contingency of infecting the Earth with possible microorganisms from the Moon. The astronauts were put into a stainless-steel cabin (a modified trailer) aboard the aircraft carrier *Hornet,* and were flown in this from Pearl Harbor to Houson, where they entered a 2-week quarantine.

Although scientific results were secondary in the Apollo-11 mission, there was great interest in the lunar samples and in three experiments set up by Armstrong and Aldrin. NASA established a Preliminary Examination Team (PET) of about 50 geologists, chemists, and other scientists at MSC, Houston, to catalog, photograph, weigh, and make the first rough analyses of the 22 kg (45 lbs) brought back from Mare Tranquillitatis. The precious rocks and soil were removed from sealed aluminum boxes and plastic sample bags first in a vacuum, then handled and stored in an atmosphere of pure, dry nitrogen, by technicians outside glass cases, their hands thrust into arm-length gloves in the sides of a case, at the Lunar Receiving Laboratory at MSC.[1] Quarantine rules were strict; if anyone's glove was torn, or if someone got a whiff of gas from the sample case, he was isolated, like the astronauts, for two weeks.

The few dozen rocks ranged in size from "golf balls to cobblestones"; some were igneous (crystalline basalt) and others crumbly and heterogeneous (breccia). The soil contained "dust" of fine glassy spherules (Fig. 32), as well as sandlike fragments of the rock material. Over-all chemical composition showed minerals familiar on Earth—quartz, feldspar, and compounds of iron, magnesium, and aluminum—but with several major differences from terrestial rocks. The content of titanium, chromium, zirconium, and yttrium is much higher, and the rock structure shows evidences of shock and repeated melting-solidification. The exposed rock surfaces are rounded by meteorite erosion and pitted with minute glassy craterlets. Many more details are given in the following articles. No biological (organic) matter and no water was found.

One of the first three experiments performed on the lunar surface was returned to Houston—a sheet of thin aluminum foil hung in the open

[1] Security of the lunar samples is maintained by closed-circuit TV, special combination locks, and round-the-clock armed guards—measures instituted in 1973 after one Moon rock on display in California was stolen. Also, since there is some risk of catastrophe at any one location, NASA moved 1/7 of the samples (about 843 lbs at the end of the Apollo program) from Houston to Brooks Air Force Base in San Antonio, where another safe repository had been built by 1976.—TLP

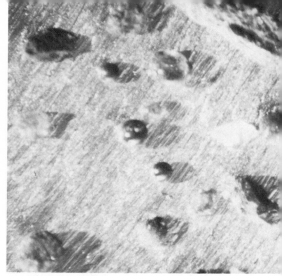

FIG. 32. Typical glassy spherules from the lunar soil range from colorless through gray green, green, brown, and wine red, to black. The large one at the top is ½ mm in diameter. The background is an aluminum dish. (NASA photo.)

FIG. 33. Apollo-11 landing site, with LM in background and the deployed instruments in the foreground. Aldrin stands behind the seismometers, which have a solar panel on each side—one facing east, one west—for power, and a radio antenna for telemetering the seismic data to Earth. Farther back and left is the laser retro-reflector facing the Earth. Left of the American flag is the TV camera on a small tripod. (NASA photo by Armstrong.)

like a handkerchief drying on a clothesline to collect ions from the solar wind (which were adsorbed in the foil and measured in accurate laboratory analyses). The other two experiments were left on the lunar surface (Fig. 33): a seismometer to measure moonquakes (telemetering to Houston data on long-period seismic waves reaching Tranquillity Base), and a "laser retro-reflector" (LRR). The latter is a block of 100 corner-cubes ("cats' eyes") of quartz, accurately cut to return a beam of light precisely *in the direction from which it came. This and two other LRRs left on later Apollo missions are illuminated by lasers in terrestrial telescopes, the round-trip time of the returned light pulse giving very accurate measurement of Earth-Moon distances.*—TLP

Findings from a Sample
of Lunar Material

RAYMOND N. WATTS, JR.

(Sky and Telescope, March 1970)

A team of Smithsonian Astrophysical Observatory (SAO) scientists headed by John A. Wood hurried to prepare their laboratory and to perfect their experimental techniques before the Apollo-11 flight. Under NASA contract, they were to make mineralogical and petrological (rock-history) analyses of an unsorted sample of lunar material brought back by the astronauts. They had guessed that soil might be the most promising available material for a try at unraveling the Moon's tangled history, since it would probably contain a rich variety of lunar substances.

Expecting to study tiny grains, the scientists acquired a microfocus generator for x-ray diffraction work and devised a method for mounting single grains on fine glass fibers.

On Sept. 17, 1969, Wood received the lunar material at MSC in Houston, and the next morning it was already under study in Cambridge, Massachusetts. It consisted of 16 gm taken from the bulk sample collected on Mare Tranquillitatis by Neil Armstrong. This material turned out to be coarser, easier to manipulate, and even more interesting than the scientists had hoped. About 11 gm of it was in the form of particles between a millimeter and a centimeter in size, and the remainder smaller.

The material was sifted and washed in an acetone bath in an ultrasonic (high-frequency) vibrator. Then 1676 of the larger pieces were embedded in epoxy, cut, and polished into thin sections for a survey of rock textures and for electron-probe analysis. Meanwhile, the finer materials were sorted magnetically and in heavy liquids, in preparation for a study of the minerals and glasses by optical-immersion and x-ray diffraction techniques. Several hundred specimens were mounted on fibers.

In sorting the tiny pieces, the SAO investigators noted that there were some white or light-gray fragments, quite unlike the others, which were very dark. (See Figs. 34 and 35.)

FIG. 34. A representative set of particles from the SAO lunar-soil sample returned by Apollo 11 from Mare Tranquillitatis. Identifications given in Figure 35. (SAO photo.)

TABLE 5. RELATIVE ABUNDANCES OF APOLLO-11 ROCK TYPES, IDENTIFIED IN THIN SECTIONS

Type	Percent
Soil breccias	52.4
Basalts	37.4
Glasses	5.1
Anorthosites	3.6
Others	1.5

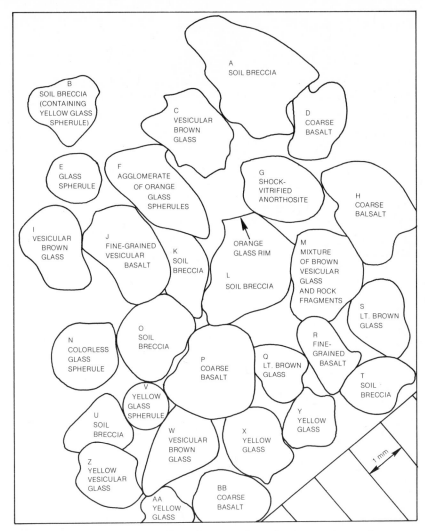

FIG. 35. Key to Figure 34. The colors are those later seen in a microscope view of thin sections. The mineralogical terms are defined in the Glossary. (SAO diagram.)

Classification of the thin sections revealed that the dark samples consisted of soil breccias, basalts, and glasses mixed with a small amount of meteorite fragments, shock debris, and other miscellaneous matter. The light-colored specimen unexpectedly turned out to be anorthosite (see Glossary). The anorthosites are distinguished by low titanium content and are rich in calcium and aluminum. Their densities are between 2.8 and 2.9 gm/cm³.

The darker basalts are rich in titanium and get their color from finely divided ilmenite. They have a density of about 3.3 gm/cm³.

There is no close genetic relationship between the anorthosites and the more common, titanium-rich basalts. If both were formed together by fractional crystallization near Tranquillity Base, there should be a variety of intermediate rocks also. But these were lacking in the SAO sample. The anorthosites seemed to be out of place.

TABLE 6. COMPOSITION OF APOLLO-11 ANORTHOSITES COMPARED WITH SURVEYOR-7 ANALYSIS OF HIGHLAND MATERIAL

Element	Apollo 11	Surveyor 7
C	—	under 2
O	61.2	58 ± 5
Na	0.3	under 3
Mg	2.1	4 ± 3
Al	12.9	9 ± 3
Si	16.0	18 ± 4
Ca[1]	6.3	6 ± 2
Fe[2]	1.2	2 ± 1

[1] Including S, K, and P.
[2] Including Cr, Mn, Ti, and Ni.

Table 6 compares the chemical compositions of these anorthosites with the results of the analyzer aboard Surveyor 7 that landed in the highlands north of Tycho (p. 96).

The agreement is startling. Did the anorthosites come from the highlands?

Tranquillity Base is only 41.5 km NNE of the highland called Kant Plateau. It seems quite likely that the anorthosites are highland material.

If the Moon is, by and large, in isostatic equilibrium, then the anorthositic highlands material (density 2.9) can be visualized as "floating" on the basaltic gabbro (density 3.3). The density differences would have to persist to a depth of about 25 km to support the height difference of roughly 3 km between highlands and maria. The densities could not increase very much at greater depths because the over-all density of the Moon is only about 3.35.

From this evidence, Wood and his associates have proposed a simple model of the Moon that is consistent with the observations and can account easily for the required layerings of rock types. They postulate that the Moon was at least partly molten at some time in the remote past. Anorthositic material began to crystallize and float toward the surface of the gabbroic melt. Meanwhile, heavier olivine crystals would have been sinking, leaving a titanium-rich magma between.

On Earth, when feldspars crystallize from a silicate melt, the first crystals are poorest in sodium, which explains why the highlands are poorer in sodium than the mare basalts.

As it formed, the crust of the Moon would have been punctured by very large meteoroids. Following a puncture, the molten surface material would well up to a level of hydrostatic equilibrium. Then, cooling, it would contract and subside by about 10 percent. A later lava flow from another puncture could flow out on top of this subsided surface, making the mare more massive than an isostatic model would predict. Such gravitational anomalies, called mascons, have been observed in association with the circular maria (see p. 97). An unanswered question is how the interior of a hot Moon could have cooled to the currently accepted temperature of 1000° C.

Oldest Moon Rock

(Sky and Telescope, July 1970)

Perhaps the strangest object yet brought from the Moon is a lemon-sized rock weighing 83 gm, picked up by an Apollo-12 astronaut on Oceanus Procellarum on Nov. 20, 1969. It is called 12013, being the 13th rock removed from the Apollo-12 sample box. It looks somewhat like granite, but is very complex mineralogically.

It contains 20 times as much uranium, thorium, and potassium as any lunar rock previously studied.

Gerry Wasserburg of CalTech announced at the end of May that this rock has an apparent age of 4.6 billion years—the oldest rock yet found on the Moon. Other rocks from the Apollo-11 and -12 sites appear to have crystallized from a magma that was formed 3.3-3.7 billion years ago.

The 4.6-billion-yr age of 12013 was determined by measuring its content of strontium-87 and rubidium-87. Because the former isotope is produced by the radioactive decay of the latter at a known rate, their abundance ratio indicates the time elapsed since the rock solidified. The age determination by Wasserburg was made easier and surer in this case by the high abundance of rubidium.

The age of the solar system, indicated mainly by isotope-ratio studies of meteorites, is also slightly less than 5 billion yrs. It would therefore appear that this rock dates back almost to the time when our solar system condensed.

As the studies of lunar samples progressed from Apollo 11 through Apollo 17, astronomers had to study up on mineralogy and other aspects of geology to understand such terms as breccia, anorthosite, vesicular basalt, and gabbro. Summaries of all the Apollo findings are presented on pages 121-132 and 142-153; we will not attempt here to cover the detailed geochemical reasoning of the lunar-sample investigators. A complete library of their results is maintained at the Lunar Science Institute, 3301 NASA Road 1, Houston, Texas 77058, a university organization supported by NASA near the Johnson Space Center (JSC, formerly MSC).

Many other approaches to lunar science have been developed from the high-resolution photographs taken on Orbiter and Apollo missions. One puzzle that intrigued many amateur astronomers has not been resolved: the peculiar brief brightenings generally noted by eye (not photographs or electronic detectors), and thought by some to be fluorescence of lunar surface materials. NASA organized a special study; a telephone reporting system was set up by the Smithsonian Astrophysical Observatory (SAO) in Cambridge, Massachusetts; and a special telescope was built in Arizona (with NASA funds) by Northwestern University. Although visual reports of brightening events, mostly in the crater Aristarchus, continue, none of these efforts has produced an explanation—or even verification—of the "events."

Another kind of event on the Moon, landslides, leaves a definitive record, but no one has yet seen one in action.—TLP

Lunar Landslides[1]

PHILIP JAN CANNON

(Sky and Telescope, October 1970)

Using high-resolution Lunar Orbiter and Apollo photographs, we can, in effect, make geological field trips on the Moon. Thus, in the last few years there has been increased understanding of the ways in which the Moon's surface features are evolving.

Landslides and rockfalls, debris-slides, slump, and creep are important lunar geological processes. Under the general category of *mass wasting*, they are the major agents of degradation that continually modify the lunar craters. Though the five processes differ in detail, in each one gravity causes a general movement of large and small masses of rock debris downward and inward toward the center of a crater.

The lunar landforms are thus modified mainly by landslides and related processes, as well as by subsequent impacts. Many features on the

[1] This study was performed in part under National Aeronautics and Space Administration contract R-66. Publication authorized by Director, U.S. Geological Survey.

Moon attest to the magnitude of mass wasting, the most conspicuous being the terraces inside large craters, such as Copernicus. They are caused by the slumping of rim- and wall-material into the crater. The pair of photographs of Jansen B in Figure 36 show a small terrace along the lower part of the inner wall, whereas the slumping of a large block of the eastern rim has given this crater in Mare Tranquillitatis an irregular shape.

In many places, blocks that have rolled downslope and left tracks are obvious instances of mass wasting. The tracks attest to the unconsolidated nature of the lunar materials. There is also the unconsolidated debris (talus) deposited at the base of a slope.

What starts downslope motion under the influence of gravity on the Moon? One cause is seismic vibrations from a meteoroid impact; another is thermal heave, illustrated in Figure 37. The fragment of material expands from position 1 to position 2 when heated during the long lunar day. During the ensuing night it cools by hundreds of degrees and contracts into position 3, and each month the cycle is repeated.

The sliding material leaves debris stripes or radial bands on the walls of relatively young primary craters.

FIG. 36. This stereoscopic pair of high-resolution photos taken from Lunar Orbiter 4 shows the effects of slumping in the crater Jansen B. On the right-hand (eastern) wall, a large slump block has increased the crater's diameter from 16 km to 18 km. (To get a three-dimensional, stereoscopic view, hold a card or sheet of paper vertical over the middle of the figure and place your face over it so that each eye sees only one of the two crater photos.) (NASA photo selected by the author.)

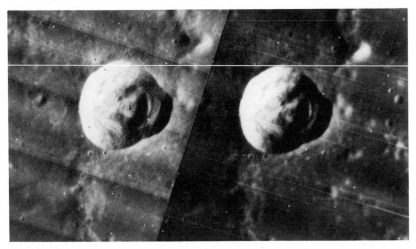

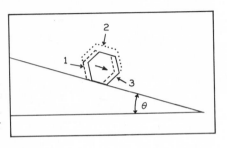

FIG. 37. "Thermal heave" is shown schematically by this block on a slope (angle θ). The outlines 1 (dashed lines), 2 (dotted), and 3 (solid) show the effects of thermal expansion and contraction of the block during lunar day and night. (Diagrams adapted by the author from Sharpe, *Landslides*, 1938.)

Lunar-impact craters range from sharp-rimmed, steep-walled pits to rimless shallow depressions. Immediately after the formation of an impact crater, slumping occurs along the walls, resulting in terraces. Although these lend temporary stability to the walls, they eventually break down and the resulting talus spreads over the crater floor. As the rim material is degraded, the scarp of the rim retreats outward. The result is partial filling of the crater as its diameter increases.

Meanwhile the space race continued, the Soviet Union now concentrating on unmanned lunar landings. They achieved technological success in returning lunar samples (without a man to select them) on Luna 16, launched Sept. 12, 1970, which landed in Mare Fecunditatis. An electric drill, guided by ground controllers using TV aboard Luna 16, obtained soil samples from depths up to 34 cm (13.5 in.) and placed them in a sealed container. After 24 hrs on the lunar surface, Luna-16's ascent stage blasted off and returned the samples by parachute landing in Kazakhstan on September 24.

The Russians also developed a lunar vehicle ("Lunokhod") that can move about slowly and make several experiments on the lunar surface. The first was carried by Luna 17, which soft-landed at 35° W, 38° 17' N, in Mare Imbrium on Nov. 17, 1970. A folding ramp was lowered so that Lunokhod 1 could roll out on the surface under its own power (electric motors on eight wheels), carrying a pointable TV camera, solar cells, a retro-reflector like LRR (p. 128), magnetometer, and other instruments. A computer and inclinometer stop the motors if Lunokhod tips dangerously, since the controlling crew (at a computer center near Moscow) are always 2.5 sec behind time due to the radio travel time from Moon to Earth and back. The vehicle moves slowly, averaging about 600 m/hr (33 ft/min), but can climb 20° slopes. It is used only when Moscow's radio antennas can "see" the Moon, and must be inactivated during the 2-week lunar night for lack of power. It lasted for a year but very little information about Lunokhod 1 was released by the Soviet Union.

In January 1973, Luna 21 carried Lunokhod 2 to the eastern edge of Mare Serenitatis, where it followed a similar, unhurried procedure. This time, Sovfoto released diagrams (Figs. 38 and 39) showing the design.

FIG. 38. The diagram shows the last three stages of Luna 21 landing with Lunokhod 2. Top: the retrorocket slows the lander's orbital speed and fall; at 70-ft altitude (center) it is turned off and small thrusters level the lander, allowing it to fall gently to the surface. Bottom: the ramps are lowered so that Lunokhod 2 can be driven off Luna 21 onto the lunar surface. (Sovfoto —N.Y.—diagram.)

After 2 days of battery charging, Lunokhod 2 was driven 1.15 km in 6 hrs, getting TV views of lunar mountains and making other measurements. These operations continued to mid-1975, with TV pictures and other data reported in the Soviet press. Cessation of these reports and lack of

FIG. 39. Lunokhod 2, the Soviet unmanned lunar explorer, which began operation near the Taurus Mountains on Jan. 18, 1973. Numbers indicate: 1, directional antenna for radio communications with Earth; 2 and 8, three TV cameras; 3, photosensor to detect sunlight; 4, solar cells on inside of (opened) cover; 5, magnetometer; 6, retroreflector for laser beams from Earth; 7, photometer to measure sky brightness; 9, obstacle sensor (for steering); 10, telephoto lenses for distant views; 11, "travel-capability evaluator" (to stop motors before Lunokhod tips over). The eight wheels are independently driven by electric motors powered by rechargeable batteries in the body of the vehicle. (Sovfoto—N.Y.—photo.)

laser returns from the Soviet LRR indicate to Western experts that Lunokhod 2 turned over in a deep crater, despite its inclinometer.

As was often the case, the Soviet use of a new development preceded the NASA use. Lunokhod was more complex, but NASA's 6 ft x 10 ft Lunar Rover was ideal for manned missions, allowing accurate navigation and much longer traverses than on foot. With a man at the controls, it traveled much faster than Lunokhod (13 km/hr, or 8 mi/hr) and carried fewer instruments. Figure 40 shows its load of cameras and tools, and room for sample bags and two astronauts in bulky space suits.

There were hazards in lunar driving, of course: boulders and craters had to be avoided, and the dust kicked up by the aluminum-mesh wheels in the lower lunar gravity (about 1/6 of the Earth's) required efficient "mudguards." Seatbelts were required, and several of the astronauts have

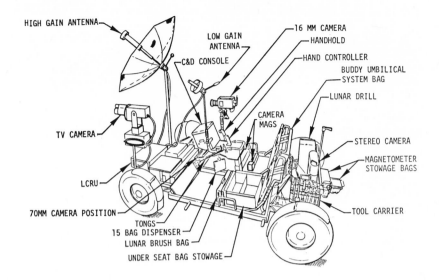

HIGH GAIN ANTENNA
LOW GAIN ANTENNA
C&D CONSOLE
16 MM CAMERA
HANDHOLD
HAND CONTROLLER
BUDDY UMBILICAL SYSTEM BAG
LUNAR DRILL
CAMERA MAGS
TV CAMERA
STEREO CAMERA
MAGNETOMETER STOWAGE BAGS
LCRU
70MM CAMERA POSITION
TONGS
15 BAG DISPENSER
LUNAR BRUSH BAG
UNDER SEAT BAG STOWAGE
TOOL CARRIER

LRV STOWED PAYLOAD INSTALLATION

FIG. 40. NASA's Rover, or Lunar Roving Vehicle (LRV), first used at Hadley Rille, the Apollo-15 landing site. Two astronauts sit side by side facing left (forward), the right-hand one controlling speed and turns with the hand controller. The C & D console includes an on-board computer that displays distances and direction back to the trip's starting point (LM). The high-gain antenna transmits TV pictures from the camera (front, left), which can be pointed by the controllers at MSC, Houston. The four wheels are independently driven by electric motors powered by two batteries at the front end. (NASA diagram.)

remarked that their ride was very bumpy. Development of the Rover was largely due to Anthony J. Calio, and use of its TV camera was controlled by Ed Fendell (popularly known as "Capt. Video"), both at JSC. The 480-lb Rover folded up for stowage in an outside compartment on the LM. It was first used on the Apollo-15 mission to Hadley Rille at 3° E, 25° N.—TLP

Apollo-15 Pictorial

JOSEPH ASHBROOK

(Sky and Telescope, October 1971)

On these pages are a few of the many excellent photographs obtained during the fourth visit of men to the Moon, from July 31 to August 2, 1971. Some were taken on the lunar surface by Scott and Irwin; others were taken by Worden as he circled the Moon alone in the Command Module at a height of approximately 70 mi.

The photographic equipment used by astronauts on lunar missions has grown during the past 2 years. On Apollo 11 a Hasselblad camera and Ektachrome film were used to record the first human footprints on the Moon. On the Apollo-12 mission to Oceanus Procellarum, stereo cameras with built-in electronic flash were used to take close-ups of rock and soil before their removal. On the Apollo-14 flight a movie camera was added to the chest pack of one of the astronauts—the first time movies were made on the Moon.

On Apollo 15, Scott and Irwin had seven different cameras—three 16-mm movie, three 70-mm still, and a 35-mm still—with various accessories.

More significant, perhaps, was the great improvement in the photographic equipment used in the Command Module for lunar mapping from orbit. As H. J. P. Arnold of Kodak Limited explained:

"To make accurate maps and to interpret surface characteristics accurately and in detail, photogrammetrists require the precise altitude, position and attitude of the vehicle from which photographs have been taken. In addition, it is of great assistance to the photo-interpreters if a large-format camera is used with as long a focal length lens as possible."

FIG. 41. Part of the large, far-side crater Tsiolkovsky, photographed from about 70 mi. above by Apollo-15's mapping camera. Forward motion of the spacecraft was compensated by moving the camera film during exposure so that 100-ft craterlets can be seen on this 70-mi. expanse of crater floor. Note the slumped crater wall and crooked rilles (top right and above "island"). (NASA photo.)

FIG. 42. The deep, 28-mi. crater Aristarchus (upper left), shallow crater Herodotus, and Schröter's Valley (right)—a southward view recorded by Apollo-15's mapping camera. Note volcanic dome-crater above Herodotus. (NASA photo.)

FIG. 43. Apollo-15 CSM photographed from the LM, showing the mapping cameras on the open side. The Command Module is facing down, the Service-Module rocket nozzle up. (NASA photo.)

Previous Apollo missions had not satisfied these needs, but Apollo 15 had in its Service Module a camera bay (Fig. 43), with three cameras (mapping, stellar orientation, and panoramic) and a laser for measuring the spacecraft's altitude.

The mapping camera had an $f/4.5$ lens of 3-in. focal length and a magazine·with 1800 feet of Kodak 5-in. black-and-white film. This camera was equipped with a reseau plate, illuminated fiducial marks, and a forward-motion compensator—devices designed to aid photogrammetrists and cartographers. Each frame covered an area about 100 mi. square (Figs. 41, 42), and at 70-mi. height was expected to resolve 60-ft objects.

Rigidly attached to the mapping camera but pointing almost perpendicular to it was the $f/2.8$ stellar-orientation camera, which has the same focal length. It obtained a photograph of a star field simultaneously with each exposure of the mapping camera. Because the angle between the optical axes is known, the pointing direction of the mapping camera at the moment of exposure can be reconstructed.

Since the three cameras were outside the pressure-tight skin of the spacecraft, it was necessary during the return journey from the Moon for Worden to recover the exposed films on EVA.

Weighing 336 lbs and as large as a desk, the panoramic bar camera had an 8-element $f/3.5$ lens of 24-in. focal length. During each exposure, the camera was swung across the flight path from the lunar horizon on the left to the horizon on the right, the film being moved to match the image precisely. Each sweep photographed an area about 12½ x 180 mi. of the lunar surface on a film strip 4½ x 47 in. In all, 1650 such exposures were taken on about 6500 ft of film. The Army Map Service used a special printer to correct for distortion of the pictures. (See p. 142).

FIG. 44. Four-fifths of a panoramic-camera film, showing a strip 12½ mi. x 140 mi., including Hadley Rille and the Apollo-15 landing site near top center. North is to the left. (NASA photo.)

FIG. 45. Between LM (left) and Rover (right—with Irwin in the passenger's seat) is St. George Crater on the slope of Hadley Delta Mountain, the goal of the first Rover excursion. (NASA photo by Scott.)

Figures 45, 46, and 47 are photographs taken on the lunar surface by Scott and Irwin.

These are some of the most beautiful photographs taken on the Moon, partly because the scenery (Hadley Rille and the Hadley Mountains) is so spectacular. The Apollo-15 mission was an exciting and enjoyable TV spectacular, with long Rover traverses, views down into Hadley Rille

FIG. 46. Close-up of a small, rock-strewn, "fresh" crater near the Apollo-15 landing site, with Hadley Delta Mountain in the right background. (NASA photo by Scott.)

FIG. 47. Scott and the Rover, parked near the edge of Hadley Rille at the "elbow" (top center on Fig. 44). The view is northward, showing the near-vertical walls of the 1100-ft-deep rille and the talus material at the bottom. (NASA photo by Irwin.)

showing layered beds in the lunar crust (see p. 124), and the pixie voice of Joe Allen, CapCom in Mission Control at Houston, who did all the talking with astronauts Scott, Irwin, and Worden. It was the first Apollo mission to give high priority to scientific goals, and the geologists assembled at MSC (now JSC) in Houston had a heyday in requesting pre-photographed samples (photographed before they were picked up), deep drill cores (samples of lunar soil and rock obtained in tubes drilled down more than 6 ft), and seismic data showing the nature of the rocks far below the surface.

Analysis of the data obtained on one of these missions takes several years. Soil is analyzed for age, chemical composition, surface condition (meteoroid pits), previous history, etc.; photographs are combined in various ways; statistics are collected on seismic data, and so on. The lunar samples from Apollo 11 onward are still under study. The results are discussed each year in a Lunar Science Conference.—TLP

The Third Lunar
Science Conference

THORNTON PAGE

(*Sky and Telescope*, March and April 1972)

Every year, a special conference is held by NASA at MSC near Houston, to discuss the scientific results obtained from its missions to the Moon. The first of these conferences, in 1970, was highlighted by studies of the lunar samples returned by Apollo 11. The 1971 meeting, dealing mainly with findings from the Apollo-12 mission, filled three volumes of proceedings.

At the Third Lunar Science Conference on Jan. 10-13, 1972, an audience of 1000 listened to almost 200 investigators present papers about the Apollo-14 mission to Fra Mauro, with a few preliminary reports on the Apollo-15 visit to Hadley Rille.

Apollo 14 brought back 43 kg (95 lbs) of rock- and soil-samples, while Apollo 15 returned with 76 kg that included one cylinder extending 2 m. (6½ ft) below the lunar surface. These and the smaller samples obtained by Apollos 11 and 12 differ markedly in chemical composition from terrestrial materials (see p. 103). Lunar rocks have radioactive ages (see Glossary) near 3.7 billion yrs, whereas the Moon's soil is about 4.7 billion yrs old, which is probably close to the age of the solar system.

Of course, the composition of other regions on the Moon may be different. Moreover, the surface may have been heavily contaminated by meteoroids that have hit it during several billion years. So we are interested in studying drill samples and material ejected from deep craters, as well as making seismic, thermal, and magnetic measurements of subsurface layers.

The more we learn about the Moon, the more puzzling it becomes. There are lava flows on the surface, yet the interior is rigid (cold) enough to have deep moonquakes and to keep the mascons (p. 97) from sinking in the course of billions of years. The radioactive content of the interior (if the same as at the surface) should have caused melting and allowed a dense liquid core to form. Nevertheless, the Moon has a low average density, hardly any magnetic field (such as would be expected from dynamo action in a fluid core), and an asymmetrical shape that is "frozen in" against gravitational settling.

Curiously, while the general magnetic field is only 0.0001 that of the Earth, some lunar rocks are magnetized as if they had solidified in a much stronger field, and the increase of temperature with depth implies a hot interior. These and other puzzles added to the excitement at the Third Lunar Science Conference.

SOLAR WIND

The solar wind streams around the Earth, which is shielded by its magnetic field, but it hits the sunlit side of the Moon. A magnetometer on the 80-lb subsatellite, left in circumlunar orbit by Apollo 15, is recording magnetic fields in the solar wind and in the cavity (shadow) behind the Moon. When certain craters are at or near the terminator, they produce magnetic anomalies in the solar wind due to differences in the craters' magnetic properties there, as reported by P. J. Coleman (UCLA).

In a related effect, the spectrometers left by Apollos 12 and 15 have recorded bursts of solar wind at sunrise and sunset. The average wind is about 2,000,000 ions/cm^2 sec, but the bursts get up to 150 times more, according to C. W. Snyder (JPL).

PHOTOGRAPHY AND SELENODESY

Over 4500 photographs were taken from the orbiting CSM (Fig. 43) on the Apollo-15 mission, as reported by F. J. Doyle of the Army Topographic Command. The panoramic camera resolution is about 0.7 m. on the lunar surface, and the overlap between photographs allows stereo viewing to detect relative heights of ±3 m. Measurement of the panoramic negatives is done with an instrument called Unimace that plots contour maps at a scale of 1:10,000, corrected for viewing angle, and accurate to about 1 m.

A separate camera located accurately a set of control points on the Moon's surface relative to its center of mass, the interior point about which the spacecraft orbits. A pulsed-laser altimeter measured the distance to the lunar surface every 20 sec with an accuracy of ± 2 m., but after the first few orbits the ruby laser failed.

Nevertheless, since the CSM was tracked accurately by Earth-based observers from the Doppler effect in the radio transmissions, its orbit is accurately known relative to the Moon's center of mass. The many control points in a ring around the lunar equator were therefore faithfully located (by laser only on the far side), allowing the shape and size of the Moon to be accurately determined.

W. R. Wollenhaupt (MSC) found that the best-fit sphere has a radius of 1737 km (1080 mi.), with the center of mass 2 km closer to the Taurus Mountains, 28° N, 43° E, northwest of the Moon's center as we see it.

Near the center of the far side there is a "dimple" 6 km deep—the Van de Graaf depression—which means that the figure of the Moon is actually somewhat shorter along the Earth-Moon line than across it.

SURFACE OF THE MOON

The great majority of papers at the conference concerned surface features: maria, lava flows, craters, layerings, rilles, and highlands. It is now generally accepted that the maria were formed by large impacts, that many of them (but not all) were later filled with lava flows, and that most of the highlands were pushed up by these impacts several billion years ago. However, scientists disagree as to where the lava came from, what the layering is due to, and what the internal structure of the Moon is.

The amount of material vaporized or melted by a meteoroid impact depends on the particle velocity and the densities of the meteoroid and its target. In a major impact on porous rock, there may be much melting, allowing the separation of minerals, according to T. J. Ahrens (CalTech).

While huge impacts may have released much lava in the early history of the Moon, they do not account for recent lava flows. Most of the meteoroid flux striking the Moon today consists of particles of 10^{-6} gm or smaller. Over a time scale of many million years, these micrometeoroids act like a sandblast to round and smooth the lunar rocks. Meteoroids as large as 1 gm are much less frequent, producing one impact/m²/10⁶

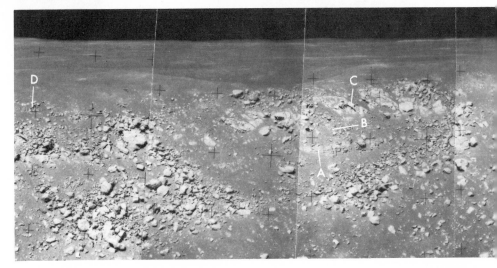

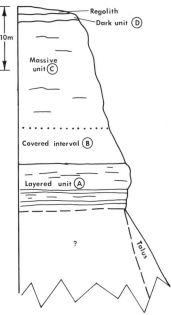

FIG. 48. Panorama of Hadley Rille's southwestern wall from photos taken by Scott and Irwin, and a vertical section of the top layers sketched by Gordon Swann. The photos show rock layers that are generally horizontal. A and B are about 8 m. thick; C is a massive, light-colored basalt, probably an ancient lava flow; D is irregular, just below the thin regolith at the surface. Talus (landslides) cover everything below A. (NASA photo and U. S. Geological Survey diagram.)

yrs, but they serve to plow up the regolith (the surface layer of soil and rock fragments), which is about 1 m. deep near the Hadley Rille.

Layering beneath the regolith was well observed on the Apollo-15 mission (see Fig. 48), as discussed by Gordon Swann (U.S. Geological Survey). Scott and Irwin's photographs show horizontal layering in the sides of the 350-m.-deep Hadley Rille. More surprisingly, the photographs of Mount Hadley, 16 km away, showed layers in its southwestern wall, 4000 m. above the flat mare surface. The stratification high on Mount Hadley probably shows that flat-lying beds were pushed up that high by the catastrophic Mare Imbrium impact.

Frank Low (Lunar and Planetary Laboratory) reported scans of the night side of the Moon with far-infrared (10-50 micron wavelength) detectors, observing hundreds of thermal anomalies—small regions that cool off more slowly than their surroundings during lunar night or eclipse. Some of these warms spots are 100°C warmer than their surroundings.

A new observing technique is *bistatic radar*, in which an Earth-based station compares radio signals beamed directly from an orbiting Command Module with a second beam reflected earthward from the lunar surface. Such observations indicate topographical irregularities and radio reflectivity at some depth, since the radio waves penetrate about 20 wavelengths into the lunar ground. Bistatic measurements at 115-cm wavelength show smooth electrical properties about 23 m. beneath the surface, whereas 13-cm measures reveal submerged highlandlike roughness.

By contrast, the x-ray fluorescence studied by I. Adler (Goddard Space Flight Center) refers to the top millimeter of rock and dust. In effect, he used the hot solar corona as an x-ray source, measuring differences in surface composition below the orbiting spacecraft by changes in the energy distribution of the secondary "fluorescent" x-rays produced by the interaction of the solar x-rays with the lunar surface. Among other things, his results show no horizontal transport of the dust (such as had been proposed by T. Gold), and a strong correlation between surface albedo and local abundance ratio of aluminum to silicon. Surface material with more aluminum in it is generally lighter in color (higher albedo) and shows the x-ray fluorescence characteristic of aluminum.

LUNAR ATMOSPHERE

The Moon lost any significant atmosphere long ago because its noon surface temperature is so high (400°K) that even the heaviest gas molecules in time exceed the low velocity of escape, 2.4 km/sec. Three cold-cathode ionization gauges (detectors of gas ions) left on the Moon have shown that the daytime atmospheric pressure there is normally about 10^{-10} torr (roughly 10^{-13} of sea-level barometric pressure on Earth). That is a fairly "hard" vacuum.

But short intervals of higher pressure occur (up to 10^{-6} torr) when cabin gas is released from a nearby Lunar Module before the astronauts step out, and near an astronaut (his space-suit backpack exudes water vapor). Other brief pressure increases recorded after the astronauts left the Moon are believed to be "burps" from tanks on the abandoned Lunar Module, but some may be from volcanic gases seeping out of rock crevices. Radioactive disintegrations yield helium, argon, and other rare gases that must be escaping from the lunar surface, while on the sunlit hemisphere solar-wind gases are coming in toward the Moon.

On the Apollo-15 CSM, a sensitive alpha-particle detector recorded ionized helium and a little radon, mostly emanating from radioactive surface materials. The maximum effect was near the large crater Tsiolkovsky on the Moon's far side.

F. S. Johnson (University of Texas at Dallas) told about three months of observing by a cold-cathode ionization gauge left near Hadley Rille. In the lunar daytime there were about 1,000,000 atoms of hydrogen and helium/cm^3 and 60,000 neon atoms/cm^3, with about a tenth these amounts during the night, plus 10 times as much in contaminants from the abandoned Lunar Module.

The suprathermal ion-detector experiment (SIDE) left at Fra Mauro by Apollo 14 measures high-speed gas molecules coming in from a direction about 15° E of the zenith. For several hours on March 7, 29 days after the astronauts had left, bursts were detected, consisting almost entirely of ions of mass 20—probably water vapor.

Since the analysis of lunar rocks and soil has shown no trace of water, the observed water ions probably came from the Command Module when it orbited the Moon a month earlier. Records show that 47 kg of waste water were dumped through a nozzle as the spacecraft traveled westward at 1.6 km/sec, 100 km above the surface. Although some of the spray droplets probably froze into snowflakes, the water would soon be in the form of individual molecules, each in orbit around the Moon. This swarm must have been influenced by solar radiation, interaction with the solar wind, and the Earth's tidal action. During the first month, the molecules probably spread out into a ring around the Moon. Containing some 30 ions/cm^3, the ring would become 1000 km wide and would reach from the surface to a height of 2000 km.

As the ring developed and many ions began to hit the Moon's surface, some came in at the appropriate angle to be detected by the SIDE experiment at Fra Mauro. We cannot prove that all this happened, and it is uncertain how long the ring of molecules will last in orbit around the Moon—perhaps several months or even years.

Richard R. Vondrak of Rice University points out that if the lunar atmosphere were 10,000 times denser that it is, the solar wind would be unable to sweep it away, and the much slower process of thermal escape would dominate. The lifetime of such an atmosphere would be thousands of years for atoms as heavy as oxygen or neon.

Each Apollo mission has added nearly 10 tons of rocket exhaust to the lunar environment. Vondrak notes that extensive lunar exploration and colonization could well release enough gas to produce a man-made lunar atmosphere.

SURFACE MATERIAL

Lunar surface materials returned by the Apollo missions fall into two broad categories—mare-type and highland-type (p. 108). These names are somewhat misleading, since the highland-type material that forms the bright mountains is now also believed to underlie the mare-type in the large dark lunar seas.

Most rock samples show evidence of multiple origin; that is, they have undergone several epochs of freezing, shattering, remelting, and refreezing. The highland materials represent original lunar crust, modified only by small impacts, but with much variability in composition. Samples from the sites of Apollos 12, 14, and 15 show it to be enriched with potassium (K), the Rare-Earth elements (of intermediate atomic weight), and phosphorus (P)—a combination lunar scientists name by the acronym "KREEP."

Many of the chemical analyses were summarized by Paul Gast, a Columbia University geochemist at MSC. He points out that in the outer 200 km of the lunar crust the abundances of calcium, aluminum, the rare earths, uranium, and thorium are more than five times higher than in the sun and meteorites. However, in order to account for the Moon's interior density, there must be a normal solar abundance of calcium and aluminum at deeper levels. This raises the question of how this differentiation came about. Gast thinks that the materials falling into the Moon near the end of its accretion period differed from the earlier infalling materials.

Although the abundances of the rare-earth elements are only a few parts per million, as compared to 5-10 percent by weight for calcium and aluminum, the rare earths provide a sensitive index of the lunar composition, for they range from 10 to 300 times the solar-system (meteoritic) abundances. Thus, in mare-type samples the rare-earth elements are about 20 times overabundant, while in highland-type material they are 100 to 250 times overabundant. Therefore, the KREEP content of both types cannot be meteoritic, and Gast concludes that the original composition of lunar material 4.5 billion yrs ago was different from the sun's and Earth's.

 He thinks that the mare material contains less of the KREEP elements because the earlier infall contained less than the final infall, which made the outermost "highland" crust. Later, the mare material percolated upward from a 300-km depth by a series of partial meltings. This required temperatures near 1200°K, possibly achieved by localized radioactive heating within the upper 300-km layer.

A partial confirmation of this radioactively heated layer was provided by the Apollo-15 heat-flow measurements, and detection of radioactive emanations from the surface (p. 126). Also, as mentioned on page 125, Adler's x-ray-fluorescence experiment indicated that the aluminum-to-silicon abundance ratio is greater in the highlands than in the maria, consistent with the Gast model.

THE INTERIOR OF THE MOON

The exploration of the Moon's interior is by no means confined to seismic observations, such as those mentioned on page 102. Four other procedures were reported, involving measurements of magnetic fields, electrical properties, thermal gradients, and the Moon's libration ("wobbling").

The last method is purely astronomical. It has long been known that minute irregularities in the Moon's rotation are caused by the Earth's attraction on the slightly ellipsoidal body of the Moon. As reported by Derral Mulholland (University of Texas), the laser retroreflectors (see p. 104) placed by Apollos 11, 14, and 15 at three widely separated sites can be used to measure the orientation of the Moon's three axes with high precision.

Laser ranging errors are now estimated at ±15 cm, and they may be reduced to ±3 cm next year. [This was achieved in August 1975, when a very powerful laser was installed at the observatory on Mount Haleakala, Hawaii.] The changes will be used to determine the Moon's moments of inertia around its three axes, as well as a dissipation parameter that shows the rate at which the physical libration is damped. The moments are related to the distribution of mass inside the Moon, and the dissipation tells us something about how rigid the Moon is.

The best values currently available for the three moments, as deduced from ground-based observations of libration, were discussed by Zdenek Kopal (University of Manchester). He noted that the differences between the moments are much too large for a body in hydrostatic equilibrium. Hence, he argued, the Moon must be rigid, and its interior temperatures cannot be near the melting point of the material, except in a relatively small core.

If there were a large fluid core, a magnetic field should be produced by dynamo action there. The steady general magnetic field of the Moon is now only about 5 gammas (10^{-4} of the Earth's).

The situation may have been different in the remote past. David

Strangway (MSC) has measured a weak but appreciable remanent magnetization of lunar rocks, which indicates a field of several thousand gammas at the time the rocks solidified, about 3.8 billion yrs ago. Either the Moon had a large liquid core then, or there were strong local fields produced by pools of molten rock near the surface.

Measurements on the surface do indicate present local fields. Thus, at the Apollo-12 site in Oceanus Procellarum, the field was about 40 gammas downward toward the southeast, and the solar wind undergoes irregular deflections at the limb of the Moon, as if the local magnetic field were irregular. However, no one at the conference seemed sure that liquid magma from a major impact could produce by dynamo action a field strength of 1000 gammas at the surface, to cause the remanent magnetization observed in rock samples.

Abundant data on moonquakes is being provided by the seismometers left at the sites of Apollos 12, 14, and 15—more sensitive than the ones left by Apollo 11 (p. 102). One moonquake was recorded at 800-km depth—$\frac{2}{5}$ of the lunar radius under 20° W, 20° S, in the vicinity of Bullialdus. At this location and 10 others, moonquakes occur frequently —evidence of repeated slips in solid rock.

One of the earliest peculiarities noted in lunar seismic records was the long "ringing time," which for a strong quake or impact could produce oscillations lasting one or two hrs (Fig. 49). It now appears, according to Gary Latham and his collaborators at Columbia University, that the great difference between lunar and terrestrial seismic signals can be explained by a heterogeneous layer that blankets the Moon to a depth of a few kilometers. Seismic waves are highly scattered in this zone, but below it wave propagation is highly efficient.

The arrival times of various types of seismic waves have been used by M. N. Toksöz (MIT) to calculate seismic-wave velocities at various depths in a large region south of Mare Imbrium (Fig. 50). Two layers are indicated by a change in velocity from 5 to 7 km/sec at a depth of about 20 km. The upper layer is almost certainly basalt, whereas the lower has the characteristics of eclogite (a gabbro-type rock containing garnet) or anorthosite. At a depth of 60 km the wave velocity again increases, to about 9 km/sec. The basement material below 60 km may be olivine, which has a density of 3.2 gm/cm^3.

Marcus Langseth (Columbia University) reported on the measurements of heat flow from the lunar interior performed during the Apollo-15 mission. First of their kind, these measures aroused much discussion because they indicated a larger thermal gradient and a higher internal temperature than expected for the Moon. On July 30, 1971, astronaut Scott drilled two 6-ft holes for two sets of thermometers, which since

FIG. 49. Seismic records telemetered to MSC, Houston, from seismometers at the Apollo-12 landing site just after the known impacts of the Apollo-14 LM and S IV-B rocket at known locations on the lunar surface. X and Y are horizontal movements of the surface, Z vertical, and the sensitivity (scale in the records) was reduced by a factor of 25 for the S IV-B records (expected to be much more violent). Note how the "ringing" lasted over 50 min. (Columbia University diagram by Latham.)

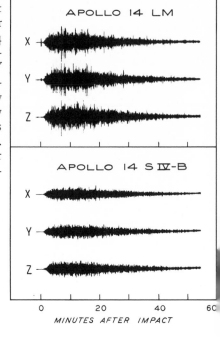

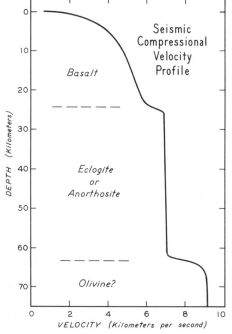

FIG. 50. Velocity of seismic pressure waves at depths from o to 70 km below the lunar surface derived from seismic records by M. N. Toksöz.

then have radioed back to Houston the temperatures at 16 different depths in the lunar regolith near Hadley Rille.

The measurement of the heat flow is more complicated than might be expected. The thermometers are accurate to $\pm 0.001\,^{\circ}C$, but they are inside a tube that partially insulates them from the lunar material and conducts heat itself. Also, the conductivity of the regolith had to be measured, and corrections had to be applied for the initial heat of drilling and for the change in surface temperature during the lunar "day."

The conductivity of the regolith amounts to about 200 microwatts/cm²/ degree for a 1-cm layer, and is probably due to radiation between soil particles separated in a vacuum.

When all the data are combined, the vertical temperature gradient is found to be 1.7°C/m., corresponding to heat output of 3 microwatts/ cm²—about ⅓ that in the Earth's crust—with an uncertainty of ±15 percent. At 1-m. depth, the temperature averages 252°K, or about 20° higher than the average at the surface.

If this lunar gradient continued downward, the temperature would be about 1953°K at a depth of 1 km! Although the higher conductivity of solid rock can be expected to reduce the gradient, a calculation of the Moon's central temperature based on heat conduction comes out very high. The English physicist D. C. Tozer (University of Newcastle upon Tyne) offered an explanation. He suggests that the Moon's interior is in *convective* equilibrium; that is, the energy is carried outward by large-scale motion of the material. This idea was suggested for the Earth's mantle several years ago.

Tozer points out that solid-state creep of about 10^{-7} cm/sec takes place in rocks if the temperature is above 700°K. The "convection currents" in the Moon's solid interior need not be any faster than that to keep the interior temperature below 1000°K everywhere. In fact, he has calculated that if the Moon at first had a hot core, it would cool and "forget" its earlier high temperature in 10^8 yrs. With a reasonable amount of radioactive material releasing energy in the interior, the Moon could have a roughly uniform temperature from its center out to about 90 percent of the lunar radius.

If these ideas are correct, then the Moon can have little rigidity, and it would be expected to have the prolate-spheroid shape of hydrostatic equilibrium. However, Tozer points out that stable convective currents might modify the shape, producing upthrusts in some places and dips in others.

ORIGIN OF THE MOON

It is natural to suppose that the Earth and Moon were formed near each other in the solar nebula about 5 billion yrs ago. Why, then, should the original lunar composition differ from the Earth's? We might imagine that the Moon was formed at a different distance from the sun, where the composition of the solar nebula differed from that where the Earth was formed. How, then, could the Earth capture the Moon?

To escape this difficulty, we might suggest that the Moon was formed from the proto-Earth, taking only the surface material of low density, after the denser materials had sunk toward the center of the rotating and evolving Earth. But how could the ring of material spun off the proto-Earth condense into one solid object?

These basic questions have been discussed (in somewhat different ways) at all of the Lunar Science Conferences.

The idea that the Moon originated by the accretion of cold material goes back to about 1900, when T. C. Chamberlin and F. R. Moulton proposed their famous "planetesimal hypothesis" of the origin of planets and satellites. (See Vol. 3, *Origin of the Solar System.*) The problem of temperatures in a Moon formed by accretion has now been studied by A. E. Ringwood, a geochemist at the Australian National University in Canberra. He finds that the energy of infall might melt most of the Moon initially.

Recent calculations by M. N. Toksöz at MIT show that the outer half would then remain molten for 1-2 billion yrs, allowing radioactive elements to concentrate near the surface in what is now the highland-type material. Such a model (based on thermal conduction, not convection) is consistent with the basaltic rock samples' measured ages, all between 3.2 and 3.9 billion yrs, as reported by D. A. Papanastassiou (CalTech). The soil samples are generally older—up to 4.6 billion yrs—which implies that there was a lunar crust on top of molten material, less than a billion years after the sun itself was formed. According to Gast's chemical ideas (page 127), the Moon's outer composition has changed little since then, except for large impact-produced lava flows of mare-type material depleted in KREEP and iron.

For the last 3 billion years, the Moon has probably not changed much physically as it spiraled farther from the Earth, decreased its spin, and acquired a small amount of meteoritic material that made fresh craters and some new lava flows.

If the uncertainties can be smoothed out, these lunar studies will contribute much to theories of the origin of the solar system.

A First Report on
Apollo 16

RAYMOND N. WATTS, JR.

(*Sky and Telescope,* June 1972)

An ambitious program of exploration and collection was performed by the crew of Apollo 16. While John W. Young and Charles M. Duke, Jr., visited the lunar highlands near Descartes (a change from the smooth maria where the first four Apollo landers came down), Command-Module pilot Thomas K. Mattingly II remained in orbit above the Moon.

The NASA staff in Houston is busy sorting out 213 lbs of samples and processing 12,446 ft of film. Only brief preliminary results have been announced, and it will be months before the detailed analyses are completed. As Mattingly flew over the far-side crater Guyot, he saw a distinctive lava-flow formation similar to the volcanic lava fields of Hawaii, while Young and Duke, on the rocky, undulating plateau north of Descartes, found none of the expected ancient lava flows. Instead of crystalline rocks of volcanic origin (frozen lava), these explorers found breccias composed of broken rock fragments later consolidated.

Another surprise came from the magnetometer readings made in the vicinity of Descartes. Ranging from 121 to 313 gammas (see App. V), they were higher than had been measured elsewhere on the Moon. Unexpectedly, the magnetic polarity was reversed in places (different directions to magnetic north). Does this mean that continuing meteoritic impacts have scrambled the magnetic pattern long ago "frozen" into the rocks of the Descartes area? The regolith is as much as 56 ft deep—evidence of intensive plowing by meteorite bombardment.

Observations from the CSM in lunar orbit revealed that most lunar radioactivity is concentrated in the basins on the western side of the Moon. Why?

The launch of Apollo 16 went well on April 16, 1972. However, a later problem with the main engine of the CSM made a lunar landing seem impossible. The trouble began behind the Moon, soon after the LM had separated. Mattingly was making required checks on the gimbal-mounted main engine when he found that operating the backup yaw control caused the engine to wobble sideways.

Standing NASA orders forbid any use of the engine unless both its primary and backup controls are working, except in an emergency. (If the engine could not be used, then the only way to bring the astronauts back to Earth would be for the LM to use its descent engine for thrust.) It was finally decided from prelaunch tests that the observed wobble could be tolerated and the LM was allowed to land, 6 hrs late, on April 20.

Once on the Moon, Young and Duke spent more time exploring than any previous astronauts. On their first EVA, they set up surface experiments about 250 ft southwest of the LM. Unfortunately, the accidental breaking of a wire rendered the Heat-Flow Experiment inoperative. The two men had no difficulty in keeping the schedule during their 71 hrs on the Moon, 20¼ of which were spent outside the LM. They traveled

15.6 mi. in the Lunar Rover, and ascended to the top of Stone Mountain, the highest point yet reached by Moon explorers.

Prior to returning to Earth, the astronauts released an 80-lb subsatellite to study particles and fields in the lunar neighborhood. (Apollo-15's subsatellite ran into trouble on Feb. 3, 1972, with failure of telemetering circuits on the magnetometer and particle sensor.) The Apollo-16 subsatellite ceased broadcasting on May 29, when it probably crashed on the far side of the Moon.

On May 23, radio commands from Earth caused three mortar shells to be fired from a launching device that Young and Duke had erected at their landing site a month earlier. The three shells traveled 3000, 1000, and 500 ft westward and exploded on impact. The seismic shock waves were recorded by an array of three sensitive geophones, and preliminary analysis shows that the regolith or topsoil in this vicinity is at least 100 yards deep—more than twice as deep as expected.

Just before Young and Duke boarded the LM for departure from the Descartes site, they parked the Rover upsun from the LM so that its TV camera could view the LM ascent-stage take-off, shown in Figure 51. It

FIG. 51. Views of the Apollo-16 LM before and during blast-off from the Descartes site, as seen from the Rover's TV camera. First, the LM stands ready, with Young and Duke inside (above). Then the initial rocket exhaust scattered pieces of the LM's gold-coated mylar heat shield (facing, above), and the ascent stage slowly rose, revealing its downward-pointing rocket nozzle (facing, right). (Photos from a TV screen by Dennis Milon.)

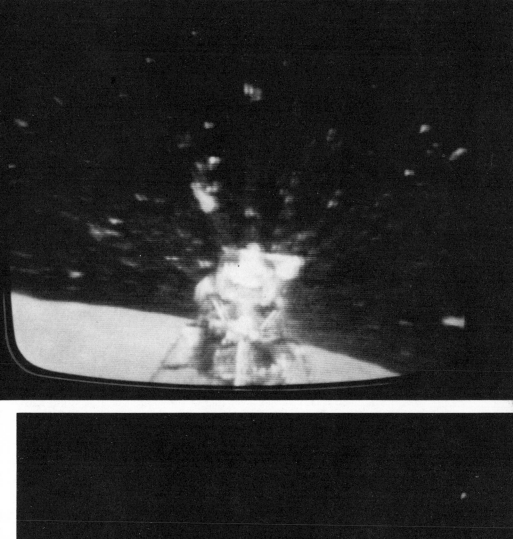

was also planned to use the TV for other views after the astronauts de-
parted, but the batteries failed in the hot sun.

The Apollo-16 CM splashed down in the Pacific on April 27, bringing
astronauts, lunar samples, and exposed film safely back to Earth. Among
the 12,446 ft of film were 34 ft with 180 far-ultraviolet photos of the
Earth, stars, nebulae, and galaxies taken with the S201 Far-UV Camera
shown in Figure 52. Some of its results, worked out by George Carruthers and Thornton Page (NRL) are given in Chapter 8. The camera was
set up by John Young in the shadow of the LM, and pointed by him at
eight different targets in the sky, taking 10-40 photos and spectra of each
in wavelengths shorter than 1600A, about half of them limited to the
range 1250-1600A. The other half record Lyman-alpha, 1216A, showing
atomic hydrogen in some of the targets, including an extensive corona
around the Earth (Fig. 134).

The camera was left on the Moon, but a duplicate was later modified
for use on Skylab (Chap. 7). (Before launch, I reminded John Young
that he would be matching the first optical astronomer, Galileo—the first
person to look at the Moon through a telescope from Italy in 1608, 364
years ago. Young was not impressed, but he pointed the camera accurately at the Earth from the Moon.)

The Apollo-16 rock samples were found to be richer in aluminum and
calcium than mare samples from earlier missions, and many of the results

FIG. 52. The far-UV Car-
ruthers camera-spectro-
graph in the shadow of
the LM at the Apollo-16
landing site, with Rover
and Astronaut Young in
the background. The in-
strument and tripod are
gold coated for tempera-
ture control. The cylinder
at the top is the camera's
aperture, facing right; the
square opening below it
holds an objective grating
for spectroscopy. The turn-
table above the tripod was
leveled by Young and al-
lows pointing at a specified
azimuth. The camera also
swings from horizontal up
to a specified altitude.
(NASA photo.)

are given on pages 142-150. One sample collected by Duke near the northernmost traverse on April 23 seems to be the remains of a comet which impacted the Moon, leaving small amounts of water, carbon dioxide, methane, hydrogen cyanide, and hydrocarbons detected by E. K. Gibson (MSC) in a vacuum evaporation chamber (see Science for Jan. 5, 1973).

We now turn to Apollo 17—the lunar scientists' last chance. We all regret that no Apollo 18 was sent to confirm or deny the theories developed from data obtained through Apollos 11 to 17.—TLP

Visit to Taurus-Littrow

ROGER W. SINNOTT

(Sky and Telescope, February 1973)

The last flight of the Apollo series may well be the best remembered. Apollo 17 landed the first geologist—Harrison H. Schmitt—who roamed the surface with mission commander E. A. Cernan for 10 times as long as the crew of Apollo 11 had in 1969. They brought back 249 lbs of geological samples, including the mysterious "orange soil" first seen on this mission.

They also set up four new experiments, and succeeded in deploying the troublesome Heat-Flow Experiment. Overhead in the Command Module, Ronald E. Evans operated a powerful radar sounding device expected to penetrate ¾ mi. into the Moon to map subsurface layers, perhaps outlining the lunar mascons (p. 97).

Ironically, part of Apollo 17's success must be attributed to financial cutbacks in the space program; the schedule was crammed with all the best projects originally planned for Apollos 17, 18, and 19. There was an air of finality about Apollo 17 when Schmitt and Cernan, after the last excursion, were told to keep their 40-lb excess of lunar rock samples.

Cernan and Schmitt landed on the Moon south of the crater Littrow (Fig. 53) on Dec. 11, 1972. During the first hour out, they readied the Rover and its television camera, and had a chance to look around.

The region was highly pitted with small, shallow craters. Cernan noticed tracks of rocks that had rolled down nearby hills. On the valley floor, Schmitt saw what appeared to be "boulders covered by some of the dark mantle." This dark material, possibly volcanic and young, was one reason for selecting this landing site.

As on the five earlier manned landings, high priority was placed on setting up the Apollo Lunar-Surface-Experiments Package (ALSEP).

FIG. 53. Artist's reconstruction of the view across the Taurus-Littrow Valley toward South Massif, showing the intended Apollo-17 landing site, the three planned traverse routes, I, II, and III, and the planned stopping points (stations) numbered 1 to 10. (NASA painting by Jerry Elmore.)

About 300 ft east of the lunar module, Cernan and Schmitt deployed the five experiments in a spokelike pattern centered on the plutonium-powered radio transmitter (Fig. 54).

The mass spectrometer, intended to analyze any trace of lunar atmosphere, a device for studying the speed, direction, and mass of micrometeorites, the Seismic-Profile Experiment, and the much-publicized Heat-Flow Experiment (which had failed on two previous Apollo missions) were successfully deployed. (See p. 129.)

Only the gravity meter seemed defective. Joseph Weber (University of Maryland) had hoped to confirm the existence of the controversial gravity waves predicted by General Relativity Theory to come from binary stars and other oscillating masses. Weber thought that one of the wires suspending a weight in the apparatus must have broken. Nevertheless, the instrument was functioning as a conventional seismometer.

Because of difficulties in placing the ALSEP instruments, the planned traverse I, shown on Figure 53, in the Rover was somewhat curtailed. The ride was plagued by flying dust, because Cernan had accidentally knocked off part of the Rover's right rear fender. A makeshift was clamped on for the second EVA, as shown in Figure 55.

FIG. 54. ALSEP instruments deployed near the Apollo-17 landing site around the dark central station with rodlike radio antenna pointed at Earth. To the right is the RTG plutonium power supply, and to the left (background), the surface gravimeter covered with a curved sunshade. The foreground instrument is the meteorite experiment, connected to the central station (as other instruments are) by a cable. (NASA photo.)

On December 2, at 6:17 p.m., the cabin was depressurized again, to begin the longest interval that astronauts have ever spent on EVA— 7 hrs, 37 min. During a round trip of some 12 mi., Cernan and Schmitt followed closely the prescribed route (II on Fig. 53) that took them to the foot of 7000-ft South Massif mountain.

While driving, the men encountered many shallow, dish-shaped craterlets, some with glassy linings, instead of the blocky, deep, rimmed variety attributed to recent impacts. At the foot of the mountain, the explorers saw blue-gray and tan-gray conglomerates (consolidated soil and rounded rock fragments), while on the slopes were blue-gray outcrops, possibly of extremely old rocks.

Here, the traverse gravimeter carried by the Rover indicated lower gravity than elsewhere. Apparently the massif consists of a block of low-density material extending as deep as a mile below the valley floor.

But the biggest surprise of the mission came at station 4, Shorty Crater, when Schmitt happened to kick at some dark material covering the crater's rim. "There is orange soil!" he shouted, and Cernan came over to confirm it. The geologist suggested that the orange material might be from a fumarole (vent releasing volcanic gases), while his companion likened its color to "oxidized desert soil." Preliminary tests at MSC in Houston revealed that at least 90 percent of the orange soil consists of tiny glass spherules, brownish to burnt orange (see p. 141).

At 5:31 p.m. on December 13, Cernan and Schmitt began their last lunar excursion. In their four-wheeled, battery-powered vehicle they were able to follow closely the assigned route. They visited the foot of mile-high North Massif and collected samples of boulders that may have rolled down its slopes. The travelers reached the safety of the LM after 7¼ hrs.

For three lonely days, astronaut Ronald E. Evans circled the Moon by himself, conducting experiments, and exposed some 2 mi. of film with his 3-in. mapping camera and 24-in. panoramic camera.

In the lunar sounding experiment, radio pulses were reflected from the Moon, received by antennas on the Command Module (including one extendable to 80 ft long), converted to light signals, and photographed. Stronger reflections were expected from subsurface layers of the lunar crust which conduct electricity better, such as those containing iron ore or water. In this way, it was hoped to map subsurface layers almost a mile deep, using signals at 5, 15, and 150 megahertz.

FIG. 55. Unneeded maps were taped and clamped in place to repair the Rover's broken right-rear fender so that dust would not be thrown all over the vehicle. (NASA photo.)

FIG. 56. At Station 6 on Traverse III, Schmitt examines a large boulder. (NASA photo by Cernan.)

Other remote-sensing devices aboard the CSM were successful: the infrared radiometer made approximately 100,000,000 surface-temperature measurements in a belt covering a third of the lunar surface.

The far-ultraviolet spectrometer worked perfectly, but seemed to find only insignificant traces of any lunar atmosphere. William E. Fastie of Johns Hopkins University commented, "The Moon is not degassing— it simply has nothing left to create an atmosphere."

The safe return of the Apollo-17 astronauts concluded the last, longest, and most successful of six manned lunar landings.

The orange soil got more attention than it merited, partly because direct evidence of a fumarole or recent volcanic vent on the Moon would be highly significant. It obviously excited geologist Jack Schmitt, who dug a trench down into it and took several photographs. Color contrast with the dark-gray lunar lands must have been striking. Within four months, the age of the orange-glass spherules was measured as 3.71 billion years ± 60 million by Oliver Schaeffer at Stony Brook, N.Y., using the argon-40/potassium-39 ratio, and chemical analysis showed them similar to the Apollo-11 samples.

The large boulder (Fig. 56) near Station 6 at the base of North Massif showed an interface between two kinds of rock: blue-gray breccia and greenish-gray basalt, an important record of an early lava flow.

It is just possible that another Apollo mission to the Moon could take place before 1999. There is a Saturn 1B booster in reserve at Kennedy Space Center, and the Command Module, LM, and experiment hardware could be built for less than $500 million. The scientific goal would be partly to check on previous Apollo data and all the conclusions drawn from them (see pp. 142-153). Another goal might be to prepare for the space colonization proposed by O'Neill (see p. 418). Although NASA administrators have given up hope, they agree that another Apollo mission is "highly desirable."

More of the Apollo scientific results were presented in a 1973 conference.—TLP

Notes on the Fourth
Lunar Science Conference

THORNTON PAGE

(*Sky and Telescope*, June-August 1973)

At the fourth Lunar Science Conference, held March 5-8, 1973, at the L. B. Johnson Space Center (JSC) in Texas, scientists were faced by changes in basic conclusions about the Moon:

> *The Moon is now believed to have a hot core. (Last year* [1972] *it was considered cold.)*
>
> *There is strong evidence that many large impacts about 3.95 billion yrs ago splashed material all over the Moon's surface, especially on what is now its far side.*
>
> *The puzzle of magnetized lunar rocks may now be solved by Harold Urey's theory of how a onetime strong lunar magnetic field was later lost by internal thermal changes.*
>
> *Last year* [1972] *there were two major types of lunar surface materials known; now there are four.*
>
> *Cosmic-ray tracks in the lunar soil are giving new data on the early history of the solar system and nearby regions of the Galaxy.*

These are but a few highlights from the enormous body of results of Apollo missions 15, 16, and 17 reported at the conference, organized and sponsored jointly by Anthony J. Calio, director of science and applications at JSC, and Joseph W. Chamberlain, director of the Lunar Science Institute (LSI).

THE LUNAR SURFACE

Accurate mapping in the equatorial zone between latitudes 32° N and 32° S was reported by Donald Light and James Hammack of the Defense Mapping Agency in Washington, D. C., who described how

they combine all the data from the mapping cameras (p. 118) to minimize errors. Distances between any two mapped points will be reliable
to about 25 m., on two sets of lunar maps soon to be published, with
scales of 1:250,000 and 1:20,000.

The *mantling* of extended areas of the lunar surface with ejecta from
craters received a great deal of attention at this conference. The most
recent crater ejecta and rays seem to be white (except the man-made
dark splashes where booster rockets and jettisoned ascent stages have
crashed into the Moon). Somewhat older ejecta (probably volcanic)
apparently formed a dark mantle that has been traced over a large elliptical region on the Moon's near side, from longitude 20° W to
65° E and from latitude 20° N to 15° S. Farouk El-Baz (Smithsonian
Institution, Washington) described this dark mantling as volcanic, and
calls this region "the ellipse of fire."

Using data from an infrared scanning radiometer on Apollo 17, Frank
Low (Rice University) and W. W. Mendell (JSC) again surveyed
the night side of the Moon. This time they found a number of *negative*
thermal anomalies—localities where the surface temperature is 5°-10°C
lower than their surroundings. These "cold spots" must be low-density,
highly porous material probably not of impact origin. The more usual
hot spots are places where rocks (that conduct heat well) protrude above
the soil, perhaps having been uncovered by impact. There are far fewer
hot spots on the far side of the Moon than on the near side.

SOME ANCIENT HISTORY

The general idea of heavy mantling by large-scale ejecta from the
Moon's near side seems to have caught on among selenologists. In a
most interesting paper which combined seismic and gravity data with a
"computer game" of impacts on the Moon, John A. Wood (SAO) made
a good case for an ejecta mantle twice as thick on the back side as on the
front.

This would explain gravity measurements,[1] the 2-km displacement
earthward of the Moon's center of mass relative to its center of figure
(p. 123), and the measured moments of inertia around the polar axis
and the Earth-Moon line. In Wood's model of the battered, mantled
Moon, the low-density crust (2.9 gm/cm^3) is about 70 km thicker on the
far side, and the large maria on the near side are supposed to be of
basalt (density 3.35) about 30 km thick.

The question remains: Why should one side of the Moon have been
battered by more impacts than the other hemisphere? Wood's calcula-

---

[1] For the most recent data, see A. J. Ferrari, *Science*, June 27, 1975.

tions show that this would happen because of the Moon's orbit around the Earth. When large-sized planetesimals moving under solar gravitational attraction encounter the Earth-Moon system, some collide with the Earth, some with the Moon, and most pass by. Wood's computer-generated charts show the general pattern of collisions on the two halves of the primitive Moon when it was only 40,000 miles from the Earth.

The leading face would have received four times as many impacts as the trailing one. In the following 4 billion yrs or more, the Earth's gravitational torque (twist) on our battered and reshaped satellite would have twisted it around to place the heavily cratered side toward us, while lunar gravitation consolidated the ejecta and mare basalt lava flows into rough isostatic equilibrium (with the weight of all layers supported evenly underneath).

Other evidence of this cataclysm 4 billion yrs ago has been collected by G. J. Wasserburg and his colleagues at CalTech. They have measured the ages of lunar rocks by several methods, and find that all the ages fall into two groups with a half-billion-year gap between. The basalts are about 3.95 billion yrs old, whereas the "ancient rocks" of the lunar crust are about 4.47 billion. This age of the Moon, estimated to be accurate to ±20 million yrs, is less than the age of the Earth.

Wasserburg and his colleagues reason that the "planetesimal battering" of the Moon took place 4 billion yrs ago, remelting the original crust. After the cataclysm, they reckon that liquid magma flowed out of the Moon's front surface for 600 million yrs or so, forming the mare basalt we find today.

History of the Moon's Magnetic Field

At last year's conference, one important puzzle was the *remanent magnetism* of lunar rocks that implies they crystallized in magnetic fields of 1000 gammas—about 100 times stronger than the Moon's present very weak magnetic field (see p. 128). This year David Strangway (JSC), who has measured the remanent magnetism in many lunar samples, connected the phenomenon with an idea of Harold Urey's.

In essence, Urey considers that the first step in planet formation from the primordial solar nebula was the condensation of *Moon-sized objects,* rotating balls of gas and dust having about the Moon's mass. [From this point on, Urey's theory has been modified somewhat.—TLP] It is likely that there were magnetic fields in the solar nebula, and when the spinning gases condensed to form the primordial Moon, the magnetic field trapped in the hot plasma increased to over 100,000 gammas (several gauss) in the hot core. Then the sun began to shine, blowing away the outer gases from the Moon, in which most of the solid materials had already condensed. (More matter was collected later by impacts.) Cooling of the lunar core proceeded past the Curie point of iron (780°C, the temperature below which iron is magnetic) when a field of some 20 gauss was "frozen" in the core material. Strangway notes that this provided a surface field of 1000 to 2000 gammas.

The outermost 200 km of crust was still hot from impacts and differentiation (lighter materials rising from the interior), so it was not magnetized. In the second stage, the crust cooled and became magnetized in the 1000-gamma field, but because of its low iron content (1 percent) its magnetization was less. During this stage "breccia flows" took place, where loose rocks and soils were compacted in the 1000-gamma field, and basaltic lava flows crystallized in the field, too.

Finally, in the third stage, the Moon's core heated up due to the radioactive decay of uranium, thorium, and plutonium, becoming hotter than the Curie point of iron, causing the "frozen" magnetic field to disappear. This lowered the surface field to a few gammas, as observed today.

Strangway's calculations show that this history can account for all the magnetic measurements made to date, although he warns that other sequences of events might have been possible. Of course, this goes for any theory of the Moon's or Earth's early history.

The Moon's Interior

By measuring subsurface temperatures at several depths, it is possible to find the rate at which heat is flowing outward from the lunar interior (see p. 129). Marcus Langseth of Columbia University reported further measurements of this kind from the Taurus-Littrow site.

The Apollo-17 heat-flow probes were placed in two drill holes 233 cm (92 in.) deep. One probe shows peculiar subsurface temperatures, probably caused by a nearby buried boulder, but the over-all heat flow is 3.2 microwatts/cm². The other gives around 2.8. (The Apollo-15 results gave 3.0.) These values indicate high internal temperatures in the Moon, consistent with other evidence of a hot core.

Moonquake data continue to be telemetered from the seismometers left by Apollos 12, 14, 15, and 16, as reported by Gary Latham and his co-workers in the University of Texas at Galveston. At Station 16 (with the most sensitive equipment), 3200 moonquakes are recorded each year, 1600 per year at Station 14, and 650 per year at 12 and 15. Many recur repeatedly at one or another of 41 known "active source zones," each up to a few kilometers in extent and located from 600 km to 1000 km below the lunar surface. The quakes vary in frequency during each month, showing that they are triggered by tides in the Moon. Their very low intensity proves that the accumulation of strain in the Moon's interior is extremely small. That is, the material is probably semiviscous.

Near-surface moonquakes are much less frequent (5/yr) and much more violent, apparently caused by settling of the Moon's crust at about 0.1 mm/yr.

The S-waves (transverse oscillations) come through the top 1000 km from all moonquakes on the Moon's front side, demonstrating that down to that depth the material is rigid, not molten. But four moonquakes and one meteoroid impact on the far side gave P-waves (longitudinal) and no S-waves, indicating a semimolten core of radius about 700 km. If the materials there are mostly silicates, the temperature would be about $1500°C$.

About 300 meteorite impacts are detected each year—easily distinguished from moonquakes. The recorded seismic energy allows estimates of meteoroid masses. Latham finds that most of the impacting bodies detected have masses between 0.5 and 1000 kg (15 ounces-1 ton). These numbers imply a bombardment rate of a one-kg meteorite or larger on each 40,000 km^2 (15,000 mi^2) of the lunar surface each year. This is lower than estimates based on counts of "recent' lunar impact craters. The latter discrepancy may indicate that most of the meteoroids have already been swept up by the Moon, Earth, and other planets during the last few million years. [However, there was a 7-day period of high impact rate in June 1975.—TLP]

M. N. Toksöz and his associates at MIT have continued to analyze seismic-wave velocities in the Moon, comparing them with laboratory measurements on lunar and terrestrial rocks under pressure. Modifying their conclusions of last year (see Fig. 50), these researchers find that the velocity between 25-km and 60-km depth is best fitted by lunar gabbro. Below 60 km, there is some uncertainty as to whether the P-wave velocity is 7.7 or 9 km/sec. The latter would require garnet-rich layers. Deep moonquake records show that the velocity is roughly constant as far down as 900 km.

For the outermost layers, more detailed data, discussed by Robert Kovach, Stanford University, were obtained with the active seismometers at the Apollo-14, -16, and -17 sites. At the last two (p. 134), eight explosive charges were fired at various distances from the seismometer.

At both Fra Mauro and Descartes, Kovach finds two similar layers. The upper is regolith soil and boulders, extending down to about 12 m., with a P-wave velocity near 110 m/sec. In the lower layer, probably breccia, the velocity is 250-300 m/sec.

At the Apollo-17 site, Taurus-Littrow, the top layer is 19 m. thick. Below 100-m. depth, the seismic velocity rapidly increases to 4 km/sec at a depth of 5 km, indicating increasingly rigid, solid rock.

Lunar Lava Flows

The interesting features known as *wrinkle ridges* are prominent in Mare Imbrium, Mare Serenitatis, and the Aristarchus plateau. They may be due to compression of the surface after lava flows caused the mare basins to sink, according to W. B. Bryan, Woods Hole Oceanographic Institution.

In a similar investigation on accurate Apollo photographs, Gerald G. Schaber of the U. S. Geological Survey traced the materials of three lava-eruptive periods northward across Mare Imbrium, all from a single source region in the south-southwestern corner of the Imbrium basin. About 3.0 billion yrs ago, the first and largest of these flows spread some 1200 km across the basin, and deposited tens of thousands of cubic kilometers of mare basalt. Somewhat later, the second flow complex extended 600 km from the source and about 2.5 billion yrs ago the third eruption spread lavas out for 400 km over the first two. These flows undoubtedly came from magma pools below the surface, indicating that the outer 200 km of the Moon's crust were hot at that time.

Another source of lava is the melting of surface rocks in great impacts such as produced the crater Tycho, studied by K. A. Howard and H. G. Wilshire, U. S. Geological Survey. Orbiter photographs show that nearly every fairly fresh crater larger than 40 km across has some lava ponds on the rim opposite to the direction from which the impacting body came.

Apparently some of the ejecta were molten, though most of the impact melt presumably had solidified within 10-20 min. of the fall. Large crater floors seem to have been completely flooded by liquefied material. In Copernicus, an estimated 700 km^3 of rock were melted.

Chemical Composition

Despite the emphasis on variety in lunar surface composition, there is really a remarkable uniformity in the abundances of the commonest elements, as Table 7 shows. It was compiled by Anthony Turkevich (University of Chicago), who made the first determination of lunar chemical composition after Surveyor 7 soft-landed in 1968 (see p. 96).

TABLE 7. AVERAGE CHEMICAL COMPOSITION OF THE LUNAR SURFACE
(percent of atoms)

	O	Na	Mg	Al	Si	Ca+K	Ti	Fe
Maria (eight missions)	60.6	0.4	5.3	6.6	16.8	4.7	1.0	4.5
Highlands Surveyor 7	61.8	0.5	3.6	9.2	16.3	6.9	0	1.6
Luna 20	60.3	0.4	5.2	9.7	16.0	5.9	0.15	2.1
Apollo 16	61.1	0.3	3.0	11.6	16.3	6.1	0.15	1.6
average	61.1	0.4	4.0	10.2	16.2	6.3	0.1	1.8
Over-all average	61.0	0.4	4.3	9.5	16.3	6.0	0.3	2.3

The results serve to define what we mean by lunar surface material, and they confirm that Turkevich's Surveyor-7 work was remarkably accurate.

These numbers show that the Moon's surface, compared with the solar system as a whole, is about 6½ times richer in aluminum (Al), calcium (Ca), and titanium (Ti), whereas it is four times poorer in sodium (Na), magnesium (Mg), and iron (Fe).

At last year's conference, much was made of the differences between mare and highland materials (see pp. 108 and 127). This year, Paul Gast (JSC) and his associates identified four broad categories of lunar rocks: 1, mare-type basalts; 2, so-called KREEP basalts, rich in potassium, rare earths, and phosphorus; 3, so-called VHA basalts, containing 20-24 percent aluminum oxide; 4, anorthositic rocks containing more than 24 percent aluminum oxide. It seems likely that all the rocks in each group had a common origin and similar history.

From their mineral composition, Gast reasoned that outer layers of the Moon were once molten, allowing low-density materials rich in aluminum oxide to flow to the top. An early crust formed about 4.6 billion yrs ago, when the highland-type, group-4 rocks crystallized. After that, the VHA and KREEP basalts (groups 3 and 2) flowed out to harden on the surface. The mare basalts (group 1) apparently were remelted several times just below the crust, which accounts for their complex composition. Dating of the mare basalts shows that large amounts of magma flowed on mare floors 3.6-3.9 billion yrs ago, possibly triggered by the cataclysmic impacts discussed by G. J. Wasserburg (see page 144).

LUNAR ATMOSPHERE STUDIES

Several investigators reported on the data radioed back by instruments left at Apollo sites on the Moon. A Rice University group has observations by the suprathermal ion detectors at Fra Mauro and Hadley Rille designed to study ionized gases liberated on the Moon and reaching the Moon from the sun (the solar wind).

Ion concentrations strong enought to detect occurred nearly every time the sun rose or set at one of the sites. The Rice investigators believe that these ions are from an extremely rarefied lunar atmosphere rather than from local sources.

At the University of Texas at Dallas, R. R. Hodges and others point out that the lunar atmosphere is fed by the solar wind, by gases baked out of the surface materials, by the vaporization of meteoroids, by products of cosmic-ray absorption and natural radioactive decay, and possibly by some volcanic venting. The atmosphere is kept low by the daytime solar wind.

A mass spectrometer on the Apollo-15 CSM detected molecular nitrogen near Mare Orientale on the Moon's far side, and this may have been volcanic in origin. A small mass spectrometer left behind by Apollo 17 shows significant quantities of argon-40 that probably come from the radioactive decay of potassium-40 in lunar materials.

Astronaut John Young made 10 pointings with the Apollo-16 far-UV camera, described on p. 136, four near the lunar horizon. In the lower half of each of the 1-min. exposures near the horizon was recorded over 100 rayleighs of Lyman-alpha emission in excess of the sky background —evidence for atomic hydrogen in the lunar atmosphere. But William Fastie and his co-workers at Johns Hopkins University had an ultraviolet spectrometer on the Apollo-17 CSM. When it was pointed across the Moon's terminator, viewing the sunlit atmosphere against the dark lunar surface, the spectrometer showed less than two rayleighs of Lyman-alpha emission, indicating only 50 hydrogen atoms/cm^3, less than 1 percent of the value expected from solar-wind protons losing their charges on impact. The disagreement with the Lyman-alpha photographs has still to be resolved.

History of the Solar System

Lunar science has recently developed methods of examining lunar materials that give direct evidence of events far from the Moon in space and time. Meteorite impacts on the Moon indicate past meteoroid densities in the solar system, and nuclear-particle tracks in lunar materials give a wealth of information about the sun's past activity, and possibly about nearby supernova explosions.

The group at the University of California at Los Angeles working with P. A. Baedecker finds meteoritic material to comprise about 3.5 percent of Apollo-16 surface soils, about 2 percent of Apollo-14 soils (also from a highland area), and about 1 percent of the mare-type soils prevalent at other Apollo landing sites. The material is identified as meteoritic by comparing the abundance of its trace elements with those in type C-1 chondrites found on the Earth. Figure 57, a plot of the meteoroid flux against the soil samples' estimated ages, indicates that the flux must have been much larger 4 billion yrs ago than it was half a billion years later—a result that also comes from crater counts.

From this plot Baedecker proposes two meteoroid components: one that produced the ancient high flux, half of which was swept up in about 45 million yrs, and a slowly waning flux with a half-life of 2 billion yrs. To some extent, J. W. Morgan and his collaborators at the University of Chicago confirm this by their analyses of lunar meteoritic material, which also shows two components.

J. Borg and others at Orsay, France, have studied etched nuclear tracks in lunar grains, etching the glass, pyroxene, and feldspar with hydrofluoric acid. They identify a class of old "fossil tracks" that are not erased by exposing the grains to a million-electron-volt beam, possibly because the lunar glass has been annealed by billions of years of mild temperature variations. The fossil tracks provide a 4-billion-yr record of solar-system history as the sun moved in the Milky Way Galaxy, spending about 50 million yrs in a spiral arm and 150 million yrs between arms (see Vol. 7, *Stars and Clouds of the Milky Way*). The particle-track record would vary with these changes in environment. For example, if inside a spiral arm a supernova exploded within 15 light years of the sun, it would blow its ejection shell into the solar system at a velocity greater than 3000 km/sec. Ions moving at such a speed would

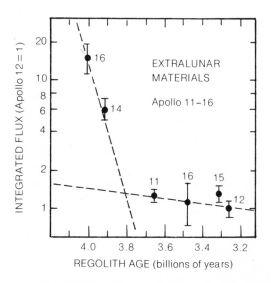

FIG. 57. Plot of meteoroid flux (logarithmic scale) *vs* age of the soil (regolith) where the meteoritic material is found at five Apollo landing sites and Luna-16's site. The steeper dashed line shows a much higher but rapidly dwindling rate of meteorite impacts about 4 billion yrs. ago. Since 3.7 billion yrs. ago, there has been a lower rate of impacts, halving every 2 billion yrs. (Diagram by P. A. Baedecker.)

INTEGRATED FLUX (Apollo 12 = 1)

EXTRALUNAR MATERIALS

Apollo 11–16

REGOLITH AGE (billions of years)

leave a high track density at the surface of an exposed lunar grain, with a steep gradient in the first 5 microns of depth in the grain.

Some of the lunar dust examined by Borg revealed as many as 10^{11} fossil tracks/cm^2, with a steep gradient, but their densities are only a tenth as large as expected from a supernova ejecting 10^{33} grams of material 15 light years away. Obstruction by interstellar dust and gas may account for this discrepancy. Also, the 10-percent proportion of lunar dust grains that show such steep gradients of fossil tracks would imply that one supernova erupts within 15 light years of the sun every 10 million yrs. This is more frequent than expected from Fritz Zwicky's counts of supernovae in distant galaxies (see Vol. 8, *Beyond the Milky Way*).

The source of the true cosmic rays (not those originating in the sun) is even farther away, and these very-high-energy particles were studied by Narendra Bhandari and his colleagues at the Tata Institute, Bombay, India. They chose to examine lunar olivine crystals, in which the tracks of iron nuclei and other abundant elements are reduced in number. Over 100 heavy-nuclei tracks were found 7 cm deep in a lunar rock. These help to establish the existence of very heavy particles in galactic cosmic rays.

Notes on Lunar Research

THORNTON PAGE

(*Sky and Telescope*, August 1974)

Recent research reported at the Fifth Lunar Science Conference, held in Houston, Texas, during the third week in March 1974, extends and alters the results reported for the Fourth Conference. Some 565 scientists from 15 nations met at JSC and LSI, under NASA sponsorship.

A few highlights from the lunar papers will be mentioned here. The conference also heard two accounts of Skylab accomplishments, several reports on Comet Kohoutek, and on the fifth day a number of topical summaries.

As soon as it was learned from the Apollo samples that the Moon's chemical composition differs significantly from the Earth's, the origin of the Moon became a more difficult scientific problem. If formed near each other in the primordial solar nebula, the two bodies should have similar chemistry. But if the Moon originated far from the Earth, how was it captured? J. V. Smith of the University of Chicago and, independently, H. E. Mitler, SAO, discussed ways in which chemical differentiation might take place during the capture process.

It is currently supposed that the forming Earth (proto-Earth) and the material of which the Moon is made coalesced from planetesimals at roughly our present distance from the sun. Once it became relatively large and massive, the proto-Earth would capture many small bodies through collisions. Those bodies which did not collide, but passed within the Roche limit, would have been broken up into many smaller pieces. (The Roche limit is the distance from the Earth's center—2.88 Earth radii—within which the tidal effect of terrestrial gravity would disrupt a passing body with little or no cohesive strength.)

Such proto-moonlets—loosely packed material, several hundred miles in diameter—would have been originally formed between 0.8 and 1.2 astronomical units from the sun, not in some more remote region. Smith believes a number of these moonlets passed close to the proto-Earth, and he emphasized the difference between the Roche limit for a moonlet's high-density, consolidated core and that for its low-density silicate mantle. Shattering of the mantle could take place, resulting in the accumulation of low-density debris in Earth orbit, which later coalesced to form the Moon. Much of the dense core material would go on into solar orbit, leaving the Earth-Moon system. It also may be possible, with repeated fragmentation of moonlets after the Moon had formed, to explain lunar mascons and the major maria impacts by later collisions with the sun-orbiting core material.

Mitler pointed out that it is normally difficult or impossible to capture material that passes by the Earth. E. Öpik has recently shown, however, that in passing within the Roche zone a proto-moonlet coming in on a parabolic orbit would have about half of its mass torn off and captured.

In extending this idea, Mitler shows that different incoming orbits for the proto-moonlets result in different and well-defined fragments of crust and mantle being captured. Thus the Moon would have coalesced from iron-poor material. This theory goes far toward explaining the formation of the Moon with peculiar chemical composition.

Of course, many other topics were discussed and measurements reported—far too numerous for this chapter. The main theme of the Sixth Lunar Science Conference in March 1975 was that impacts have played a prominent role in the history of Moon, Earth, Mars, Mercury, and Venus. C. H. Simonds (LSI), with W. C. Phinney, J. L. Warner, and G. H. Heiken of JSC, showed from experiment and theory what happens to large volumes of molten rock produced by an impact on the Moon. Mixed with surrounding cold material in the lunar regolith, it cools suddenly, forming a finely crystalline rock with chunks of unmelted lunar soil—just like many of the lunar-rock samples.

Herbert Zook (JSC) calculated the proportions of meteoric and lunar materials melted and vaporized, using the estimated distribution of impact velocities from Earth observations of meteors. He concluded that meteoroids melt about 20 times their mass and vaporize over twice their mass of lunar-surface material, but that little of the vaporized material escapes. The impact debris, which covers much of the Moon's surface, contains about 20 percent meteoroid material. Later, John Lindsey (University of Texas) and L. J. Srnka (LSI) found evidence of cosmic (interstellar) dust in the subsurface lunar soil. Refining Figure 57, they noted marked increases of the micrometeoroid flux every 100 million yrs, an interval that matches the estimated travel time of the solar system across the spiral arms of the Milky Way Galaxy (see Fig. 145 and Vol. 8).

Continuing this theme, the conferees discussed craters on Mercury and Mars photographed in recent fly-bys—discussed in the next two chapters.—TLP

Trips to Venus
and Mercury

Figure 58 is a diagram of the solar system (from Vol. 1, Wanderers in the Sky) showing some of the regularities: nearly circular orbits (except for the comets, not shown), all nearly in the same plane, "direct" motions —clockwise as seen from the south—and the range of planet sizes, with the largest (Jupiter) in the middle of the sequence from Mercury to Pluto. The distances between the planets are very large, so that space probes take months to reach nearby planets, instead of the 2½ days required to reach the Moon. The sun-Earth distance, averaged over the year (93 million mi., or about 150 million km) is called the astronomical unit (AU), long used by astronomers to save zeros when listing distances in the solar system. (For interstellar distances they use the light year, about 63,000 AU, or 5880 billion mi., or 9.46×10^{12} km. Note that 1 AU is about 500 light seconds; that is, it takes sunlight about 8 min. 20 sec to reach the Earth.)

The Bode-Titius law is an easy way to remember the approximate sun-planet distances: to the series of numbers, 0, 3, 6, 12, 24, 48, 96, 192, 384 (each double the preceding one) add 4 and divide by 10. The results, 0.4, 0.7, 1.0, 1.6, 2.8, 5.2, 10.0, 19.6, and 38.8, are roughly the distances of

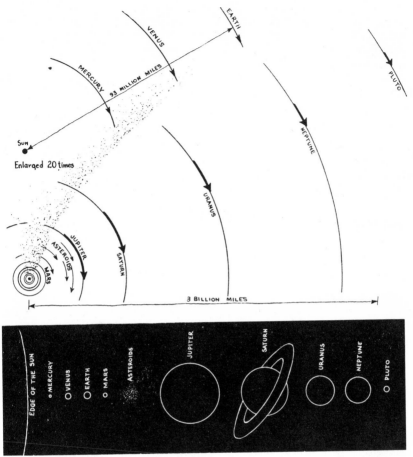

FIG. 58. Diagram of the solar system to scale, with an enlargement of the central section (upper left) and the planet sizes (below). (Yerkes Observatory diagram.)

Mercury, Venus . . . Neptune from the sun in AU. Neptune's distance by Bode's law is too high by 8 AU, and the formula is almost twice too large for Pluto. As shown in Volume 3, *Origin of the Solar System*, this roughly geometric progression of orbit sizes is another regularity to be explained by theories of the origin. In 1945 a German physicist, C. F. von Weizsäcker, postulated eddies in a rotating nebula of dust and gas from which the sun and planets were formed. As modified by the American astronomer G. P. Kuiper, the theory goes like this: some 5 billion yrs ago the rotating "primordial solar nebula," about 100 AU in diameter, cooled and contracted. Eddies caused material to collect in large rotating clouds; most of the mass in the center was later to become the sun. As the sun condensed, it heated up, nuclear reactions started in its center, and it became bright enough to "blow away" (by radiation pressure) most of the nearer unconsolidated clouds, leaving the "terrestrial" planets, Mercury, Venus, Earth, and Mars. The next (fifth) cloud was broken up, leaving the asteroids—a swarm of thousands of "minor planets." The next four clouds were too far from the sun to lose their light gases (mostly hydrogen) and became the giant planets: Jupiter, Saturn, Uranus, and Neptune. The Jupiter- and Saturn-clouds were large enough to have sub-systems of eddies which formed miniature solar systems of almost a dozen moons each (see Chapter 6). The origin of Pluto is uncertain, but the outermost shell of the primordial solar nebula condensed into billions of small frozen chunks, a few of which were disturbed each year so that they either escaped from the sun's gravitational attraction or else fell in toward the sun to become comets in long, elliptical orbits (see Glossary).

More recently, Harold Urey and other astrochemists computed the chemical compounds that would condense in the solar nebula at the various planetary distances and thus explain in more detail the high densities of the terrestrial planets and the low densities of the planets farther out (see Table 8). The whole process was a complicated one, with eddies, temperature, density, chemical affinities, and over-all chemical composition of the solar nebula playing significant roles. Space-age astronomy has made some progress in explaining comets (see Chapter 7), the giant planets (Chapter 6), and the Moon (Chapter 3), but leaves several unanswered questions about the terrestrial planets.

Venus, most similar to planet Earth in many respects (Table 8), is an enigma because its surface is entirely obscured by clouds, first thought to be water, then estimated to be various molecular combinations, including strong acids in the form of dust and droplets supported by carbon-dioxide gas (CO_2). As noted in Volume 2, Neighbors of the Earth, recent radar observations show mountains and basins on Venus' invisible surface, and a very slow retrograde rotation (east to west, opposite to the Earth's), with a 243-day period relative to the stars (119 days relative to the sun). Starting in 1967, several Soviet and American space probes visited Venus.—TLP

TABLE 8. PLANET CHARACTERISTICS

Name[1]	a^2[2] (AU)	Diameter[3] (km)	Average Density[4] (gm/cm³)	Surface Temp.[5] (°K)	Atmosphere[6] major gases (minor)	Rotation[7] Period (days)	Space-probe visits[8]	Condensation Temp.[9]	Expected composition[10]	Satellites[11]
Mercury	0.39	4,880	5.4	(100-600)	— (He)	59	Mariner 10, 3 passes	1000	$(Fe,Mg)SiO_3$ Fe-Ni Core No FeS	0
Venus	0.72	12,104	5.3	750	CO_2 (HCl,HF,CO, H_2O,H, O,He)	243R	SU Venera 1-10 Mariner 5, 10 (2 Pioneers)	800	Oxides Fe-Ni Core No FeS	0
Earth	1.00	12,756	5.52	300	N_2,O_2,A,H_2O (H,O,He,O_3)	1.0	Many down-looking	600	Oxides(H_2O) $(Fe,Mg)SiO_3$ FeS,Fe-Ni core	1
(Moon)	1.00	3,476	3.34	350	— (H,A,space-craft contam-inants)	27.3	SU Luna 1-31, Zond 3 Ranger 7-9 Surveyor 3-7 Orbiter 1-5 Apollo 8-17	600		0
Mars	1.52	6,787	3.94	250	CO_2 (CO,O_2,H_2O, A,H)	1.03	SU Mars 2-6 Mariner 4,6,7,9 (Viking 1,2)	450	Oxides(H_2O) FeS core	2

	a	diameter (km)	density	(temp °K)	constituents	rotation				sat.
Uranus	19.18	51,800	1.2	(63)	H₂,He,CH₄ (NH₃)	0.46R	None	—	(same as sun)	5
Neptune	30.06	49,500	1.7	(53)	H₂,He,CH₄ (NH₃)	0.67	None	—	(same as sun)	2
Pluto	39.44	6,000?	5.?	(43)	— (none)	6.39	None	(low)	?	0?
(comets)	(10–10,000?)	10?	1.?	(5–3000)	H,CN,CO,C₂, NH,OH,CH,NH₂	None? 10?	None	(very low)	Dust and frozen gases	0

[1] Column 1 includes the Moon, Ceres (largest minor planet in the Asteroid Belt), and the large satellites of Jupiter and Saturn. Comets are not planets, but typical approximate values are listed for them, too.

[2] a is the semi-major axis of the planet's orbit, or average distance from the sun, known to 1 part in 10^5. (For satellites, the parent-planet's a is listed.)

[3] Equatorial diameter in km. (1 km = 0.62 mi.) The polar diameter of a rapidly spinning planet is shorter (133,220 km for Jupiter).

[4] The density of water is 1 gm/cm³. Note that Saturn's average density is less than water's.

[5] Average surface temperature in degrees Kelvin. Water freezes at 273°K and boils (on Earth's surface) at 373°K. Values in parentheses are uncertain.

[6] The major constituents are given for each atmosphere in the first line, minor ones in parentheses below.

[7] The sidereal rotation period (relative to the stars). Because of orbital motion, the period from noon to noon (relative to the sun) is very different for Mercury and Venus, one minute longer for Earth, and the same for outer planets and their satellites.

[8] Space-probe visits include planned ones in parentheses. The listing is incomplete for Earth and Moon.

[9] The probable temperature (°K) in the solar nebula where the planet or satellite condensed about 5 billion yrs ago.

[10] Layers from surface (top line) to core (lowest line) expected to condense from the solar nebula and differentiate (separate) after radioactive heating melted the planet's interior. The theory (John S. Lewis' "Equilibrium Model" explained in *Scientific American* Magazine for March 1974) gives average densities and surface layers in agreement with observations. Gravity dominated the condensation process—for Jupiter, Saturn, Uranus, and Neptune, gravity held on to *all* material, regardless of its "freezing temperature."

[11] The number of known satellites in orbit about each planet. At least 2 or 3 of Jupiter's and Saturn's satellites were captured long after these planets and their "miniature solar systems" condensed from the solar nebula.

TABLE 8 (Continued)

Name[1]	a^2 (AU)	Diameter[3] (km)	Average Density[4] (gm/cm³)	Surface Temp.[5] (°K)	Atmosphere[6] major gases (minor)	Rotation[7] Period (days)	Space-probe visits[8]	Condensation Temp.[9]	Expected composition[10]	Satellites[11]
Ceres	2.77	785	3.?	(200)	— (none)	0.3	None	350	FeS	0
Jupiter	5.20	142,800	1.33	(123)	H_2,He (CH_4,NH_3,H_2O)	0.41	Pioneer 10,11	—	(same as sun)	13
(Io)	5.2	3,660	3.5	(100)	H,Na (No H_2O)	1.76	Pioneer 10,11	250	$(Fe,Mg)SiO_3$ $Mg_3Si_2O_5(OH)_4$	0
(Europa)	5.2	3,100	3.3	(110)		3.5	Pioneer 10	190	$(Fe,Mg)SiO_3$ $Mg_3Si_2O_5(OH)_4$	0
(Ganymede)	5.2	5,270	1.9	(110)		7.2	Pioneer 10,11	170	Water ice $(Fe,Mg)SiO_3$ $Mg_3Si_2O_5(OH)_4$	0
(Callisto)	5.2	5,000	1.7	(110)		16.7	Pioneer 10,11	140	NH_3-H_2O $Mg_3Si_2O_5(OH)_4$ FeS core	0
Saturn	9.54	120,000	0.69	(93)	H_2,He (CH_4,NH_3,C_2H_4)	0.42	(Pioneer 11)	—	(same as sun)	10
(Titan)	9.5	4,950	2.2	(105)	H_2	15.9	(Pioneer 11)	90	CH_4-H_2O NH_3-H_2O $Mg_3Si_2O_5(OH)_4$-FeS	0

Preliminary Results from the Venus Probes

RAYMOND N. WATTS, JR.

(*Sky and Telescope*, December 1967)

Oct. 18-19, 1967, saw two important advances in space exploration. Only 34 hrs apart, a Soviet probe plunged deep into the atmosphere of Venus, and an American craft sailed past the planet for closeup measurements.

On October 16, the president of the U.S.S.R. Academy of Sciences, Mstislav Keldysh, asked Sir Bernard Lovell, director of Britain's Jodrell Bank radio observatory, to help in monitoring signals from the Soviet spacecraft Venera 4 as it approached our sister planet.

The 250-ft radio telescope at Jodrell Bank soon picked up faint signals from Venera 4 that indicated it was on course. Two days later, the Soviet craft released to the pull of the planet's gravity a 35-in. egg-shaped capsule coated with an ablative heat shield that burned away as it entered the atmosphere. Once it had been sufficiently slowed by atmospheric friction, the package released a parachute and began a 90-min. drift toward the surface.

Soviet scientists released results from Venera 4 after a delay of only 7 hrs: the atmosphere consists of about 98 percent CO_2, the remainder being water vapor and oxygen. The temperature some 16 mi. above the surface of the planet was reported to be $+104°F$, but it rose to $+536°F$ as the capsule descended. The pressure or density is 15-22 atmospheres (sea-level values on Earth). It is not certain whether the final readings were made at the planet's surface. In the intense heat, the electronic equipment aboard the capsule may have been destroyed before it reached the ground.

The American Mariner 5 began its 4-month voyage to Venus on June 14, 1967.

On October 19, Mariner 5 flew past Venus some 2500 miles from its surface, while its instruments viewed the planet and sampled its environment. It sent its radio signals through the Venus atmosphere as it looped behind the planet and then reappeared on the other side 21 min. later.

Then began its long task of radioing the observations back to Earth, about 49 million mi. away. Operating on only 10 watts of power,

Mariner's tiny transmitter required 34 hrs to relay all the data stored on the tape recorder during the 2-hour observing session.

Preliminary analysis has resulted in several significant findings: no radiation belt analogous to the Van Allen belt encircling the Earth, a weak magnetic field, probably no stronger than 0.03 of the Earth's, and an atmosphere 75-85 percent CO_2, with small amounts of hydrogen but no recognizable oxygen. It was also discovered that the night side of the planet emits a faint ultraviolet glow of unknown origin.

The problems of landing on Venus get worse as we learn more. In 1965 they looked so formidable to NASA engineers that plans for a U.S. landing have never gotten very far.—TLP

Surface Conditions on Venus

GEORGE S. MUMFORD

(*Sky and Telescope*, March 1965)

Robert B. Owen, a NASA scientist at the George C. Marshall Space Flight Center, in Huntsville, Alabama, has made a critical comparison of several theoretical models of the atmosphere of the planet Venus. After weighing the evidence, Owen concludes:

"The surface of Venus appears to be a forbidding place, with surface temperatures comparable to a red-hot oven and pressures that exist on Earth only at ocean depths greater than 350 m. Since the melting points of aluminum, lead, tin, magnesium, zinc, and bismuth might be reached, pools of molten surface material could cover much of the bright side. The high pressures may produce clouds of exotic materials that would ordinarily be gases at such temperatures.

"The temperature of the dark pole has been estimated by Frank Drake (Cornell University) to be 540°K, and with the high surface pressures, several possible constituents of the lower atmosphere may condense out in that region. Such polar seas may contain liquid benzene, liquid acetic acid, liquid butyric acid, liquid phenol, and if the pressure exceeds 60 atmospheres (see App. V), perhaps a bit of liquid water. . . . If this model is valid, a surface landing presents an engineering problem of a magnitude never encountered before."

The Exploration
of Venus

RAYMOND N. WATTS, JR.

(*Sky and Telescope*, February 1971)

Thirty-five minutes' worth of telemetered data was the payoff for the 4-month mission of the Soviet spacecraft Venera 7. Launched on Aug. 17, 1970, the 2600-lb vehicle had traveled some 200,000,000 mi. to perform its brief sampling of Venus' atmosphere.

At 8:02 a.m. Moscow time on December 15, Venera 7 entered the atmosphere of the cloud-covered planet and the spherical instrument-carrying capsule was separated from the spacecraft. When atmospheric drag had slowed the entry package to about 800 ft/sec, a large parachute was deployed.

Although there was considerable speculation that this improved version of previous Venus probes might reach the planet's surface, it apparently did not do so. (But see p. 164).

The next NASA flight now scheduled is for 1973, a "slingshot" that will pass close to Venus on its way to Mercury, but there will be no encounter with the Venus atmosphere (p. 168).

The National Academy of Sciences has recently published a 79-page report, *Venus: Strategy for Exploration*, which reviews the recommendations of a 21-man study panel convened in June 1970 by the Space Science Board in cooperation with the Lunar and Planetary Missions Board of NASA. Co-chairmen of the panel were Richard M. Goody of Harvard and Donald M. Hunten of Kitt Peak National Observatory. The panel strongly urges direct exploration by means of orbiting satellites and entry probes, such as the 850-lb "universal bus," capable of carrying orbiters, entry probes, balloons, and landers.

Information about the extensive cloud cover blanketing Venus and the makeup of its turbid carbon-dioxide atmosphere could contribute to understanding atmospheric systems in general, including the Earth's. The several conflicting theories about Venus could be tested with relatively simple measurements. For example, determining the concentration of hydrogen atoms and compounds could shed light on whether or not oceans once covered the planet's surface.

The universal bus, its experiments, and the regions of Venus that it would explore are already well planned by GSFC and the Systems Di-

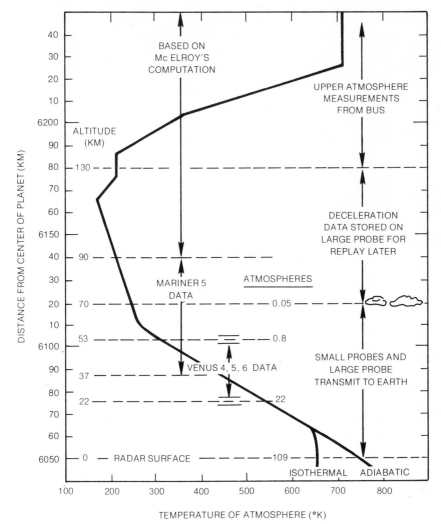

FIG. 59. Temperature *vs* height in Venus' atmosphere. The heavy line shows the best estimates in 1971, and the vertical arrows (center) show the basis for estimations. The vertical arrows at right show plans for the 1977 mission. (NASA diagram by J. E. Ainsworth.)

vision of Avco Corporation of Massachusetts. The bus consists of a 12-sided prism with a fixed array of solar cells. The craft's width is only 48 in., its height 45 in., but it can deploy booms for stabilization and to carry its antennas and experiment sensors. Eight small hydrazine thrusters (rockets) are symmetrically arranged around the bus to permit maneuvering or spinning it as the mission requires.

Planned for 1977, the multiple-probe mission would measure Venus' upper and lower atmosphere by means of the bus itself, a large entry probe, and three small probes.

The observed deceleration of the large probe as it falls through Venus' upper atmosphere will aid in determining pressure-temperature-density profiles 130-70 kilometers above the surface. Like the small probes, the large one will descend by parachute from the upper cloud layer, at 70 km, to about 52 km, then be released for free fall again.

Meanwhile, instruments on the bus itself will make observations in the extreme upper atmosphere, as indicated on Figure 59. Tying all the results together will permit a new temperature profile to be drawn of far greater accuracy than is now possible.

A geometrically favorable opportunity will occur in 1978 for sending a universal bus to Venus to become a satellite of that planet.

Future missions in this proposed Planetary Explorer series should include measures of the planet's atmospheric circulation by balloons floating at preset levels; radar, thermal, and optical mapping of the planet and its cloud layers from orbiters; study of the interior by seismic apparatus lowered to its surface; and TV examination of that surface by a landed probe.

Soft Landing
on Venus

(*Sky and Telescope*, March 1971)

It now appears that Venera 7 did actually land. On Jan. 26, 1971, the Moscow newspaper *Isvestya* published an announcement that computer analysis of the radio transmissions revealed weak signals continuing 23 min. more, although at only 1 percent of the previous intensity. Moreover, the Doppler shift in the signals changed in a manner indicating that the fall had ceased and that the ship was lying on the surface.

Telemetry data after the landing revealed a temperature of 475° $\pm$ 20°C, and a barometric pressure of 90 $\pm$ 15 atmospheres, according to the summary in the New York *Times*.

This is the first occasion on which scientific observations have been made from the surface of another planet. Three earlier Soviet space probes that entered the Venus atmosphere all ceased to transmit while still descending. Venera 7 was generally similar to these three forerunners, but it was constructed to withstand temperatures as high as 530°C and pressures up to 180 atmospheres.

Soviet Venus Probe

RAYMOND N. WATTS, JR.

(*Sky and Telescope*, May, September, and November 1972)

The flight of Venera 8 was announced by the Tass news service in a release datelined March 27, 1972:

"The Soviet Union launched the automatic station Venera 8 at 7:15 Moscow time today. It is planned to conduct research on the physical characteristics of interplanetary space, in particular, on changes of concentration of neutral hydrogen and fluxes of solar plasma."

After a flight lasting 117 days, Venera 8 reached its destination on July 22. At 7:40 UT, the instrument-carrying descent capsule was separated from the spacecraft. As it rushed down through the planet's dense atmosphere, the capsule was slowed by atmospheric drag to 250 m/sec. Then a parachute opened and allowed the package to make a soft landing at 9:29 UT, at a point on the sunlit side of Venus.

During the capsule's descent and for 50 min. after landing, it transmitted telemetry signals to Earth.

From these signals, much valuable information has been gained about Venus and its atmosphere. From a preliminary report which appeared in the Moscow newspaper *Pravda* on September 10, here are some highlights:

As the 1090-lb descent capsule parachuted through the dense atmosphere of Venus, it measured the ammonia content (by means of a reagent that changed from yellow to blue in the presence of this gas), indicating 0.01-0.1 percent ammonia at elevations 46-33 km.

Winds of 110 mi/hr were observed above 45 km, slackening to less than 4 mi/hr in the lowest 10 km of the atmosphere. These winds, blowing in the direction of the planet's rotation, were determined from the lateral motion of the capsule.

During the descent, a photometer monitored the decreasing light intensity. "A certain amount of sunlight does penetrate to the surface so that there is a detectable difference in illumination between the planet's night and day sides," says the Soviet report.

After the capsule had settled on the surface, its gamma-ray spectrometer measured the radioactivity of nearby rocks. Calibrated against samples of terrestrial rocks, the spectrometer found that material at the landing site contained about 4 percent potassium, 0.002 percent uranium, and 0.00065 percent thorium. These abundances are more like those in terrestrial granites than in lunar rocks, but Soviet scientists warn about generalizing from a single site.

The Venera-8 spacecraft landed on the planet's sunlit side, near the terminator that divides day and night. The surface temperature was measured at 470°C (878°F). When the Venera-7 ship landed on the night side on Dec. 15, 1970, it reported 475°C. Both probes determined the surface pressure as 90 atmospheres. Thus, there seems to be surprisingly little difference between the day and night hemispheres. On each side of Venus it is hot enough to melt lead and zinc.

The remarkably similar surface temperatures on both sunlit and dark sides of Venus were at first a puzzle until it was noted (by R. M. Goody) that refraction of infrared rays in the dense CO_2 atmosphere is extreme; heat radiation from the surface can barely escape from the atmosphere and equalizes the temperature over most of the surface at about 750°K (477°C or 890°F). Higher up, as Figure 59 shows, the atmospheric temperature drops to about 200°K, then rises again to about 600°K in the very low-density upper layers.

FIG. 59-A. Shortly after touchdown at 8:13 A.M. Moscow time on Oct. 22, 1975, the Venera-9 lander transmitted this panorama of slablike rocks as large as 15 in. across, one rounded rock, and a view of the distant horizon in the upper right corner. (The white arc in center foreground is the lander's shockproof base, and the vertical stripes are gaps in the picture transmission while other data were radioed back *via* the Venera-9 orbiter.) (Tass photo.)

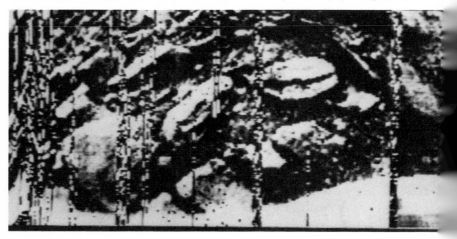

In June 1975, the Soviet Union launched two more Venus probes, Venera 9 and 10, with improved landers. As Venera 9 orbited the planet, its chilled lander descended on October 22 through the hot atmosphere, was slowed by a French-designed parachute, landed, and transmitted data for 53 min. on the surface before the heat "killed" it. The temperature was measured at 485°C (905°F), pressure as 92 atmospheres, and the first TV pictures from another planet's surface were received in Moscow (Fig. 59A). Surprisingly, the pictures show sharp-edged, slablike rocks lying on a gravel surface well illuminated by sunlight that casts shadows, despite Venus' heavy clouds. On October 25 the Venera-10 lander reached the surface on lower land where the temperature was 465°C and the wind some 7 mi/hr. This lander lasted 65 min. and transmitted more photographs of the surface—another triumph for the Soviet Tyuratam Cosmodrome. (NASA plans for two Pioneer space probes to Venus with landers have been held up by lack of funds. This missed chance to "beat the Soviets" roused criticism of space scientists, NASA executives, and the U.S. Congress.)

Other questions about Venus concern its magnetosphere (see Chapter 2), the composition of the atmosphere, and wind motions as revealed by cloud patterns. NASA's Mariner-10 mission was designed to learn more about all of these.—TLP

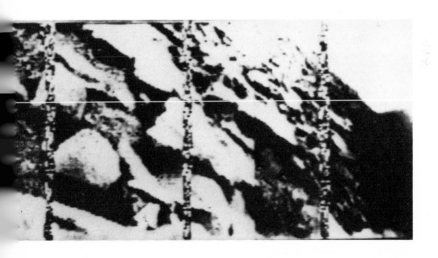

News of Mariner 10

RAYMOND N. WATTS, JR.

(*Sky and Telescope*, January 1974)

Mankind's first closeup look at Mercury could come March 29, 1974, when a 1108-lb spacecraft with two television cameras aboard passes within 700 mi. of the solar system's innermost planet. Launched Nov. 3, 1973, from Cape Canaveral, Florida, Mariner 10 will be the first space vehicle to study two planets in turn. As it sweeps by Venus on February 5, that planet's gravitational attraction will help propel the spacecraft on toward Mercury.

To use Venus' gravitational field to advantage for this "slingshot" effect, Mariner 10 must pass the planet at exactly the right distance. Thus, four in-course trajectory corrections have been scheduled: one in November to correct launching errors, one just before and one just after the encounter with Venus, and one about four weeks before reaching Mercury. Thereafter, Mariner will enter an orbit around the sun with a period just twice that of Mercury, so that another encounter is possible on Sept. 21, 1974.

In addition to its television systems, Mariner 10 is equipped to measure magnetic fields, the solar-wind intensity, and infrared radiation from the planets' surfaces. An ultraviolet spectrometer will be used to analyze the planetary atmospheres. As the spacecraft goes behind each planet, the extinction of radio transmission will yield data about the atmosphere and the planet's size.

Television coverage of Venus will extend from ½ hr before closest approach—some 3300 mi. above the surface—to about 17 days later. Plans include ultraviolet photography of cloud patterns and a search for possible small natural satellites of our sister planet.

Then, 2 months later, Mercury will come under intense scrutiny in a 32-hr interval. Pictures of the sunlit face are expected to have a resolution of better than 1 mi., with views of selected regions attaining 300 ft.

On Nov. 3, 1973, an Atlas-Centaur rocket carried Mariner 10 into a temporary orbit around the Earth. After a 25-min. coast in orbit, the Centaur engines were restarted and Mariner left with initial speed over 7 mi/sec westward relative to the Earth. This impulse, opposite to the Earth's motion around the sun, was necessary for Mariner's inward motion, slowing behind the Earth, to "fall" toward the orbits of Venus and Mercury.

Once in space, the delicate, spiderlike craft (Fig. 60) came alive, unfolding the booms and panels that had been compactly stowed inside the booster's protective nose cone. All went well, except that the heaters designed to warm the barrels of the two TV cameras failed to turn on.

The television experiment's principal investigator, Bruce C. Murray of CalTech, expressed concern that the cold, $-18°F$ to $+20°F$ (50° lower than planned), would warp the lenses and spoil the pictures. However, early test views of Earth, Moon, and stars showed no ill effects. It was decided to leave the cameras turned on all the time, to eliminate the chance of cracking the cold vidicon (TV) tubes if they were to be suddenly turned on as Mariner approached Venus.

In design, Mariner 10 is much like the earlier Mariner spacecraft. As it draws closer to the sun, the solar panels can be tilted as much as 76° to keep them from getting too hot.

FIG. 60. Mariner 10 has two solar panels, each 106 in. x 47 in., which normally face the sun but can be tilted. (JPL-NASA diagram.)

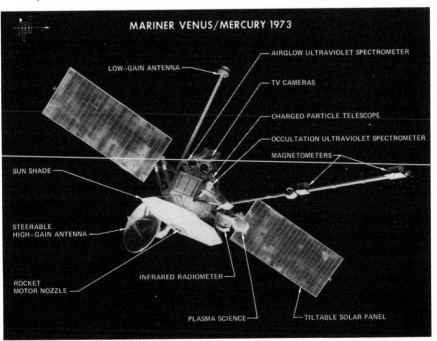

Two identical Cassegrain telescopes (Fig. 63), of 1500-mm (59-in.) effective focal length and 0°.5 fields of view, are on a motor-driven platform on the shaded side of the spacecraft. A pair of vidicon cameras can either use these telescopes for high-resolution photography or employ wide-angle lens systems of 62-mm focal length for views covering a rectangular 11° × 14° field. Each television frame consists of 700 scan lines, with 832 picture elements per line.

Mounted on the directable platform with the telescopes is an ultraviolet airglow spectrometer. On the main octagon of the spacecraft are an ultraviolet spectrometer, a charged-particle telescope for studying solar flares and cosmic rays, and an infrared radiometer.

A 20-ft hinged boom supports two magnetometers. Plasma detectors will be measuring the flux of positive ions and electrons that comprise the solar wind, during the entire trip in to 0.4 AU from the sun.

Television transmissions will be at 117,600 or 22,050 bits per second. At the higher rate, full-resolution TV frames take 42 sec each. The slower rate ensures separating the signal from background noise. There is a choice of two picture modes: the entire frame can be transmitted at high rate, or a strip equal to one-fourth of a frame can be sent at slow rate. It is also possible to record up to 36 TV pictures for later slow transmission.

To keep track of Mariner 10, a two-way radio link will be maintained throughout the flight so that the returned signal received back on Earth will have suffered a double Doppler shift in frequency (used to calculate Mariner's accurate radial velocity with respect to the Earth, see p. 98). Range or distance is determined by the time out and back.

NASA's Deep Space Network stations at Goldstone in California, Canberra in Australia, and Madrid, Spain, have been assigned to track Mariner 10 and relay the scientific data to Mission Control at JPL in Pasadena. Each station has 210-ft and 85-ft dish antennas. The communication frequencies to and from the craft are all in the S-band, between 2100 and 2300 megahertz. Mariner 10 also carries an experimental X-band (8415-megahertz) transmitter for use in atmospheric measurements when it moves behind Venus or Mercury. Since the Goldstone 210-ft antenna is the only one equipped to receive X-band frequencies, the encounters have been planned to occur when Mariner 10 is above the horizon for California.

Venus Observed
by Mariner

(*Sky and Telescope*, April 1974)

After a 94-day voyage inward through the solar system, the 1108-lb space probe passed about 3600 mi. from the planet on Feb. 5, 1974, just as planned.

Nearly all the observations planned for the encounter with Venus could be carried out. The initial failure of the television-camera heaters was cured, and the anomalous behavior of the roll gyro was compensated. Certain planned pictures of Venus could not be taken, however, because the camera platform would turn only through an angle of 107°, instead of the intended 122°.

Mariner 10 tested its cameras by photographing the Earth and Moon, returning excellent high-resolution pictures from ranges of up to nearly a million miles. The cloud cover of Earth imitated the conditions to be expected in photographing Venus, while the Moon's rugged surface and low albedo simulated Mercury.

For transmission, the picture elements are digitally coded into eight-bit words. Approximately 3000 pictures of Venus were taken, beginning at 16:21 UT on Feb. 5, 1974, 40 min. before closest approach. A picture was made every 42 sec and transmitted to the 210-ft antenna in California, 28 million mi. away. During 21 min. beginning at 17:07 UT, while Mariner was occulted by the planet, the pictures were tape recorded aboard the spacecraft.

In the next few hours, the cameras began transmitting high-resolution mosaics of Venus. This ended on the 13th, when the increasing distance of Mariner from Venus had reduced the resolution to about that of good Earth-based photography of Venus.

Pictures taken along the planet's illuminated limb at very high resolution (about 200 m.), show distinctly a faint stratified haze, with a variable number of layers above the very uniform top of the planet's cloud deck.

As has been known for many years, ultraviolet images show vague dark bands and patches. The ultraviolet mosaics taken by Mariner 10 (Fig. 61) show for the first time the detailed pattern of motions in the planet's very turbulent atmosphere.

Mariner-10's ultraviolet airglow photometer has given valuable information about the composition of Venus' upper atmosphere. Lyle Broadfoot (KPNO) reported that hydrogen is abundant but that there is only slight evidence of deuterium (heavy hydrogen).

If the atmosphere of Venus originated like the Earth's, from the outgassing of rocks and from oceans, it should be relatively rich in deuterium.

It is likely that Venus' hydrogen comes from the solar wind, which is mostly protons with very few deuterium nuclei. Because Venus has little or no magnetic field, the positively charged protons can enter the atmosphere and form hydrogen atoms. This process would be balanced by continual escape of hydrogen.

Broadfoot also reported that Mariner 10 has observed helium, as a trace constituent, and an important amount of atomic oxygen, revealed by the emission line at 1304A.

The Mariner-10 infrared radiometer was designed primarily for Mercury; hence only one of its two channels—wavelength range 35-55 microns—was useful for temperature measurements of Venus. About 8 min. before closest approach, the thermal scans showed a sharp cutoff at both the bright and dark limbs of Venus, but no strong change across the terminator to the day side (see p. 166).

Investigating teams at GSFC, Los Alamos Scientific Laboratory, and MIT all participated in the plasma-science experiment—measuring the properties of the solar wind and its interaction with Venus and Mercury. The solar-wind plasma acts like a supersonic gas streaming outward through the solar system, producing a shock wave on the sunward side of each planet, and an elongated wake on the opposite side. Mariner 10 approached Venus nearly along the wake, and its instruments first detected the wake some 5 days out from the planet, at a distance of about 500 Venus radii, or approximately 2 million mi.

Venus' plasma wake is only about 1/10 as wide as the Earth's, where the magnetosphere (Fig. 7) is much larger than the Earth itself. Venus seemingly has no magnetic field—a fact noted by Mariner 5 and confirmed by Mariner-10's magnetometer experiment. Consequently, the diameter of Venus' plasma wake is little bigger than the planet itself.

As Mariner neared the edge of Venus, its 2295-megahertz S-band radio signals passed through an increasing thickness of that planet's atmosphere, changing both in frequency and strength. Analysis of these changes will yield pressure- and temperature-values from the top of the atmosphere down to the surface.

The final cutoff of the signal as it grazed the solid surface provides a measure of the planet's radius. At reappearance from occultation, a similar sequence of observations was made in reverse order. In addition, an X-band radio transmitter, working at 8415 megahertz, was employed to measure the absorption by the planet's atmosphere at a second frequency.

The Doppler shift of the S-band transmissions also measured the gravitational attraction of Venus, indicating that the mass of Venus is slightly less than had been inferred from the corresponding Mariner-5 result in 1967.

The S-band and X-band transponders, rebroadcasting signals received from Earth as Mariner 10 passed behind Venus, confirmed the central part of Figure 59, and added a set of four small temperature inversions— irregularities in the temperature at heights of 6108, 6110, 6113, and 6115 km from the planet's center. The effect of solar wind on Venus' outer atmosphere seems to be a narrow, low-density tail of ionized gas extending down-sun about 2 million mi. Somewhat closer to the planet, both Mariners 5 and 10 detected atomic hydrogen by Lyman-alpha emission (see Chapter 8), extending some 30,000 km from the planet center. Mariner 10 also detected helium (He) by its 584A emission and oxygen by 1304A emission, both similar to the Earth's, and smaller amounts of He+ 304A, Ne 740A, Argon 1048A, CO 1480A, and C 1657A.

The near-ultraviolet (UV) photos in 3500-4000A light showed a remarkable circulation pattern, symmetrical about Venus' equator, recording cirruslike clouds about 80 km above the planet's surface. These were fitted into the mosaic of Figure 61 by Bruce Murray (CalTech) and his photographic team to show a full equatorial band first published in Science for March 29, 1974, and reproduced in the August 1974 issue of Sky and Telescope.

Striking as these results on Venus were, Mariner 10 went on to learn even more about Mercury which, like Pluto, had been a "mystery planet." In our view from Earth, Mercury is always close to the sun, thus difficult to observe at 50-120 million-mi. distance. Mariner 10 made three visits to Mercury, the first within 4500 mi. of the surface, the second within 30,000 mi., and the third within 200 mi. Only six of the 37 photographs in Sky and Telescope can be reproduced here, so we will augment them with some summary comments made at the Sixth Lunar Science Conference (in Houston, March 17-21, 1975) by those who had studied many more of the hundreds of photos obtained by Mariner 10. D. E. Gault (NASA Ames Research Center) explained the surprising similarity between the cratered surfaces of Mercury and the Moon. The surface gravity on Mercury is 2.3 times larger than that on the Moon, so one would expect craters to be 20 percent smaller because the material thrown

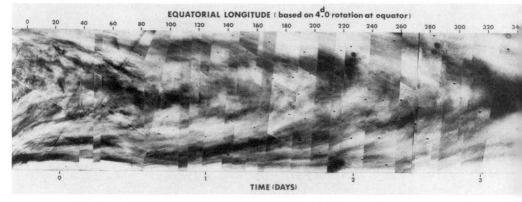

0 20 40 60 80 100 120 140 160 180 200 220 240 260 280 300 320 34

0 1 2 3

TIME (DAYS)

FIG. 61. Mosaic of Mariner-10 UV photos of Venus' cloud pattern. The subsolar point moves, as Venus rotates, from left to right. The top is at latitude 40° N, the bottom, 50° S. The large white clouds about 1000 km in extent are probably condensates in Venus' stratosphere, and indicate divergent flow around the subsolar point. (*Science* photo by Bruce C. Murray.)

out by impact falls back to the surface more quickly. Atmospheric drag on Mercury makes them smaller yet. However, Mercury's larger mass causes meteoroids to hit the planet with about twice the velocity of impacts on the Moon. The higher velocity—impact energy—produces a larger crater, just about cancelling the effects of surface gravity and atmospheric drag. The ratio of crater diameter to depth is about the same on Moon and Mercury, implying that the surface material (silicate rock on the Moon) is the same as on Mercury.

James E. Guest (University of London) noted that the most conspicuous feature on Mercury is the Caloris Basin—a huge 1300-km crater surrounded by gouged radial grooves and ejecta (see Fig. 66). On the opposite side is another interesting feature not matched on the Moon, called the "Weird Terrain"—about 100,000 sq km of obviously disturbed ("chopped up") surface. Guest suggests that seismic waves from the Caloris impact were partly focused on the opposite side of Mercury, where they produced the earthquake damage of the Weird Terrain. Mercury also has scarps (Fig. 67), not seen on the lunar surface. Their arrangement suggests thrust faults, like wrinkles in a shriveled prune, possibly indicating that Mercury's radius decreased by 2 km after its solid crust formed.

Many preliminary statistical statements about Mercury's surface may be later confirmed: there are far fewer rilles than on the Moon; one hemisphere has more plains than the other. Both speakers praised the computer-enhanced contrast of the Mariner-10 photos.—TLP

The Planet Mercury as Viewed by Mariner 10

ROBERT G. STROM

(*Sky and Telescope*, June 1974)

During the last few days of March 1974, Mariner 10 achieved the first close-range inspection of Mercury. The primary goal of the TV experiment was to photograph the surface for insight into its structure and evolution. Over 2000 frames were acquired.

As the spacecraft approached Mercury, it saw the planet's disk approximately half illuminated, and after the fly-by this was also the case. Closest approach was on the night side and also on the far side of Mercury as then seen from Earth. This aiming point precluded the TV cameras from seeing the sunlit surface at the minimum distance of about 750 km (470 mi.).

FIG. 62. Artist's concept of Mariner 10 passing Mercury, showing the sun and a star (Canopus) in the background, the TV views, and other observations made. (NASA drawing.)

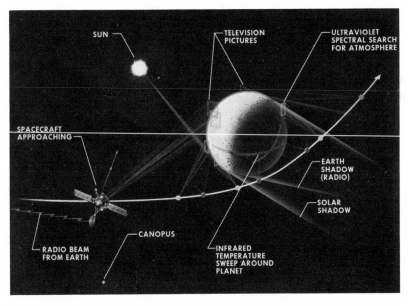

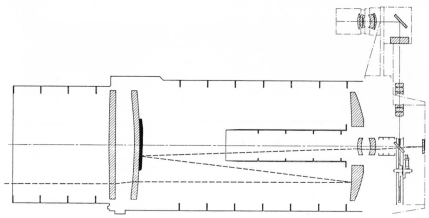

FIG. 63. Diagram of Mariner-10's TV cameras, showing the wide-angle lens (top) and the much larger Cassegrain telescope of 7-in. aperture (below). Light enters from the left and is focused on the Vidicon (TV tube) at far right. The diagonal mirror at lower right can be replaced with filters on a wheel seen edge-on. (NASA diagram.)

Because of this trajectory, Mariner-10's twin TV cameras had optical systems of 1.5-m (59-in.) focal length (see p. 170 and Fig. 63). The TV telescopes are $f/8.43$ Cassegrains with all-spherical optics, and fields of view of 0.36° by 0.48°. In each instrument, the primary mirror is made of Cer-Vit (glass of very low thermal expansion), whereas the two correctors and the field-flattening lens are of Suprasil II. A reflecting coating centered on the back of the second corrector serves as the secondary mirror (see Vol. 4). Each camera has an 8-position filter wheel, 6 positions providing different effective-wavelength responses from 3740A in the ultraviolet to 5760A in the orange.

Mariner 10 began photographing Mercury on March 23 from a distance of 5.4 million km (3.4 million mi.). Intermittent picture-taking went on daily until March 28, when nearly continuous operations commenced for the close encounter. After they ended, periodic operations were kept up until April 3, by which time the spacecraft was 3.5 million km (2.2 million mi.) past the planet.

The near-encounter picture-taking sequence, which began 16 hrs before closest approach, consisted of reading out an 18-frame mosaic of the planet every hour. The resolution of these frames ranges from about 12 to 4 km. From 3½ hrs to 35 min. before closest approach, a series of mosaics was transmitted with 3- to 0.7-km resolution. Thirty-six pictures were tape-recorded from 35 min. before closest approach to 22 min. after, with resolution from about 500 to 100 m. As Mariner 10 receded farther from Mercury, the picture-taking sequence was similar but in reverse order.

FIG. 64. Photomosaic of Mercury from 145,000 mi. on March 29, 1974, 6 hrs before Mariner-10's closest approach, showing numerous craters from the bright limb (top) at 110° W longitude to the evening terminator at 20° W. North is to the right. (JPL-NASA photo.)

Only about half of the planet's surface could be recorded on this fly-by, and only a small fraction of the pictures have yet been studied.

"Another Moon!"—has been the impression of many persons on seeing these closeups of Mercury. The resemblance was to some extent expected by astronomers, partly on the basis of radar observations.

Inspection of the photographs shows numerous craters, scarps, ridges, circular basins, and relatively smooth plains. Most of these features strongly resemble their lunar counterparts. The illuminated quarter of Mercury photographed by the approaching Mariner (see Fig. 64) has a rugged terrain similar to the lunar highlands, rich in overlapping craters and in basins. The plains are cratered to approximately the same degree as the Moon's maria.

The mercurian craters range in size from large basins down to pits barely detectable on highest-resolution photographs. They exhibit a complete spectrum of impact forms. Some preliminary measurements suggest that the depth-to-diameter ratios of mercurian craters are similar to those of lunar craters of the same size and type.

The Mariner-10 television team has proposed that one conspicuous rayed crater be named Kuiper in honor of the planetary astronomer Gerard P. Kuiper, who died on Dec. 24, 1973, and who was an original member of the TV team.[1]

[1] The 10-member team is headed by Bruce C. Murray of CalTech. It includes the author and scientists of many other colleges and institutions. The TV team published a preliminary report on the Mercury fly-by in *Science* for April 26, 1974, and results from all the Mariner-10 Mercury experiments appear in that publication.

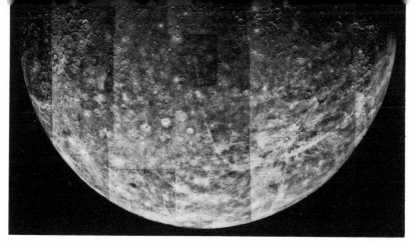

FIG. 65. Photomosaic of Mercury from 130,000 mi. on March 29, 1974, 5½ hrs after Mariner-10's closest approach, from bright limb (bottom) at 110° W longitude to the morning terminator (top) at 200° W, showing many rayed craters (bottom) and the outline of the huge Caloris Basin (center). The north pole is near the right edge, and Mercury's equator is to the left of the basin. (JPL-NASA photo.)

The largest basin is about 1300 km (800 mi.) across, and appeared on the terminator after Mariner 10 had passed the planet. It can be seen in part in Figure 65, centered at approximately longitude 195° W, latitude 30° N.[2] This huge, shallow depression has many of the characteristics of the Mare Imbrium basin and others on the Moon. Elsewhere on Mercury, many smaller basins of varying degrees of distinctness are evident.

The Mariner-10 photographs have also revealed irregular scarps, as much as 1 km high, which extend for hundreds of kilometers, cutting across large craters. Several inferences can be drawn from these photographs:

(1) There is no evidence of atmospheric erosion on Mercury. Hence the planet has not had any significant atmosphere since the end of the bombardment that produced its craters.

(2) The presence of flooded craters indicates that Mercury experienced internal heating and volcanism after the heavy bombardment. The extensive flooding implies a silicate composition for the outer layers of Mercury. However, the high mean density (Table 8) requires

[2] Longitudes and latitudes of the mercurian features are specified in a system adopted by the International Astronomical Union in 1970. The prime meridian on the planet is that which contained the subsolar point when Mercury was at perihelion on Jan. 10, 1950. As Mercury rotates with a period of 58.64617 days (two-thirds of its orbital period), the longitude of the central meridian, as seen from a fixed direction in space, increases from 0° to 360°. Latitudes are specified on the assumption that the planet's equator lies in its orbital plane.—Ed.

that the deep interior must consist of very much denser material, probably a large iron core. Hence Mercury is probably chemically differentiated, like the Earth.

The Mariner-10 mission has been an unqualified success, advancing our understanding of the early history and evolution of the inner planets.

Mariner 10 and Mercury

(*Sky and Telescope*, September 1974)

On July 2, 1974, scientists at JPL turned on Mariner-10's engines for 19 sec, placing the probe on the selected course for its second rendezvous with Mercury, on September 21. The spacecraft will pass about 47,360 km (29,930 mi.) from the planet's surface, the closest approach being over the sunlit side.

This trajectory was selected to extend the photographic coverage of Mercury's surface by Mariner-10's television cameras and will permit stereo mapping of the planet's south-polar regions, which were not accessible during the first fly-by in late March. Better viewing angles for some of the regions already photographed should yield better data on heights, slopes, and albedos of surface features.

A rich harvest of new scientific results from the March encounter has been gathered; computer "laundering" of the 2300 TV pictures has provided much clearer and sharper views of the planet. Figure 66 shows half of the 1300-km circular feature tentatively named Caloris Basin. Its broad floor is laced by fractures and sinuous ridges resembling those in the lunar maria. The fractures are believed to have resulted from the sinking of the depression's central part.

Mercury appears, like the Moon and Mars, to have an asymmetric surface—half of it heavily cratered, and the other half relatively smooth.

It is covered everywhere with a fine-grained material analogous to the lunar regolith, according to the TV team. The spacecraft's infrared radiometer, by measuring thermal radiation at wavelengths near 4.5 microns, showed a minimum brightness temperature of 100°K on the night side. The manner in which the temperature declined from sunset to local midnight implies that Mercury is covered by a porous soil, very similar in thermal properties to that of the Moon.

A very tenuous atmosphere was detected by the ultraviolet spectrometers on Mariner 10. Neutral helium is the principal constituent; hydrogen, argon, and oxygen could not be found. The total surface

FIG. 66. Mosaic of computer-processed Mariner-10 photos taken March 29, 1974, showing half of the Caloris Basin, bisected by the morning terminator (left) at about 200° W longitude. The top of the picture is 50° N latitude, the bottom 10° N. (JPL-NASA photo.)

pressure of Mercury's atmosphere is estimated at about 2×10^{-9} millibar—so low that the atmospheric atoms follow ballistic trajectories, their motions being determined by the planet's gravitational field rather than by mutual collisions.

The magnetometer- and plasma-experiments showed that Mercury has a very well-developed, detached, bow-shock wave, resulting from interaction of the planet's magnetic field with the solar wind. Deep inside this bow shock, the magnetometer detected fields up to 98 gammas as Mariner passed closest to the planet (704 km).

From the Mariner-10-radio-tracking data, the mass of Mercury has been very precisely determined. The ratio of the sun's mass to that of Mercury is 6,023,600 ± 600, agreeing well with the best previous values.

The radio-occultation experiment indicates that Mercury is very nearly spherical, for its radii in latitudes 2° N and 68° N are 2440 ± 2 and 2438 ± 2 km, respectively. Its diameter is therefore 3021 mi. The new values for the mass and size of Mercury correspond to a mean density of 5.44 gm/cm³.

Mercury does not have any satellites as large as 5 km in diameter (assuming the same albedo as the planet), a conclusion based on Mariner photographs, which show stars of visual magnitude 8.0 (see Glossary).

Mariner-10's Second
Look at Mercury

J. KELLY BEATTY

(*Sky and Telescope*, November 1974)

On Sept. 21, 1974, Mariner 10 swept past Mercury at a distance of 29,814 mi. (47,981 km), on its second visit. Within 49 hrs the pair of television cameras aboard Mariner 10 obtained about 500 high-quality images of the surface of Mercury.

Transmitted over a distance of 105 million mi. to Earth, the images were received by the 210-ft (64-m.) antennas of the Deep Space Network. Since the onboard tape recorder was inoperable during the fly-by, the high-resolution pictures had to be transmitted "in real time" (with no delay), one frame every 42 secs. Upon reception, the signals were taped so that the photographs could later be computer processed to clarify and enhance their details.

The real-time transmission rate was 117,600 bits of information per second, but the Canberra and Madrid receivers could accept only every fourth data bit. Goldstone received all 117,600 bits per second with one 64-m. and two 26-m. dishes, adjusted to bring the three signals into phase, providing a greatly improved signal-to-noise ratio.

To relay an image to Earth, each of its 582,400 picture elements (pixels) was assigned a brightness value from 0 (black) to 255—an eight-bit "word" for transmission—which was taped at the receiver and reconverted into a photograph by computers at JPL.

Such "raw" photographs have some background noise and gaps that lower the resolution. To bring out the rich detail of the pictures printed here, the taped picture data were run through special computer "stretch" programs by technicians at JPL's Image Processing Laboratory, in which brightness values in low-contrast areas were reassigned to increase the contrast. For instance, brightness values of 8, 10, and 12 in a low-contrast part of a picture, would be "stretched" to 8, 20, and 30.

The new pictures extend Mariner-10's useful photographic coverage from about 25 percent of the planet's surface last March to about 37 percent, most of the gain being in the southern hemisphere. In the 176 days between the two fly-bys, Mercury had completed almost exactly three rotations, so substantially the same half of the planet's globe was sunlit on the two occasions.

FIG. 67. Region near Mercury's south pole viewed by Mariner 10 from 40,000 mi. about 1 hr after closest approach on September 21, 1974, showing a scarp 185 mi. long. This and others are considered evidence of the planet's contraction. (JPL-NASA photo.)

The remarkable similarity of Mercury's surface to the Moon's indicates that the two bodies have similar histories. Nevertheless, there are several important differences between the mercurian and lunar surfaces, some directly attributable to the planet's larger surface gravity. Mercury's craters are shallower than lunar ones of the same size, and their ejecta cover only a fifth as much area. Terraces along the inner walls and central peaks are commoner in Mercury's craters.

Perhaps the most significant features of the photographs are the huge scarps that run for hundreds of kilometers across the planet's face (Fig. 67). Possibly as high as 3 km, these cliffs generally have sinuous outlines and cut across both craters and intercrater areas. Their lobed form suggests they may be thrust or reverse faults, possibly caused by compressive stresses while the surface was cooling early in the planet's history.

The photographic mosaic (Fig. 68) shows the "Weird Terrain," characterized by a rough, scabby surface and greatly degraded crater rims, many with downslope gouges like landslides. This appearance is also found on the Moon in the areas antipodal (directly opposite) to the large impact basins Mare Imbrium and Mare Orientale.

FIG. 68. Mosaic of computer-enhanced Mariner-10 photos taken March 29, 1974, showing the "Weird Terrain" with rough hills and ridges cutting across many degraded craters. The crater with the flat floor near center is about 90 mi. across. The top and bottom edges are at 19° and 45° S latitude, the left and right ones at 43° W and 10° W longitude. (JPL-NASA photo.)

Since Mercury's "Weird Terrain" lies at the antipodal point of the 1300-km Caloris Basin, there is a possible connection between the two features.

Sometime after lava flooded the impact area, the central part of Caloris Basin gradually subsided until it was about 2 km lower than the outer part. This subsidence produced a pattern of tensional fractures.

How do all these observations affect theories of the planet's history? Here are some tentative interpretations by the TV experiment team:

- Mercury underwent a period of early heavy bombardment, including formation of huge basins, followed by widespread volcanism.

- On Mercury (and on the Moon as well), an ancient, heavily cratered terrain has been preserved in extensive regions without major modification.

- Planet-wide scarps and ridges are suggestive of compression. . . . An obvious speculation is that an iron-rich core underwent shrinkage, resulting in compression of the outer layers, during the terminal phase of heavy bombardment but not throughout the remaining history of the planet.

Mariner 10 will make a third rendezvous next February.

Names on the Planets
and Satellites

(*Sky and Telescope*, October 1974)

In recent years, numerous surface features have been mapped on both sides of the Moon, on Mercury, Venus, Mars, and the martian satellites. As space exploration advances, by the end of this century there may be 30-40 members of the solar system whose surface features will need naming.

To deal with this problem on a unified basis, last year the International Astronomical Union formed a working group for planetary-system nomenclature, with Canadian astronomer Peter M. Millman as president. It contains five task groups having these chairmen: Moon, D. H. Menzel; Mercury, D. D. Morrison; Venus, G. H. Pettengill; Mars and satellites, B. A. Smith; Jupiter and satellites, T. C. Owen.

At its first meeting, in Ottawa, Canada, on June 27-28, 1974, the working group approved three names already provisionally used on Mercury: Caloris Basin, Hun kal, and Kuiper.

Two new crater names on Mars were also approved: Kuiper and Vishniac. The former honors the late planetary astronomer Gerard P. Kuiper. Wolf Vishniac was an American microbiologist at the University of Rochester who planned the life-detection experiment for the Viking Mars lander, and died prematurely while exploring in Antarctica on Dec. 10, 1973.

Mercury Revisited
by Mariner 10

(*Sky and Telescope*, May 1975)

Mariner 10 made its third and last useful fly-by of Mercury on March 16, 1975, skimming to within 200 mi. of the surface at 22:38 UT. The fruits of this encounter include more than 300 additional high-resolution

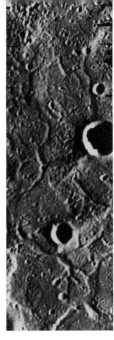

FIG. 69. High-resolution Mariner-10 photo of a strip in Caloris Basin from about 300 mi., 34 min. after closest approach on March 16, 1975. North is to the left, and the largest crater is at 31° N, 183° W. Details as small as 50 m. can be detected. (JPL-NASA photo.)

photographs of the cratered surface and new information on Mercury's magnetic field.

This field had been detected during the first encounter on March 29, 1974. However, there was uncertainty as to whether it was intrinsic to the planet or due to an interaction with the solar wind.

Using first-encounter magnetometer data and assuming that the field originates in Mercury's core, Norman F. Ness of NASA's GSFC predicted field strengths as 350 and 700 gammas at the equator and poles, respectively (1 percent of the Earth's), and a tilt of 7° between magnetic and rotational axes. A high-latitude pass was chosen for the third fly-by to permit new magnetometer observations that would check this model.

The observations of the bow shock and magnetopause agree well with Ness' predictions.[1] This inherent field is probably caused by either dynamo action in a fluid core—or by a permanent magnetization of materials in the planet's crust.

Because Mariner 10 passed much closer to the surface than in previous encounters, the two television cameras gave increased resolution—possibly as small as 50 m. The improved resolution is shown in Figure 69.

Mariner 10 had suffered a series of misfortunes, including loss of its on-board tape recorder and near exhaustion of its attitude-control gas.

The craft failed to respond when technicians attempted to turn on

[1] Later in 1975, Ness joined with Siscoe and Yates of MIT to say that at least part of the measured magnetic field was due to a violent magnetic storm on Mercury, caused by solar-wind gusts.—TLP

its scientific equipment two days before the encounter, apparently because the high-gain antenna had drifted and could not receive signals from Earth. A complex and time-consuming series of maneuvers finally oriented the probe correctly for the fly-by. German scientists helped this rescue operation by giving up time on the Deep Space Network antennas, which were following their spacecraft Helios (p. 326).

Problems at the Canberra, Australia, receiving station forced use of an 85-ft antenna instead of a 210-ft one, reducing the Mariner data-reception rate from the normal 117,000 bits per second to only 22,000. Rather than sacrifice picture quality, the spacecraft was ordered to transmit only one-fourth of the area of each photograph, preserving the excellent resolution of the images.

5

Preparations for
Landing on Mars

Public interest in Mars was intense during the first 30 or 40 years of this century, after barely perceptible linear features were named "canals" and Percival Lowell founded an observatory in Flagstaff, Arizona, to confirm the existence of humanoid life on Mars. As mentioned on page 73 of Volume 2, Neighbors of the Earth, the Mariner-4 space probe telemetered to Earth close-up photos of Mars that showed no canals—only craters and mountains. The following articles confirm this, but the interest in life on Mars is well founded. Conditions there, unlike any other planet except Earth, are suitable (although not favorable) for life as we know it. The major deficiency is water, upon which terrestrial life is heavily dependent.

Popular writers, and a few serious scientists, have pointed out two possibilities: (1) there was more water on Mars in the past, and (2) other forms of life, not requiring water, may exist there. There are still other possibilities to be checked on the surface of Mars. But the NASA Mariner and the Soviet missions to Mars have narrowed them; in 1966 the Harvard astronomer Carl Sagan (now at Cornell) wrote an article "Is There Life on Earth?" showing that photographs of the Earth with the same resolution (about 2 km) as the Mariner-4 photos of Mars would show little if any evidence of human life on Earth (although a spectrogram would show the red spectrum line indicating chlorophyll in tree leaves and foliage). This chapter will show how Mariners 9 and 10 have obtained 100-m.-resolution photos which show winding (dry) river beds, but no evidence of abandoned cotton fields or town houses.

The Mars exploration program started at NASA's Jet Propulsion Laboratory (JPL) somewhat earlier than the Venus probes. Note the more primitive picture transmission from Mariner 4.—TLP

Mariner-4 Photographs of Mars

(*Sky and Telescope*, September 1965)

In a brilliant technological achievement that came off almost exactly as planned, Mariner 4 flew past Mars at a distance of 6118 mi. on July 14-15, 1965, after a 7½-month voyage of 325 million mi. This gave man his first close-up look at the solid surface of another planet, greatly changing some widely held ideas about Mars.

In less than half an hour, Mariner was scheduled to take 22 pictures of the red planet, commencing at a range of 10,500 mi. However, the last three frames show nothing, as the camera was pointing at the night side.

As transmitted by Mariner, each complete picture consisted of 200 lines of 200 dots each, somewhat like a halftone engraving. The dots were sent as numbers (in binary notation) that indicated light intensity on a scale of zero (white) to 63 (black). Transmission of one complete frame to Earth required approximately 8½ hrs; reception of the series took about 10 days.

At JPL, processing of the radioed numbers to reconstruct the photographs has been long and complicated. The pictures have been subjected to repeated operations to improve contrast and to suppress electronic defects ("noise").

When the space probe began taking pictures at 5:18 p.m. PDT on July 14, its camera was pointing toward the planet's limb, at a region nearly vertically illuminated by the sun.

Although greatly foreshortened, Frame 1 shows a richly variegated pattern of light and dark, whose distinctness is added evidence of the thinness of the martian atmosphere, for the camera's line of sight passed through it very obliquely.

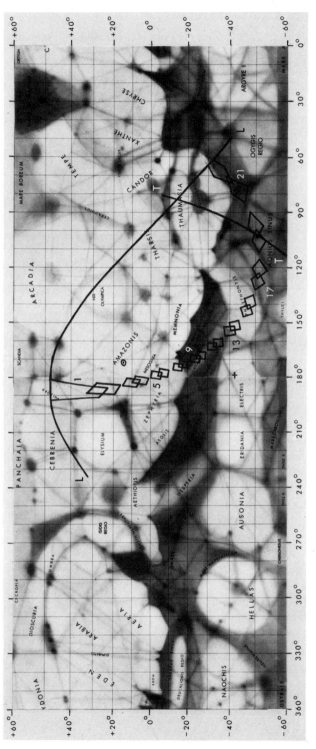

FIG. 70. A map of Mars, with conventional names and the "canals" seen by ground-based observers. Areas 1 through 21 outline Mariner-4's photos from the bright limb (line LL) to the evening terminator (line TT). (Air Force-NASA chart.)

The map in Figure 70 shows that Frames 2 through 6 lie in the martian bright regions Amazonis and Mesogaea, where many "canals" had been plotted; the photographs give no indication of them. . . .

Bruce C. Murray of CalTech commented: "When we received Frame 7, even in its initial form . . . we began to recognize clearly many, many craters as well as some light-to-dark variations on the surface of the planet. . . ."

Other CalTech members of the Mariner study team were R. B. Leighton and R. P. Sharp. They write: "We have observed more than 70 clearly distinguishable craters ranging in diameter from 4 km to 120 km. It seems likely that smaller craters exist; there also may be still larger craters . . . since Mariner 4 photographed, in all, only about 1 percent of the martian surface. . . ."

Mariners to Fly
Past Mars

RAYMOND N. WATTS, JR.

(*Sky and Telescope*, April 1969)

On Feb. 24, 1969, an Atlas-Centaur rocket was launched from Cape Kennedy. It carried Mariner 6, the first of a pair of 910-lb interplanetary probes built by JPL to make closeup observations of Mars. This is the first try since Mariner-4's success in 1965.

The schedule calls for Mariner 7 to be launched on March 24. The craft are to fly within about 2000 mi. of the red planet on July 31 and August 5, the first passing over Mars' equatorial zone, the second over its south-polar region.

These flights are precursors of NASA missions planned for 1971, when two Mariners are to orbit Mars for several months.

Two craft are sent each time to increase the chances of making successful observations.

After Mariner-6's launch, the Atlas booster carried it to a height of nearly 90 mi. and a speed of more than 8000 mi/hr.

Then the second-stage Centaur burned for about 7 min., increasing Mariner's speed to nearly 25,000 mi/hr, for a direct ascent into space without a period of orbital coasting around the Earth. On February 28 a command was radioed from JPL that fired Mariner-6's onboard engine for 5.4 sec, slightly slowing the craft. Subsequent observations indicated a path that should carry the probe within 2000 mi. of Mars.

FIG. 71. The Mariner spacecraft is a standard octagonal structure with four solar panels (later reduced to two on Mariner 10, Fig. 60), two radio antennas on top, and a scan platform for cameras on the bottom. From the small, low-gain antenna (top) to the cameras at bottom is 11 ft. (JPL-NASA photo.)

As shown in Figure 71, each Mariner measures 11 ft from the cameras at the bottom to the top of the low-gain antenna. With its solar-cell panels deployed, the spacecraft spans 19 ft. The basic 8-sided magnesium body is 18 in. high and 4½ ft across. Its eight compartments carry power-conversion equipment; midcourse-maneuver rocket engine; central computer, sequencer (electronic timer), and attitude-control subsystem; flight telemetry and command subsystems; tape recorders; radio receiver and transmitters; scientific-instrument electronics and data-automation subsystem; power-booster regulators and spacecraft battery. The high-gain antenna is an aluminum paraboloid 40 in. in diameter that weighs only 3.3 lbs.

Mariner 6 is outfitted with instruments for making a variety of observations of Mars, particularly in the ultraviolet, visual, and infrared parts of the spectrum. Neither this craft nor Mariner 7 will study interplanetary space.

During the approach, a camera of medium resolution (about 15 mi. on Mars) will record the full disk of the planet through blue, green, and red filters; as the planet rotates, most of the martian surface will be surveyed. For the close encounter, the high-resolution camera has a Schmidt-Cassegrain optical system and a yellow haze-cutting filter. It is expected to show objects only 900 ft across, an increase in resolution of approximately 10 times over Mariner-4 photos.

JPL's 210-ft Goldstone antenna will receive high-speed transmissions from the spacecraft at 16,200 bits per second.

FIG. 72. Diagrams of Mariner's ultraviolet (upper) and infrared (lower) spectrometers, light entering from the left in both. In the UV spectrometer, a is the focusing mirror, b and d are slits, e and g are two points on a concave mirror serving as both collimator and camera in the spectrograph, f is a reflection grating that rotates to scan the spectrum, h is the exit slit (2 of them) in the focused spectrum, and i is the photometer (2) whose output is amplified by electronics, j. In the IR spectrometer (lower) A and B are focusing mirrors, C is a beam splitter, D a filter wheel (2) with filters that select certain wavelength bands, E a diagonal mirror (2), F a concave focusing mirror (2), and G a cooled detector (2) whose output is amplified by electronics, H. (NASA diagram.)

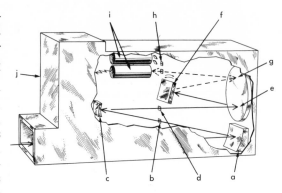

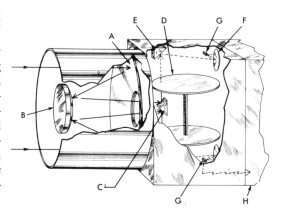

To study the lower atmosphere of Mars, an infrared spectrometer (Fig. 72) will operate in the 1.9-14.3-micron range of wavelength, searching for absorption bands of water, carbon dioxide, methane, acetylene, and ethylene—volcanic gases.

A UV spectrometer (Fig. 72) will make the first attempt at close range to identify the constituents of Mars' upper atmosphere from their ultraviolet absorption lines. This experiment will also estimate atmospheric density, temperature gradient with altitude, and the amount of ultraviolet light that reaches the surface of the planet.

Each Mariner has an infrared radiometer (photometer), so aligned with the television cameras that temperature readings of Mars' surface can be correlated with the photographs.

Temperature measures of the night side of Mars will be of particular value; they cannot be obtained from Earth because Mars moves in an orbit exterior to ours—we see only the sunlit side.

Mars Pictures from
Mariners 6 and 7

(*Sky and Telescope*, October 1969)

Two hundred pictures were obtained by the two spacecraft, 74 by Mariner 6 and 126 by Mariner 7.

With a few notable exceptions, telescopic observers have drawn Mars' "seas" with sharp, angular edges. The Mariner-6 photographs, however, show diffuse, irregular boundaries and a mottled appearance within. In place of canals, the new pictures reveal only fairly large, irregular, very-low-contrast splotches, with no physiographic detail.

The white spot familiar to telescopic observers as Nix Olympica appears on the Mariner-6 records as a large, white-rimmed crater about 300 mi. in diameter, with a bright central patch. It can be seen in Figure 73, a far-encounter picture by Mariner 7. . . .

Early Mariner-6 pictures of the south-polar cap gave the first views along the edge of the cap, showing a number of detached and partly

FIG. 73. Mars from Mariner 7 at 293,200 mi. distance, on Aug. 4, 1969, at 10:28 UT with longitude 115° centered. Nix Olympica (later renamed "Olympus Mons") is the white circle ¾ in. toward upper left from center. Bright south-polar cap at bottom. (JPL-NASA photo.)

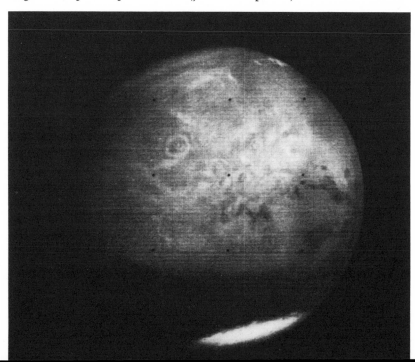

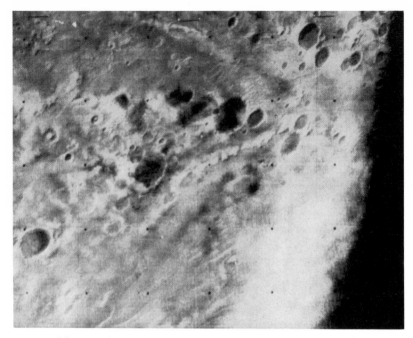

FIG. 74. Mariner-7's view of the edge of Mars' south-polar cap, an area 1200 mi. x 1000 mi. centered on 79° S latitude, 289° W longitude, including the pole at lower right. The white streaks and blotches are probably frozen carbon dioxide and water ice. (JPL-NASA photo.)

detached ovals, the smallest having diameters estimated at 20 mi. Fortunately, Mariner 7 had already been programmed to photograph the cap close up, with the striking result illustrated in Figure 74.

Even though the polar cap's irregular northern boundary is sharply defined, the cap is invisible at the south limb of the planet. This obscuration seems even more severe toward the morning terminator. Darkening toward both limb and terminator may indicate absorption of light in the martian atmosphere.

The general resemblance of the martian surface to the Moon's was stressed in the preliminary report by Robert Leighton of CalTech, who suggests that weathering and transportational processes are more effective on Mars than on the Moon. Apparently some martian craters duplicate lunar craters in having slump blocks and terraces as well as radial dry-debris avalanche chutes on steep inner faces (see Chapter 3). Another Moonlike feature is the irregularly sinuous ridges.

Differences between lunar and martian terrain lie in the seemingly more subdued relief of many martian craters. No sinuous rilles have yet been identified on Mariner-6 photos.

First Findings from the Mariner Fly-Bys

(Sky and Telescope, October 1969)

Here are some of the highlights of the six experiments that both Mariner 6 and Mariner 7 performed.

S-band occultation. Each Mariner carried an S-band radio transmitter, operating at a normal frequency of 2195 megahertz, which was monitored from Goldstone, California, and Woomera, in Australia. About 15 min. after closest approach to Mars, as the spacecraft passed behind the planet, its radio signal was cut off temporarily by the solid surface. But the martian atmosphere refracted the radio waves, and changed their frequency and strength, yielding data on the density and pressure of the atmosphere.

In the Mariner-6 experiment, the occultation of the craft by the planet's disk lasted 20 min. A. J. Kliore of JPL gave the preliminary result that the surface pressure near Meridiani Sinus was 6.5 millibars (roughly the pressure in the Earth's atmosphere at a height of 20 mi). The surface temperature at this point was tentatively reported as 260°K or +15°F.

Mariner 6 emerged from behind the disk of Mars on the night side, near the north pole, in areographic latitude 79° N, longitude 274°. Here the surface pressure was measured as about 6.2 millibars (see App. V) and the temperature as 160°K.

The gas in the upper atmosphere of Mars is ionized by ultraviolet radiation and x-rays from the sun.

Gunnar Fjeldbo of Stanford University says that the Mariner-6 occultation disappearance shows a peak ionization at 130 km above the martian surface, with about 150,000 electrons/cm^3—about 50 percent greater than measured by Mariner 4 in 1965. The increase is probably caused by two factors: in 1969 solar activity was greater than in 1965, and the sun was about 12° nearer the zenith for the Mariner-6 observation than for the Mariner-4 one.

Mariner 7 went behind the planet's limb very near the south-polar cap, at areographic latitude about 60° S, longitude 332°, near the dark marking Hellespontica Depressio, and reappeared at about 37° N, longitude 148°.

The data taken during disappearance indicates a surface pressure at that point of only about 3.5 millibars, indicating a feature about 6 km higher than the average surface of Mars.

The same run of observations confirmed the ionosphere observed by Mariner 6.

Infrared radiometer (photometer). Weighing about 8¼ lbs, the infrared radiometer contains a separate 1.0-in. objective lens for each of two channels, 8-12 and 18-25 microns. The radiation detector behind each lens is a thermopile (heat detector) with five antimony-bismuth junctions. In operation, the radiometer's view of Mars was interrupted every 63 sec by a small mirror, which substituted a view of empty space, thus furnishing a zero-radiation reference.

Both radiometers worked well. From a first inspection of approximately 600 data points obtained with Mariner 6, Gerry Neugebauer of CalTech estimated that the surface temperature at martian noon rises to $+60°F$, while during the night it falls below $-100°$. The planet's dark areas are warmer than the bright deserts and the variations in temperature from the day to night sides of Mars suggest that the surface material is a very good heat insulator.

The radiometer of Mariner 7 obtained about 200 readings on the south-polar cap, which on preliminary study yield a minimum temperature near $-190°F$. This agrees closely with the frost point of carbon dioxide under martian pressure conditions—favoring the theory that the polar caps are predominantly made of carbon dioxide rather than of water ice.

Celestial-mechanics results. In the early part of Mariner's unpowered flight, its motion is appreciably affected by the Earth and Moon, during the latter part by Mars. Because Mariners 6 and 7 were accurately tracked throughout their missions, it is possible to measure these gravitational attractions.

Monitoring the frequency of the on-board 2195-megahertz transmitter by the 210-ft antenna at the Goldstone tracking station in California gives Doppler shifts (see p. 98) and the velocity with which the spacecraft is receding from Goldstone. Also, radio signals sent to the Mariner trigger a return signal, and from the round-trip travel time, the distance of the spacecraft from Goldstone is calculated. These data give an accurate orbit from which John D. Anderson of JPL derived an accurate value for the mass of the Moon. He reports that the ratio of the Earth's mass to the Moon's is 81.3000, with an uncertainty of ±0.0015. The mass of Mars is 0.1074469 that of the Earth, with an uncertainty of ±0.0000035, agreeing very well with the Mariner-4 Doppler data (0.1074464, with an uncertainty of ±5 units in the last place).

The lesser accuracy of the Mariner-6 and -7 determination is attributed to nongravitational forces. Because the infrared spectrometers aboard Mariners 6 and 7 required cooling with liquefied gas, the recoil from the escaping gas slightly perturbed (deflected) the trajectories.

Television. The Mariner-6 and -7 photographs indicate a great advance in quality over the 22 pictures secured 4 yrs ago by Mariner 4. That probe used a single $f/5$ camera of 60-mm aperture.

This time, two cameras were carried on each spacecraft. Wide-angle camera "A," with a field of 18°, was used for the lower-resolution pictures. Its 10-mm-$f/5.2$ lens took exposures through a rotating filter wheel in the sequence red, green, blue, green.

Telescopic camera "B," used for the far-encounter photographs and alternating with camera A during closest approach, has a high-resolution optical system giving a field of 1°.8. It was designed at the University of Arizona's Optical Sciences Center by Gary Wilkerson and L. R. Baker, who adopted an $f/2.5$ Schmidt-Cassegrain configuration with an aperture of 200 mm. All camera-B pictures were taken through a yellow (minus-blue) filter to reduce the effect of any haze in the martian atmosphere.

Both cameras were mounted on a platform that could be rotated 215° and tilted 64°, for better views of the planet and to allow overlapping areas to be photographed.

Mariners 6 and 7 have 20-watt transmitters, and JPL now has a 210-ft antenna at Goldstone. Thus, it took only 5½ min. to receive one picture, made up of 665,280 pixels (945 elements in each of 704 scan lines) from the new Mariners. This represents a 2000-fold increase in the rate of data transmission over Mariner 4, and the far-encounter photographs could be transmitted back to Earth before closest approach, leaving the recorder tape clean for storing the closeup pictures.

Further, each pixel is now represented by an 8-numeral binary number, giving 256 shading variations instead of the 64 provided by the 6-numeral code used in 1965. However, the amount of tape carried on each of the new Mariners could only be doubled, to a length of 220 m., so JPL scientists had to resort to some electronic trickery to increase the effective storage capacity. The telemetered data require careful computer analysis and reconstruction.

Ultraviolet spectrometer. Intended to determine the composition of Mars' upper atmosphere, the 35-lb UV spectrometers carried by Mariners 6 and 7 were designed and built by the University of Colorado's Laboratory for Atmospheric and Space Physics (see Fig. 72). A diffraction grating (2160 rulings/mm) slowly rotates back and forth so that two exit slits scan the spectrum, each output being measured with a photomultiplier. One channel covers 1100-2150A, the other 1900-4300A.

As Mariner-6's instrument viewed the brightly illuminated limb of Mars, ultraviolet emission lines of hydrogen, oxygen, CO_2, and CO were detected, establishing the existence of these gases in the upper atmosphere of Mars. One important conclusion announced by C. A. Barth (University of Colorado) was the absence of molecular nitrogen.

And because Mariner 7 passed near the south-polar cap, its UV spectrometer could measure the abrupt increase in the ultraviolet intensity as the line of sight moved from the desert to the polar cap. This means that ultraviolet radiation at very short wavelengths, harmful to life as we know it, does penetrate to the surface.

Infrared spectrometer. The purpose of this experiment was to search for polyatomic molecules in the air of Mars. The 42-lb unit scans the planet's spectrum between 2 and 6 microns in one channel and between 4 and 14 in a second channel. Its 10-in. telescope gave a 60-by-60-mi. spatial resolution on the planet, and an infrared spectrum could be recorded every 10 sec.

On the Mariner-6 fly-by, the longer-wavelength channel did not operate, but data obtained between 2 and 6 microns showed significant thermal variations from place to place, with temperatures up to $+75°F$. The darker spots on the planet are warmer than the bright areas.

The amount of carbon dioxide varies with location on Mars. Carbon monoxide was detected, but not nitric oxide. The absorption bands of ice were also recorded, tentatively attributed to a very thin ice fog.

On Mariner 7, both channels of the infrared spectrometer behaved properly, and detected solid carbon dioxide in the region of the south-polar cap.

Recent Martian Studies

(Sky and Telescope, May 1971)

At JPL, scientists have been conducting laboratory tests of materials under simulated martian conditions in order to anticipate what the Viking soft-lander may find in 1975.

Norman H. Horowitz, J. S. Hubbard, and J. P. Hardy used soil or pulverized glass in a synthetic atmosphere of 97 percent carbon dioxide with a little carbon monoxide and water vapor. This was irradiated with ultraviolet light (wavelengths 2000A and up) of intensity equal to that measured by Mariners 6 and 7 as reaching the surface of Mars.

In these experiments, organic compounds were formed in and just below the surface of the soil or crushed glass. The compounds in question are formaldehyde (H_2CO), acetaldehyde (CH_3CHO), and glycolic acid ($HOCH_2COOH$).

"It would appear that radiation over a broad range below 3000A can

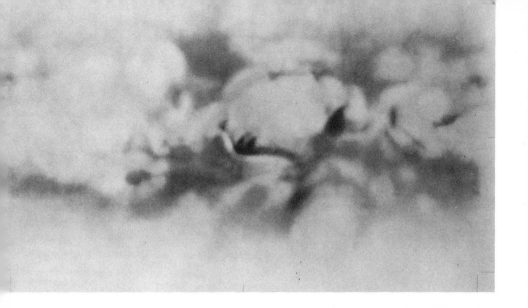

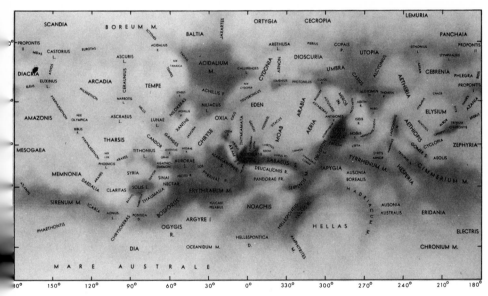

FIG. 75. Upper: International Planetary Patrol map of Mars; lower: same with 191 names of features (coordinates listed in Table 9). (Lowell Observatory airbrush painting.)

cause organic-compound formation," the JPL scientists reported. "Our findings suggest that ultraviolet presently reaching the martian surface may be producing organic matter. The rates of production would be limited by the low partial pressures of carbon monoxide and water in the martian atmosphere, but the amount of product formed could be considerable over geological time."

In 1971, a group of planetary astronomers led by William A. Baum at the Lowell Observatory published a new map of Mars based on patrol photographs taken with identical cameras at Lowell, Cerro Tololo (Chile), Mount Stromlo (Australia), Mauna Kea (Hawaii), Magdalene Peak (New Mexico), and Republic in South Africa. Distributed around the world, these observatories obtained Earth-based views of all longitudes of the red planet, and with several different color filters. The map (Fig. 75) shows visible features—not necessarily topographic features as the Mariner photos do—and shows fairly accurate positions of the 191 named places recorded by observers over the past 100 years. Positional errors are estimated to be less than 1° in latitude and longitude, and the original (without names) is 20 in. x 26 in., drawn with an airbrush.

Table 9 is a list of the 191 place names, 113 of them officially recognized by the IAU (see p. 184). More accurate topographic mapping was possible (and some names were changed) after the Mariner-9 mission.— TLP

Mariner 9 to Orbit Mars

RAYMOND N. WATTS, JR.

(*Sky and Telescope,* June 1971)

If all goes well with Mariner 9, it will be launched from KSC on May 18, 1971, or soon after[1], following the inquiry to determine the cause of Mariner-8's failure on May 8. Five minutes after launch, Mariner-8's second stage began to tumble wildly, apparently because the automatic pilot malfunctioned. This made the two Centaur rocket engines shut down when the vehicle was about 100 mi. high, and it fell back into the Atlantic Ocean.

Mariner 9 is to spend more than 6 months en route and Mars will be 84 million mi. from Earth by the time Mariner 9 reaches it on November 24.

On arrival, the spacecraft will fire a retrorocket to reduce its velocity, and will then become a 2200-lb satellite of the planet. Scientific observations and the transmission of data are scheduled to continue for at least 90 days, and it is even hoped that operations can continue on a limited basis for up to a year.

---

[1] It was actually launched on May 30.—TLP

TABLE 9. NAMES OF FEATURES ON THE 1969 MARS PATROL PHOTOGRAPHIC MAP

Italics indicate names not used on the International Astronomical Union's 1958 map of Mars (*Sky and Telescope*, November 1958, page 23).

1. *Achillis Fons*, 53°, +23°
2. *Achillis Pons*, 30°, +37°
3. Acidalium Mare, 28°, +48°
4. *Acidalius Fons*, 63°, +58°
5. Aeolis, 212°, −10°
6. *Aeria*, 310°, +15°
7. Aetheria, 240°, +40°
8. Aethiopis, 235°, +10°
9. *Agathodaemon*, 65°, −14°
10. *Albor*, 208°, +18°
11. *Alcyonius Nodus*, 268°, +35°
12. Alcyonius, 260°, +50°
13. Amazonis, 160°, +20°
14. Ambrosia, 85°, −38°
15. Amenthes, 251°, +3°
16. *Amphitrites Mare*, 322°, −58°
17. *Anian*, 228°, +48°
18. *Antigones Fons*, 295°, +20°
19. *Aonius Sinus*, 105°, −47°

20. Arabia, 320°, +28°
21. *Aram*, 12°, −5°
22. Araxes, 117°, −24°
23. Arcadia, 115°, +42°
24. *Arethusa Lacus*, 337°, +58°
25. *Argus*, 10°, 0°
26. Argyre I, 35°, −48°
27. Arnon, 337°, +50°
28. Ascraeus Lacus, 100°, +20°
29. Ascuris Lacus, 95°, +53°
30. Astaboras, 305°, +26°
31. Astusapes, 298°, +30°
32. Athos, 153°, +48°
33. Atlantis, 173°, −30°
34. Aurorae Sinus, 50°, −13°
35. Ausonia Australis, 250°, −40°
36. Ausonia Borealis, 275°, −23°
37. Australe Mare, 90°, −65°

38. Azania, 185°, +30°
39. Baltia, 40°, +63°
40. *Bathys*, 92°, −38°
41. *Biblis Fons*, 132°, +10°
42. *Bidis*, 182°, +45°
43. Boreum Mare, 95°, +65°
44. *Bosporus Gemmatus*, 63°, −43°
45. *Callirrhoes Sinus*, 3°, +50°
46. Candor, 75°, +5°
47. Casius, 275°, +43°
48. *Castorius Lacus*, 150°, +55°
49. Cebrenia, 215°, +48°
50. Cecropia Mare, 305°, +67°
51. Ceraunius, 96°, +42°
52. Cerberus, 212°, +9°
53. *Chaos*, 215°, +35°
54. Chronium Mare, 215°, −60°
55. Chryse, 32°, +8°

56. Chrysokeras, 100°, −52°
57. Cimmerium Mare, 210°, −25°
58. Claritas, 102°, −30°
59. *Coloe Palus*, 304°, +43°
60. Copais Palus, 288°, +58°
61. *Crocea*, 293°, 0°
62. Cyclopia, 218°, 0°
63. *Cydonia*, 345°, +50°
64. *Daedalia*, 120°, −34°
65. *Deltoton Sinus*, 304°, −5°
66. *Deucalionis Regio*, 345°, −18°
67. Deuteronilus, 358°, +35°
68. *Dia*, 88°, −60°
69. *Diacria*, 170°, +47°
70. Dioscuria, 315°, +54°
71. *Eden*, 350°, +28°
72. Edom, 345°, −4°
73. *Electris*, 190°, −52°
74. *Eleus*, 168°, +40°
75. Elysium, 215°, +25°

76. Eos, 37°, −15°
77. *Erebus*, 182°, +20°
78. Eridania, 218°, −45°
79. Erythraeum Mare, 30°, −30°
80. Eunostos, 225°, +15°
81. *Eurotas*, 125°, +58°
82. *Euxinus Lacus*, 155°, +43°
83. *Fastigium Aryn*, 358°, 0°
84. *Ganges*, 60°, +5°
85. *Gehon*, 358°, +15°
86. *Geryon*, 75°, −22°
87. *Gomer Sinus*, 230°, −2°
88. *Hades*, 192°, +33°
89. Hadriacum Mare, 270°, −40°
90. *Hammonis Cornu*, 316°, −13°
91. Hellas, 295°, +50°
92. Hellespontica Depressio, 358°, −58°
93. Hellespontus, 330°, −47°
94. Hesperia, 240°, −20°
95. *Hiddekel*, 347°, +18°
96. *Hyblaeus*, 228°, +30°
97. *Hydrae Pons*, 48°, −3°

98. *Iani Fretum*, 10°, −10°
99. *Iapygia*, 295°, −15°
100. Icaria, 124°, −45°
101. *Idaeus Fons*, 53°, +35°
102. Isidis Regio, 275°, +20°
103. Ismenius Lacus, 335°, +42°
104. Jamuna, 44°, +10
105. *Jaxartes*, 22°, +65°
106. Juventae Fons, 62°, −4
107. *Labotas*, 345°, 0
108. *Laocoontis Nodus*, 246°, +15°
109. Lemuria, 230°, +70°
110. *Libya*, 275°, 0°
111. Lunae Lacus, 71°, +15°
112. *Mareotis Lacus*, 96°, +32°
113. Margaritifer Sinus, 20°, −10°
114. Memnonia, 142°, −20°
115. Meridiani Sinus, 0°, −5°
116. *Meroe Insula*, 290°, +30°
117. *Mesogaea*, 168°, −2°
118. *Midas*, 165°, +56°
119. *Moab*, 338°, +10°
120. Moeris Lacus, 278°, +8°
121. *Nectar*, 60°, −28°

122. Neith Regio, 275°, +30°
123. Nepenthes, 268°, +8°
124. *Neudrus*, 4°, −15°
125. Niliacus Lacus, 32°, +27°
126. Nilokeras, 55°, +28°
127. *Nilosyrtis*, 280°, +30°
128. *Nilus*, 82°, +25°
129. *Nix Cydonia*, 3°, +40°
130. *Nix Lux*, 110°, −7°
131. *Nix Olympica*, 132°, +21°
132. *Nix Tanaica*, 55°, +52°
133. *Noachis*, 355°, −40°
134. *Nubis Lacus*, 264°, +24°
135. *Nymphaeum*, 300°, +10°
136. *Oceanidum Mare*, 35°, −60°
137. *Ogygis Regio*, 60°, −53°
138. *Ophir*, 65°, −10°
139. *Ortygia*, 350°, +65°
140. *Oxia*, 20°, +20°
141. *Oxia Palus*, 17°, +8°
142. *Oxus*, 12°, +20°
143. *Panchaia*, 205°, +62°
144. *Pandorae Fretum*, 345°, −25°
145. *Phaethontis*, 150°, −50°

146. Phison, 308°, +35°
147. *Phlegethon*, 125°, +35°
148. Phlegra, 190°, +45°
149. Phoenicis Lacus, 110°, −15°
150. *Pierius*, 310°, +59°
151. *Pontica Depressio*, 85°, −47°
152. Propontis I, 180°, +40°
153. Propontis II, 179°, +58°
154. Protei Regio, 50°, −22°
155. Protonilus, 320°, +42°
156. *Pyriphlegethon*, 140°, +20°
157. *Pyrrhae Regio*, 30°, −22°
158. Sabaeus Sinus, 335°, −12°
159. Scandia, 150°, +66°
160. *Scythes*, 75°, +64°
161. Serpentis Mare, 320°, −28°
162. *Sigeus Portus*, 335°, −8°
163. *Sinai*, 65°, −23°
164. Sirenum Mare, 140°, −40°
165. *Sitacus*, 338°, +17°
166. Sithonius Lacus, 230°, +58°

167. Solis Lacus, 85°, −30°
168. *Stymphalius Lacus*,
    205°, +54°
169. Styx, 200°, +25°
170. Syria, 90°, −20°
171. Syrtis Major, 290°, +10°
172. *Syrtis Minor*, 260°, −8°
173. Tanais, 50°, +55°
174. Tempe, 75°, +40°
175. *Tempes*, 63°, +47°

176. *Tharsis*, 103°, +8°
177. Thaumasia, 75°, −35°
178. *Thoana Palus*, 256°, +35°
179. Thoth, 263°, +15°
180. Thymiamata, 6°, +10°
181. Tithonius Lacus, 80°, −5°
182. *Tritonis Sinus*, 240°, −10°
183. Trivium Charontis,
    198°, +14°

184. *Typhon*, 322°, −4°
185. Tyrrhenum Mare,
    270°, −13°
186. Umbra, 290°, +49°
187. Utopia, 265°, +56°
188. Vulcani Pelagus, 25°, −40°
189. Xanthe, 50°, +15°
190. Yaonis Regio, 318°, −43°
191. Zephyria, 190°, 0°

SYNONYMS

1, *Craneum*; 9, Coprates; 11, *Aquae Calidae, Nuba Lacus*; 18, *Astaborae Fons*; 21, Thymiamata; 25, *Brangaena*; 27, Euphrates-Arnon; 36, Trinacria; 44, Phrixi Regio; 45, *Novem Viae*; 47, *Wedge of Cassius*; 98, *Socratis Promentorium*; 100, *Hyscus*.

107, *Daradux*; 108, *Laocoontis*; 115, *Furca Bay, Dawes Bay*; 116, Meroe; 123, Thoth-Nepenthes; 134, *Nuba, Lethes Lacus*; 146, Phison-Vexillum; 154, *Capri Cornu*; 167, Eye of Mars; 176, Tractus Albus Australis; 179, Thoth-Nepenthes; 180, *Aram*; 181, Coprates Triangle; 186, Nilosyrtis.

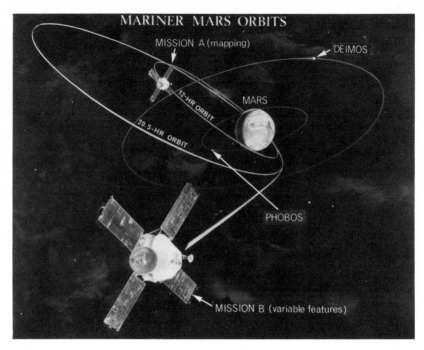

FIG. 76. Planned orbits of Mariners 8 (A) and 9 (B) around Mars relative to Deimos' 30.3-hr orbit, 14,600 mi. in radius, and Phobos' 7.65-hr orbit, at 3700 mi. above Mars' surface. (NASA sketch.)

The Mariners were to work as a team, as shown in Figure 76, Mariner 8 to map Mars' surface, while Mariner 9 would study selected areas repeatedly to observe changes in the atmosphere and on the surface.

However, the failure of Mariner 8 may change the plans for its sister ship. Fortunately, the two Mariners were equipped with identical payloads, permitting either one to carry out the six different scientific programs.

The instrumentation is mostly very similar to that of Mariners 6 and 7, and in some cases identical. Two kinds of television cameras will be used: a lens-type for wide-angle coverage (surface resolution 1000 m.) and a Schmidt-Cassegrain for high resolution (100 m.).

To make infrared observations, a new type of spectrometer for space probes replaces the simpler instrument of the 1969 missions. Its optical arrangement is shown in Figure 77. Basically, scanning is by means of a simple interferometer. Light from Mars strikes the three-position mirror, which for calibration purposes may be directed toward empty space (very cold reference) or to a special end panel of known temperature.

The filter isolates specific spectral regions, which are scanned by means of the motor-driven mirror to which one part of the light beam is sent by the beam splitter. After the combined interfering beams are focused by the collector mirror, the detector's photoelectric output is converted to pulses for coding and radio telemetry.

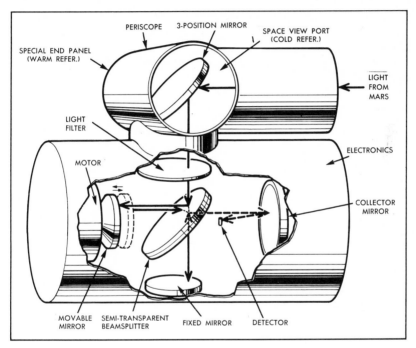

FIG. 77. Diagram of Mariner-9's compact infrared interferometer (spectrometer). Interference (adding or subtracting of light waves) between the two beams through the beam splitter changes as the movable mirror is pushed out or pulled back, thus scanning the wavelength range 5-50 microns. (NASA diagram.)

The infrared spectrometer will obtain data on gases in Mars' atmosphere, including water vapor (whose amount is an important problem); temperatures and pressures at the surface and at several heights in the atmosphere; and the composition of the polar caps (whether frozen carbon dioxide or ice or a mixture of both).

Three Spacecraft Study
the Red Planet

RAYMOND N. WATTS, JR.

(*Sky and Telescope*, January 1972)

In November 1972, the planet Mars acquired two man-made satellites: the American craft Mariner 9 on the 13th and the Soviet Mars-2 ship on the 27th. These were joined on December 2 by the Soviet Mars 3.

There was general elation that the spacecraft have been successfully placed in the orbits intended. But this enthusiasm was dampened by the great martian dust storm, which caused Mariner 9 to return nearly blank photographs of the planet. On the other hand, as some scientists pointed out, the opportunity was unprecedented to study close up one of these major storms, which seem to occur only about once in 15 years.

As Mariner 9 approached Mars, after a relatively uneventful 5½-month journey, its television cameras were turned on and successfully tested. On November 10, the first "approach" photographs, showing the bright disk of the planet, were taken from a distance of half a million miles. Telemetry indicated that everything on board was functioning well, and 31 pictures were stored on tape in the spacecraft for relay to Earth on November 11.

That day, the first pictures received at JPL showed an essentially featureless planet. However, subsequent frames did reveal some detail, particularly after computer processing to remove noise and defects caused by transmission.

On November 13 at 5:14 p.m. EST, Mariner's autopilot was turned on and tested. The spacecraft was maneuvered to aim its braking rocket in the right direction. When the spacecraft reached a position 1711 miles from the surface of Mars, the retrorocket fired for 915.6 sec, starting at 7:24 EST. Then, 2½ min, later, Mariner disappeared behind Mars in an occultation that lasted for 36 min.

After 2 days of careful tracking, a 6-sec burst from the retrorocket adjusted Mariner-9's orbit slightly to period 11 hrs 58 min., inclination 64°.36, between 868 mi. and 10,655 mi. above the surface.

From an engineering standpoint, everything seemed to be perfect, with two exceptions: a sticking valve in the gas attitude-control system with a slight leak of nitrogen, and a gradually weakening signal output from the primary radio transmitter. Although this effect is very slight, it may be necessary to switch to the backup transmitter a month or so earlier than was originally intended.

As Mariner 9 continues to follow its elliptical path around the planet, it has been relaying a wealth of information back to Earth. Doppler tracking has made possible a mapping of Mars' gravitational field, showing that the equator of the planet is elongated in the direction of longitude 110°—the Tharsis region—already known from ground-based radar to be an elevated part of the planet.

The ultraviolet spectrometer indicated provisionally that the great

FIG. 78. Changes in Mars' south-polar cap during November 1971. The black arrow in the lower-left photo shows the 60-mi. region covered by the three high-resolution photos from Mariner 9 on Nov. 19 (upper left), Nov. 28, and Dec. 1, 1971 (lower right). During the 12 days, about 1 in. of frozen carbon dioxide evaporated from the bright areas. The slow change in dark (frost-free) areas shows that they are very smooth. (JPL-NASA photo.)

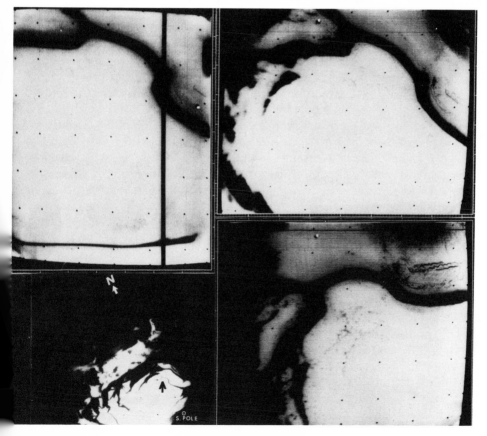

dust cloud enveloping Mars consists of fairly coarse-grained, sandlike particles, rather than ice crystals.

The infrared radiometer revealed the presence of a localized hot spot near longitude 120°, latitude −10°, where the temperature is about 7°C above that of the surrounding area.

The most spectacular early photographs were of the south polar cap, where the dust storm was abating. They show that the cap is divided into two parts, seamed with cracks that change very slowly (Fig. 78).

Pictures at close range confirm that Mars' inner satellite, Phobos, is an irregular object, about 16 mi. long and 13 mi. wide, with craters up to 4 mi. across. The profusion of craters suggests not only that Phobos is very old but that it possesses considerable structural strength.

On Nov. 27, 1971, the Soviet spaceprobe Mars 2 was inserted into a highly elliptical 18-hr orbit inclined 48°.9 to the martian equator. Highest and lowest points above the planet's surface were 15,500 mi. and 860 mi. Just before going into orbit, the ship ejected a capsule that carried a hammer-and-sickle emblem to the surface. Thus Mars became the third extraterrestrial body (after the Moon and Venus) on which a man-made object has landed.

The other Soviet interplanetary craft, Mars 3, reached the red planet by December 2, when it released a descent capsule that made a parachute landing in the martian southern hemisphere between Electris and Phaethontis, near longitude 158°, latitude −45°.

"Video signals received from the surface of Mars were of short duration and then suddenly ceased," said a Tass announcement on December 7. The signals were received and recorded aboard Mars 3, which relayed them to the Soviet deep-space tracking station in the Crimea.

Having discharged its lander, Mars 3 was put into an eccentric orbit around the planet, dipping to 930 mi. above the surface each 11 days.

Soviet Exploration
of Mars

RAYMOND N. WATTS, JR.

(*Sky and Telescope*, February 1972)

More information is now available about the twin Soviet spaceships, Mars 2 and Mars 3. Each of the nearly identical 4650-kg probes carried a special lander that would, if all went well, descend to the martian surface.

Each lander was a complex spacecraft in itself, with an instrument package, a parachute container, and a nose cone to provide aerodynamic deceleration. In the instrument package were television cameras, signal-processing equipment, timers, transmitters, and other gear, all of which had been carefully sterilized before leaving Earth, to make sure that no terrestrial microorganisms were transported to the surface of Mars.

Just before Mars 2 entered its 18-hr orbit around the planet on Nov. 27, 1971, it ejected its lander with Soviet emblem. This reached the martian surface, but the lack of further information suggests some mishap in an attempted soft landing.

The Mars-3 lander (Fig. 79) separated from its mother ship on December 2 at 9:14 UT and 15 min. later fired its descent engine. While approaching the atmosphere at an angle of 10° to the horizontal, the lander was rotated so that its heat shield was properly aligned for entry. Nearly 4½ hrs later, it entered the thin martian air, at a speed of about 6 km/sec.

Descent through the atmosphere lasted slightly more than 3 min. After atmospheric drag had slowed the craft sufficiently, a small rocket

FIG. 79. Soviet Mars-3 lander which made a soft landing on Dec. 2, 1971, but ceased broadcasting after 20 sec. The heat shield is at top, and the ring below the round lander contains the main parachute. (Tass-Sovfoto picture.)

released a drogue parachute, which was followed by the main parachute with its cupola furled. When the craft had slowed to approximately the speed of sound, the main cupola was allowed to open, giving a sharp jerk on the payload. The nose cone was jettisoned and the antennas for the soft-landing radio altimeter were deployed.

When the package had descended to 20-30 m. from the martian surface, the radio altimeter activated a braking rocket. Meanwhile, the discarded parachute was carried off to one side by a small jet motor so that it would not fall on top of the instrument package, which was encased in a hard shell and protected by shock absorbers. Within 1½ min. after its arrival on martian soil, the lander came alive under command of its time-sequence unit.

At 13:50:35 UT, a video signal from the surface of Mars was beamed toward the mother ship in orbit overhead. There, for 20 sec., the transmissions were recorded in two special memory banks for relay to Earth. But then the lander died! It sits in the southern hemisphere in the dark region Simois, between Electris and Phaethontis (see Fig. 75), near latitude 45° S, longitude 158°. So far, no one seems to know just why it stopped sending pictures.

Mars 2 and 3 remain in orbit and are still gathering data. Both spaceships carry infrared radiometers operating at wavelengths from 8 to 40 microns. The highest temperature so measured is +15°C. A hot spot was detected on Mars' night side that was 20°-25° warmer than its surroundings.

Detectors measuring radiation at a wavelength of 1.38 microns indicate very little water vapor in the planet's atmosphere—"not exceeding 5 microns of precipitation," according to the Moscow newspaper *Pravda* for Dec. 19, 1971.

The presence of atomic hydrogen and oxygen in the upper atmosphere has been detected by ultraviolet photometers. According to the same *Pravda* article, atomic oxygen can be observed to heights of 1000 km, and atomic hydrogen is detectable above 10,000 km.

Measurements of martian radiation at 3.5-cm wavelength will determine the dielectric permeability of Mars' soil and subsurface temperatures to depths of 30-50 cm. A further experiment will map surface relief by measuring the amount of carbon dioxide gas along the line of sight to various points on the planet's surface.

Like Mariner 9, both Mars 2 and 3 carry a long-focus and a short-focus TV camera for photographing the planet's surface. At year's end, however, their performance was still hampered by the great dust storm in the martian atmosphere.

Some· Mariner-9
Observations of Mars

RAYMOND N. WATTS, JR.

(*Sky and Telescope*, April 1972)

Ever since Mariner 9 began orbiting Mars on Nov. 13, 1971, many groups of scientists have been studying the huge amounts of data being obtained. Some preliminary results were published in the Jan. 21, 1972, issue of *Science*.

TELEVISION PICTURES

Because of the planet-wide dust storm, the original mapping program for the two television cameras was completely recast. The wide-angle camera (field 11° by 14°) was used to view the whole disk in search of clear areas. When a hole in the dust veil was discovered, the narrow-angle camera (1°.1 by 1°.4) was used to take two groups of four high-resolution pictures. This routine permitted 31 exposures during each 12-hr revolution of the satellite.

Nevertheless, many frames are featureless until "enhanced" by an electronic computer technique, so that a small brightness range, usually 20 percent of the original one, can be expanded into a scale of tones from black to white. (Unfortunately, this enhances other original picture blemishes, including faint vertical bands, dust specks on the vidicon light-sensitive faceplate, and afterimages from previous exposures.)

Despite these problems, much valuable information has been extracted from the pictures. Thus, typical views of the planet's edge show a gradual fading in the last 10 km out to the limb, a well-defined dark gap about 15 km wide, and a thin elevated layer of haze—sometimes several haze layers. This elevated haze is white or slightly blue. Thus, it is composed of particles different from the lower-lying dust, which shares the red color of the martian deserts. It is estimated to be about 60 km above the surface, at a level where the barometric pressure is between 0.01 and 0.1 millibar.

A number of television pictures in violet light show striations roughly parellel to the terminator. These appear to be atmospheric waves with wavelengths of about 40 km. They occur in the high blue haze, rather than in the low dust layer.

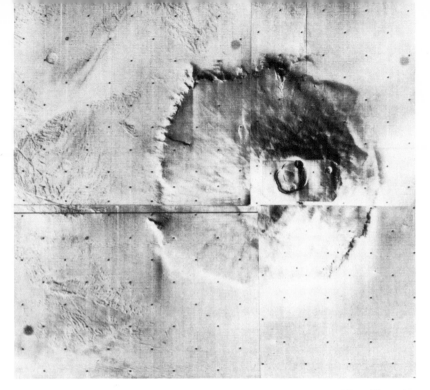

FIG. 80. Photomosaic of the enormous volcanic mountain, Nix Olympica (later renamed "Olympus Mons"), as viewed from Mariner 9 in January 1972 as the dust storm subsided. The several craters of the 40-mi. main vent (right of center) indicate that multiple eruptions occurred. At the edge of the mountain (large circle) steep cliffs drop sharply to the surrounding plain. (JPL-NASA photo.)

Four dark spots visible on even the earliest pictures have turned out to be very significant. One is Nix Olympica; the others are informally known as North Spot (longitude 106°, latitude +12°, near Ascraeus Lacus); Middle Spot (112°, 0°, near Pavonis Lacus); and South Spot (120°, −8°, near Nodus Gordii). See Figure 75.

Nix Olympica and North Spot are each composed of four or five intersecting craters with floors of different ages. Interior terrace edges are visible, indicating that these depressions are not dust filled, though the floors' apparent smoothness might be due to local dust. On high-resolution pictures, these martian craters lack the sharp rims typical of most lunar craters.

Nix Olympica, North Spot, and South Spot all resemble terrestrial calderas—craters formed by the collapse of the central parts of volcanoes when magma is withdrawn from beneath them.

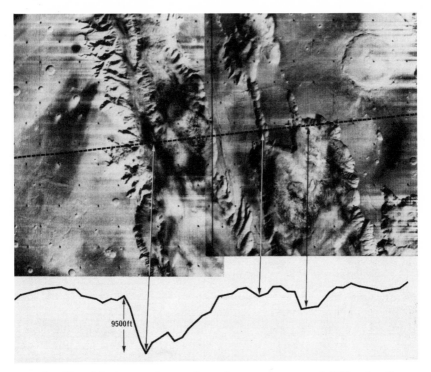

FIG. 81. Two Mariner-9 photos show the eastern part of Tithonius Lacus chasm with dashed line added to show the spacecraft pass over it while the UV spectrometer (Fig. 72) measured atmospheric pressure at the surface. These readings were converted to altitudes in feet, plotted on the graph (bottom), showing that this chasm is twice the depth of the Grand Canyon in Arizona, and about 6 mi. wide. (JPL-NASA photos and diagram.)

This impression is powerfully reinforced by Figure 80, a photograph of Nix Olympica taken as the planet-wide dust storm subsided. Here is a mountain 500 km (310 mi.) across at the base, rising well over 6 km above the surrounding plains. The main summit crater is a complex volcanic vent 65 km in diameter. Nix Olympica is even larger than the volcanic pile of the Hawaiian Islands, which is 225 km across at the base and rises 9 km from the floor of the Pacific Ocean to its summit crater, Mauna Loa.

Another surprising find is an enormous chasm with canyonlike branches, located about 300 mi. south of the martian equator in Tithonius Lacus. Over 300 mi. long and 75 mi. wide, this great trough was probably formed by subsidence along a line of weakness in the martian crust. Because of Mars' aridity, the resemblance of the branching tributaries to a terrestrial drainage system is only superficial, and perhaps wind erosion sculptured them.

The over-all impression given by the Mariner-9 photographs is of a planet that has been active geologically, with a profusion of volcanoes, crustal movements, and erosion.

ULTRAVIOLET AND INFRARED EXPERIMENTS

To study the upper atmosphere of Mars, the ultraviolet spectrometer aboard Mariner 9 observes the sunlit limb of the planet. As the spacecraft scans the limb, the spectrometer records atmospheric constituents at various heights (see p. 197).

It has also discovered that the temperature and density of Mars' upper atmosphere respond to changes in solar activity, much as the Earth's upper atmosphere has been shown to do.

Mariner-9's infrared interference spectrometer measures the bands of carbon dioxide between 12 and $18\frac{1}{2}$ microns. These appear in emission in the polar regions, but are absorption features in nonpolar areas. The reason is that near the poles the layer in which this radiation originates is warmer than the ground, whereas elsewhere this layer is cooler than the surface.

Other major infrared features are diffuse bands between 9 and 22 microns, which are attributed to silicate dust in the atmosphere and show that the dust contains about 55-65 percent silica.

Measurements of the thermal radiation from the south-polar cap yield a surface temperature of $140° \pm 10°K$. This is supporting evidence that the cap material is primarily frozen carbon dioxide, for at the martian surface any solid CO_2 in equilibrium with its saturated vapor should have a temperature of $148°K$.

The presence of a small amount of water vapor over the south-polar cap was established by high-resolution infrared spectra showing numerous rotational lines of H_2O in emission at wavelengths between 29 and 50 microns.

S-BAND EXPERIMENT AND RADIO TRACKING

During its first 40 days in orbit, Mariner 9 underwent 80 occultations by Mars. As the spacecraft curved behind the planet, the radio signal of its S-band transmitter passed through the martian atmosphere and then was cut off by the surface (see p. 195).

Preliminary results from 15 occultations reveal a curious change in the lower atmosphere. In 1969, the temperature decreased by $3°C/km$ above the planet's surface. In late 1971, however, the temperature remained unchanged for the first 15-20 km, then declined by 2°/km.

Probably the great dust storm was responsible for this partially iso-thermal condition, as fine dust suspended in the atmosphere would absorb solar radiation and warm the martian air. This suggests that dust was present to altitudes of at least 15-20 km.

The new S-band observations have also provided accurate radius de-terminations of Mars for each of the 15 locations where Mariner 9 passed behind the limb. In this way, elevation differences of 13 km have been revealed and a range in surface atmospheric pressures between 2.9 and 8.3 millibars. In particular, the floor of the bright feature Hellas was found to lie about 6 km below its western rim, while the region between Mare Sirenum and Solis Lacus is 5-8 km above the planet's mean radius.

The orbit of Mariner 9 around Mars can be deduced with high accuracy from radio tracking data (see p. 196). A detailed comparison of the observed motion of Mariner 9 with the theoretical orbit was per-formed for a 33-revolution span beginning November 16. This com-parison provided a preliminary mapping of the gravitational field of Mars. It has proportionally larger inequalities than the field of the Earth or of the Moon, and these cannot be fully attributed to surface irregularities. Thus it is possible that Mars, like the Moon, has mascons (local mass concentrations—page 97).

One by-product of the tracking-data analyses has been a new de-termination of the orientation of the planet's rotational axis. The north celestial pole of Mars lies at right ascension 21^h $09^m.2$ ± $1^m.2$, declina-tion +52°.6 ± 0°.2 (1950 coordinates), only about 0°.5 southeast of the generally accepted position.

PHOBOS AND DEIMOS

On the basis of pictures showing these satellites against the stars, im-proved orbital predictions were made. The inner satellite, Phobos, turned out to be about 2° ahead of its expected orbital position, while Deimos was about 1° behind.

Currently, Phobos is estimated to measure 25 by 21 km, Deimos 13½ by 12 km, with uncertainties in the measurements ranging from 0.5 to 5 km. When combined with G. P. Kuiper's visual magnitude values, these dimensions indicate for both moons a geometrical albedo of only 0.05. This means that they are among the darkest bodies in the solar system, comparable in reflectivity to the darkest parts of the lunar maria.

The many craters on Phobos and Deimos are probably of impact origin. One on Phobos is 5.3 km across; the impact that produced it must have been nearly the greatest the satellite could sustain without breaking up. With so many craters, the moons must be old, and the fact that Mars itself has roughly 100 times fewer craters per unit area indicates how effective erosion has been on the planet.

The New Mariner-9 Map of Mars

WILLIAM K. HARTMANN[1]

(*Sky and Telescope*, August 1972)

The earliest, crude drawings of Mars were made by Francesco Fontana at Naples in 1636. For nearly 300 years, cartography of Mars depended on the single technique of peering patiently through Earth-based telescopes and recording what faint detail was glimpsed during infrequent moments of excellent seeing. . . . Photographic charts such as Figure 75 are near the limit attainable by Earth-based observations.

A new photographic map of Mars has been prepared by the Mariner-9 project (Fig. 84). In a single step, the resolution of global martian charts has improved by two orders of magnitude (a factor of 100).

Strictly speaking, there is still no fully global Mars map, since the Earth-based ones have poor coverage of the polar regions, while the Mariner-9 map has small gaps and has not yet included all of the north-polar area (due to north-polar haze and constraints of the spacecraft's orbit). The north-polar coverage is being improved by Mariner-9 pictures taken since June 9, 1972.

The new map was compiled in two stages: a careful screening of all the televised photographs by the Mariner-9 photointerpretation team, and a major effort by the team associates at the United States Geological Survey. Once the great 1971 dust storm had cleared in mid-December,

[1] Prepared with the help of the other members of the Mariner-9 Television-Experiment Team, H. Masursky (leader), D. Arthur, R. Batson, W. Borgeson, G. Briggs, M. Carr, P. Chandeysson, J. Cutts, M. Davies, G. de Vaucouleurs, J. Lederberg, R. Leighton, C. Leovy, E. Levinthal, J. McCauley, D. Milton, B. Murray, J. Pollack, C. Sagan, R. Sharp, E. Shipley, B. Smith, L. Soderblom, J. Veverka, R. Wildev, D. Wilhelms, A. Young. Much of the article was based on a press briefing held at JPL, on June 14, 1972. Frank E. Bristow, manager of JPL's News Bureau, supplied most of the illustrations.

the wide-angle camera aboard Mariner 9 successfully covered most of the planet to as far north as latitude +50°. The first product was a photomosaic map compiled from these frames. This has the merit of displaying the original photographic data, but it has the disadvantage of resembling a montage of postage stamps.

Hence the next step was to prepare a carefully executed airbrush map, by transferring features from the original photographic format to a unified, conventionalized representation. This airbrush map presents martian physiographic features such as craters and valleys, but does not attempt to show the albedo markings (light and dark features) to which Earth-based maps like Figure 75 are limited.

INTERPRETATION

The greatly improved resolution has suddenly made available an enormous amount of geological information; we are confronted with a planet that shows unmistakable large volcanoes and tectonic canyons, together with features looking like dry river beds.

What can we now say about the planet Mars and its evolution as a result of the new knowledge from Mariner 9? Two problems inhibit detailed discussion at this time. First, scientists are still analyzing the data from the various experiments carried by this spacecraft, so many answers will not be available for months. Also, the members of the Mariner-9 Investigator Team are bound by an agreement not to publish individual interpretations of the data prior to the team's final report.

The surface of Mars is far from homogeneous. The Mariner-4, -6, and -7 missions photographed the most cratered regions, giving a hint of what came to be known as "chaotic terrain." A very coarse division of Mars into geological provinces would distinguish between two kinds of surface: heavily cratered, rough units, and smooth units which in several cases coincide with the classical deserts.

The geological map (Figs. 82, 83) suggests that these types of terrain divide Mars into hemispheres, one centered in the smooth, volcano-dominated region around Nix Olympica (longitude 132°, latitude +21°), the other centered in the densely cratered region south of Sabaeus Sinus (335°, −12°). It is well known that both the Earth and Moon show hemispheric asymmetries in crustal development. If Mars is a third case, we may be learning something quite fundamental about the evolution of a planet's crust. This question is under lively investigation.

On June 14, 1972, a fourfold classification of the martian surface was discussed by Harold Masursky of the U. S. Geological Survey,

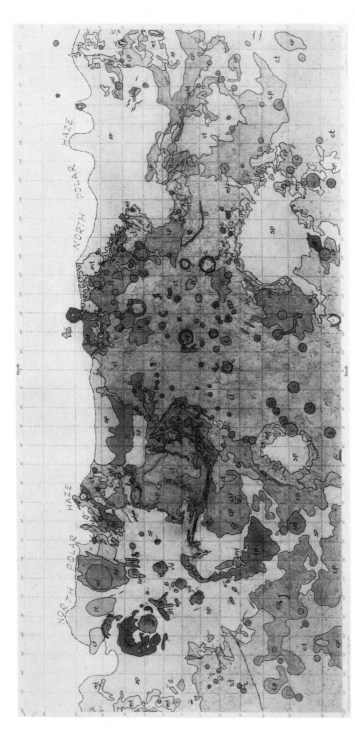

FIG. 82. Geological map of Mars' equatorial region, with 10° grid lines from 60° N to 60° S latitude, 0° longitude centered. Nature of the terrain is marked according to the key below. (JPL-NASA map.)

KEY: *c*, crater; *sp*, smooth plains; *cp*, cratered plains; *ch*, channel or canyon deposits; *ct*, cratered terrain; *lt*, lineated terrain; *mt*, mountainous terrain; *vd*, volcanic deposits; *gt*, grooved terrain; *rt*, ridged material; *pi*, polar ice; *ld*, laminated deposits; *ep*, etch-pitted plains. A graben (fault zone) is labeled by a solid straight line through a dot. Near the equator at left is a row of three giant craters (*vd*), whose positions match Ascraeus Lacus, Pavonis Lacus, and Nodus Gordii. To the northwest of these is the cone of Nix Olympica. A large dark area centered near longitude 50°, latitude −10°, marks the huge canyon pictured in Fig. 85. A great shield of mountains surrounds the smooth plain in Argyre (40°, −50°), but fewer mountains border the Hellas Basin (290°, −50°).

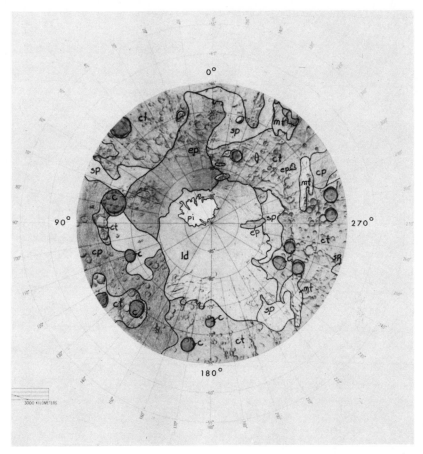

FIG. 83. Geological map of Mars' south-polar region from 65° S latitude to the pole, longitudes marked at the rim. The polar-ice (pi) region is for autumn in the southern hemisphere. Key same as for Figure 82. (JPL-NASA map.)

leader of the Television-Experiment Team. Two of the types are: **I**, the Nix Olympica-Tharsis province and **II**, cratered terrains. A third category, **III**, includes heavily faulted and lifted equatorial structures, such as the Ophir-Eos region and the enormous canyon that coincides with the broad classical canal Coprates.

The fourth type, **IV**, is the north- and south-polar terrain, which appears to be blanketed by sedimentary layers up to 100 m. thick. Scores of these layers have been identified; they seem to cover or to replace the kinds of geological structures seen at lower latitudes.

The enormous mountains of the volcanic provinces (**I**) continue to be of great interest. Not only do they imply that the martian crust was

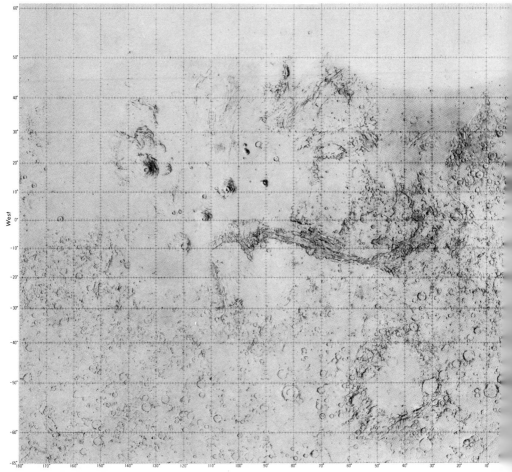

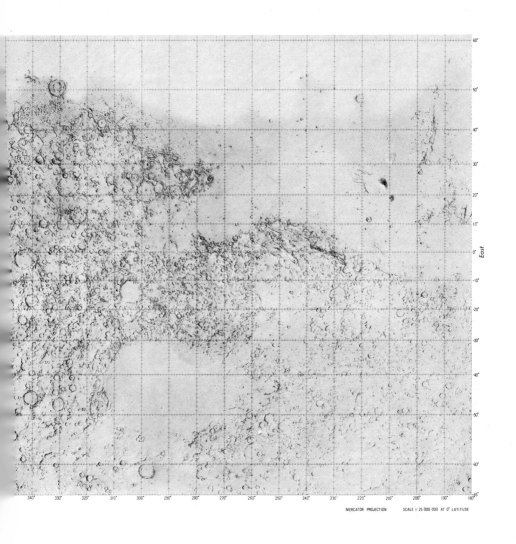

MERCATOR PROJECTION   SCALE 1:25 000 000 AT 0° LATITUDE

FIG. 84. This chart was made from Mariner-9 television image data. The positions of topographic features were taken from unrectified Mariner-9 pictures, positioned for consistency with spacecraft tracking data with some adjustment to minimize the mismatch between pictures. Positional inaccuracies as large as 60 km exist throughout the chart. It is intended to portray topography only. (USGS-JPL map.)

recently active or may still be evolving, but their sizes and forms are remarkable. These volcanic peaks rise about 8 km (26,000 ft) above the surrounding plains. If they were built of volcanic rocks that contained as much water as is found in terrestrial lavas, significant amounts of water would have been released into the martian atmosphere each time one of these mountains was formed.

In the cratered terrains (**II**), the craters range from small pits to enormous structures like the lunar maria. Thus, the Argyre-I basin has a diameter of nearly 1000 km, and the still larger Hellas basin is about 1400 km across. The latter has more than twice the area of Texas. The infrared-interferometer and the ultraviolet-spectrometer experiments indicate that Hellas lies about 17 km (55,000 ft) lower than the highest point on the planet. Its floor is about 2-3 km below the level of the surrounding plains.

The so-called canyonlands (**III**) that include the Coprates valley are seen on the map (Fig. 84) at lat. −10°, long. 70°. Near the eastern end of this canyon are some of the winding features so closely resembling river beds that it is very likely that some liquid, perhaps water, flowed in these beds, eroding them and leaving deposits. Discussion continues about the possibility that much water is locked up in the polar caps or in subsurface permafrost.

In 1966, N. H. Horowitz wrote in a *Science* article, "Water is probably the most seriously limiting factor in any martian biology." Now the question of water has been reopened by new knowledge of the morphology of Mars.

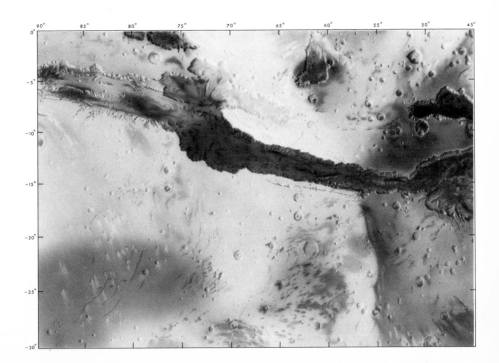

Comparison of the NASA-JPL map of Mars (Fig. 84) with Figure 75, made from ground-based photos, shows the detail and accuracy gained from Mariner-9 photos.

Another type of map, Figure 85, shows the albedo of the martian surface—the fraction of visible sunlight scattered and reflected, or the darkness and whiteness of the surface, omitting shadows. These were prepared by James Roth and Kay Walker, under the direction of G. de Vaucouleurs, at the University of Texas at Austin, using both ground-based and Mariner photos.

The Coprates canyon, 2700 km long, 150 km wide, and 6 km deep, looks like a rift in the martian crust, and two Cornell scientists (David McAdoo and Joseph Burns) think that it was caused by a shift in Mars' axis of rotation. If this happened, caused by changes in the interior, the crust would have been stretched in places as its equatorial bulge shifted to match the new rotation axis. This probably happened long ago. This canyon (renamed Valles Marineris in honor of NASA's Mariner program) is one of the very few features that match the classical "canals"— in this case, Agathodaemon in Table 9 and Figure 75. Most of the other valleys, canyons, and trenches on Figure 84 do not match Lowell's "canals," according to Carl Sagan and Paul Fox (Cornell University).

More recent changes on the surface, during the Mariner programs, can be studied by another technique—picture differencing by computer.—TLP

Mariner-9 Picture Differencing at Stanford

LYNN H. QUAM, ROBERT B. TUCKER, BOTOND G. ERÖSS, JOSEPH VEVERKA, AND CARL SAGAN

(Sky and Telescope, August 1973)

For more than a century, visual observers have been studying changes on Mars. These changes—in the boundaries between the bright and dark regions of the planet, and in their relative contrast—were first thought to be seasonal changes in vegetation. More recently it has been pointed out that windblown dust, driven by seasonal winds, offers an equally plausible explanation.

From mid-November 1971, to Oct. 27, 1972, Mariner 9 transmitted to Earth more than 7200 pictures of Mars, providing a wealth of new data to test hypotheses for the variable features of the planet. In fact,

FIG. 85. Albedo map of Coprates Canyon (recently renamed "Valles Marineris"), reproducing the white-to-gray tones of the surface (not shadows). Dark or light "tails" may be wind-blown dust. Latitudes ("-" meaning S) and longitudes are shown at the left and top edges. (University of Texas airbush painting by J. Roth.)

there is so much material that to use it efficiently requires automated data-management and data-reduction techniques.

To study surface variations, we need to examine two pictures of the same area on Mars taken at different times, but with lighting conditions and viewing geometry closely similar. (Otherwise, changes in shadows could, in certain cases, be mistaken for changes in surface albedo.)

But how do we find among 7200 or so available pictures all those that cover the selected area? How do we find the overlapping frames taken on different days? How do we gain access to the picture information itself, in order to look for differences? A unique system to perform these tasks efficiently has been developed by Quam, Tucker, and Eröss at the Stanford University Artificial Intelligence Laboratory. The system has been employed by Veverka, Sagan, and co-workers at Cornell University's Laboratory for Planetary Studies.

The Mariner-9 spacecraft had two cameras, "A" covering about 10° with resolution on Mars' surface about 1 km; the "B" camera covering 1° with resolution about 100 m. Each photographic frame consisted of an array of 832 by 700 picture elements or pixels. Each pixel described the brightness of a small region of the scene as one of 512 shades of gray described by 9 "bits"—0's or 1's. Black is recorded as 000000000, white as 111111111. This "9-bit binary code" gives $2^9 = 512$ shades from black to white. The shade was transmitted from Mariner 9 as a data number, or DN, roughly proportional to the brightness of the scene.

After the pictures were received on Earth, several corrections had to be applied: to remove transmission noise, to correct geometric camera distortions, and to convert the raw DN's to photometrically more accurate values. The last correction is necessary because vidicon responses are highly nonlinear; in general, if the brightness level of a scene is doubled, the DN value at any given pixel (picture element) is not. This photometric nonlinearity can be calibrated in the laboratory and must be corrected before the pictures are used.

Each Mariner-9 picture is stored in the file system of a PDP-10 computer, and the information can be read rapidly from magnetic tapes. Consider the steps in a typical picture-differencing sequence:

1. **Frame selection.** Suppose we are interested in Daedalia, the dusty area between Solis Lacus and Memnonia. We specify the coordinates, 25° S, 125° W (see Fig. 84), and the computer quickly generates the large display shown in Figure 86. It outlines all the A frames covering this area, identified by a number and code designation at right in the display.

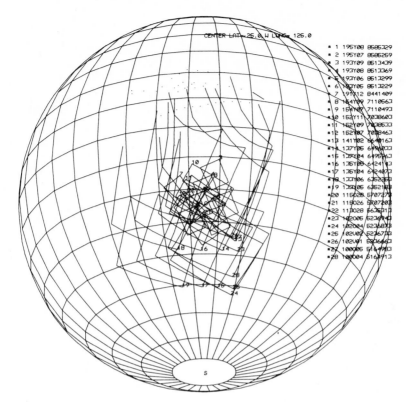

FIG. 86. Computer display showing outlines of all Mariner-9 photos of the dusty area Daedalia centered on a projected spherical grid. The key to the photo frame numbers is at the right. (Stanford University photo.)

To ensure maximum ground resolution, we eliminate all frames with large "footprints" (area coverage on Mars' surface), zoom in on the area of interest, and from this new display select two overlapping frames for further study: frame 2 (orbit 195) and frame 20 (orbit 115). On their representation, the computer draws a "window" common to both frames, as in Figure 87; this window determines the regions of the two frames that will be compared.

2. Matching the frames. At this point, comparison of the overlap areas may be difficult, since the two pictures will in general have different projections and different scales. We must take the indicated window areas out of the two pictures and project them to matching scales. The computer transforms each view to an orthographic map projection and a common scale, as if the two pictures were taken from the same distance

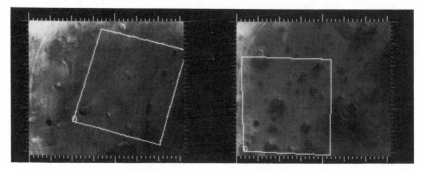

FIG. 87. Computer displays of two selected frames of Daedalia with the common "window" drawn by the computer in white. (Stanford University photo.)

FIG. 88. Results of picture differencing for the two frames of Daedalia.
Top photo is left frame minus right
in Figure 87; bottom is right frame
minus left. Some of the remaining
differences are due to changes in lighting conditions in the originals. (Stanford University photo.)

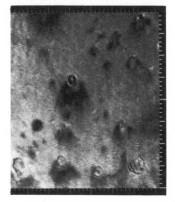

and with the same viewing geometry, relying on information about the spacecraft's orbital positions and the relative orientations of the camera during picture taking.

If this information were absolutely accurate, this process would give two views of exactly the same martian area with corresponding features in exactly matching positions. But there are always differences, caused by unavoidable uncertainties in the camera-pointing information.

3. **Registration.** To correct this, we can deform one of the pictures, pixel by pixel, to make its topographic features occur in the same location as those in the other. This process is called registration, and the displacements required to match two such areas are summarized in a distortion matrix (set of numbers), applied to one picture to make it match the other.

At this point, in principle, the two pictures could be subtracted pixel by pixel, canceling out all topography and leaving only albedo changes. The results are shown in Figure 88, a pair of prints that can be considered to be negatives of each other.

We see that traces of the craters remain in the differences, because the original frames were taken under somewhat different lighting and viewing conditions. As a result, the shadowed and illuminated portions of the craters are slightly different in the two views and do not fully cancel out in the picture difference.

4. **Interpretations.** Great changes have taken place in this Daedalia area of Mars in the 40-day interval between orbits 115 and 195. The shapes of dark crater streaks and their parallelism both suggest that they are wind-related phenomena. Are we seeing dark dust being blown from crater bottoms or bright dust being scoured away, exposing dark underlying bedrock? Since it is hard to see why scouring would be most effective in the lee of the crater walls, the first hypothesis seems more reasonable.

Picture differencing has revealed that such wind-produced albedo changes are very common on Mars, probably accounting for changes reported by Earth-based observers.

This example shows large changes; many more subtle changes stand out clearly in the "difference pictures" and would be difficult to detect without the computer differencing.

To date several hundred Mariner-9 picture pairs have been processed and differenced at Stanford. The same technique has been used to study changes on the planet over longer periods, by comparing Mariner-9 pictures with those obtained by Mariners 6 and 7 in 1969, and even with one Mariner-4 frame from 1965.

These wind effects bring us to Mars' atmosphere, which is very "thin" (surface density about 0.01 of the Earth's), but has tornado-strength winds and frequent dust storms. A question of great interest is whether it ever contained much water vapor and produced rain storms. That could account for "river valleys" like those in Figures 81 and 85, mostly near the equator, where daytime temperatures are above the freezing point of water. Many suggestions have been made, all highly speculative. An appealing one is that Mars has intervals of wet atmosphere alternating with dry periods, as at present, when the water is frozen out in the soil as permafrost under the dry-ice (CO_2) polar caps. It may be that this would account for the steplike surface near the polar caps (see p. 219). In any case, the solar wind of hydrogen ions bombarding the martian surface should produce small amounts of water that may have accumulated in the atmosphere.—TLP

Argon in Mars' Atmosphere

(*Sky and Telescope*, May 1975)

It has been widely accepted that the martian atmosphere is nearly pure carbon dioxide, with spectroscopic traces of other gases, but there is growing evidence that about 30 percent may be argon.

The first observational indication was indirect, from the performance of an ion pump aboard the Soviet spacecraft Mars 6, which crash-landed on the planet in March 1974. Soviet scientists have inferred that one molecule in three of martian air is of some inert gas, probably argon.

Other evidence comes from a preliminary study by Lewis Kaplan, University of Chicago, of high-dispersion Mars spectra taken with the 200-in. Hale telescope. Argon cannot be detected directly, as its resonance lines lie in the far ultraviolet, but its presence is inferred from a perceptible pressure broadening (fuzziness) of the martian carbon-dioxide lines.

J. S. Levine of NASA's Langley Research Center points out that radioactive decay of potassium in the martian interior should release enough argon for it to comprise about 28 percent of the atmosphere.

This outgassing, however, should have been accompanied by enormous quantities of water, some of which may still exist on the planet. The argon and water questions may be answered next year by the Viking Mission to Mars.

The two small moons of Mars, Deimos and Phobos, were long shrouded in speculation and mystery, as noted in Volume 2, Neighbors of the Earth. Mariner-9 photos dispelled the idea that they were space-craft abandoned by Martians, and now have resulted in a map of Phobos on which various features have been named in honor of past astronomers, including Asaph Hall, who discovered the satellites in 1877.
—TLP

Mapping the Martian Satellites

(*Sky and Telescope*, June 1975)

Phobos, the inner of the two tiny satellites of Mars, is visible only in large telescopes when the planet is near opposition.

The Mariner-9 spacecraft obtained over 50 high-resolution television pictures of Phobos and Deimos. Both objects were revealed to be heavily cratered and irregularly shaped (see Fig. 89). All together, 70 percent of the surface of Phobos and 40 percent of Deimos were recorded at 1-km resolution.

To construct the Phobos map, Thomas C. Duxbury (JPL) made measurements of 38 control points on nine photographs. This satellite appears to rotate in the same period (7 hrs, 39 min.) as its orbital period, keeping the same side turned toward Mars, so the sub-Mars point was adopted as the zero of longitude. The best-mapped portion of the surface lies in the southern hemisphere in the 180° of longitude west of the sub-Mars point.

Names for seven craters on Phobos have been adopted by the International Astronomical Union. The seven persons so honored are:

d'Arrest, H. L. (1822-75). German-born astronomer who searched for the satellites in 1862 at Copenhagen Observatory.

Hall, Asaph (1829-1907). American astronomer who discovered Phobos and Deimos in 1877 at the U.S. Naval Observatory.

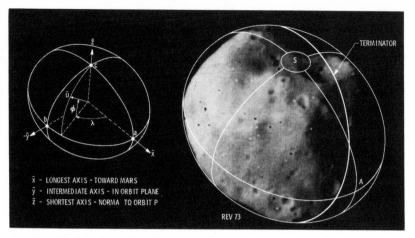

FIG. 89. Phobos from Mariner 9, with white lines added by computer to show the best-fitting spherical shape, the south pole (S), the terminator, equator, and meridians at 0°, 90°, 180° (A), and 270° longitude. To the right of S is the large crater Hall, and to the left of A is crater Todd. (JPL-NASA photo and diagram.)

Roche, E. (1820-83). French mathematician, expert on satellite dynamics.

Sharpless, Bevan P. 20th-century American astronomer at the Naval Observatory who determined the orbits of the satellites.

Stickney, Angelina. Maiden name of Hall's wife, who encouraged his search.

Todd, David P. (1855-1939). American astronomer, early observer of satellites at the Naval Observatory.

Wendell, O. C. (1845-1912). American astronomer who measured Phobos and Deimos with the 15-in. Harvard refractor.

Running southward and eastward from Stickney, Kepler Ridge is one of the most prominent features on Phobos, being at least 1500 m. (1 mi.) high. In fact, this satellite is so rough that variations of as much as 20 percent in radius occur.

If the shape of Phobos is assumed to be approximately an ellipsoid, the three principal diameters are about 27, 21½, and 19 km. The corresponding dimensions for Deimos—15, 12, and 11 km—are more uncertain.

Returning, now, to the planet itself, and the exploration of its surface, we note that NASA spent eight or nine years planning the Viking mission to land on Mars—a project that has taken on some aspects of another race with the Soviet Union.—TLP

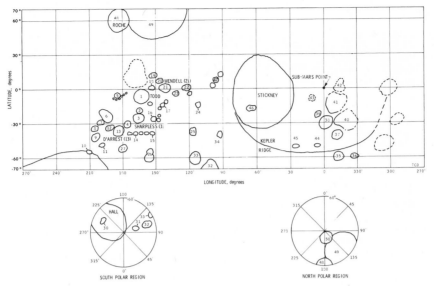

FIG. 90. Chart of craters on 70 percent of Phobos' surface. The region centered on 45° N latitude, 315° longitude has not been photographed. (JPL-NASA map.)

Soviet Mission to Mars

RAYMOND N. WATTS, JR.

(*Sky and Telescope*, September 1973)

While the United States Vikings will not be launched until 1975, three Soviet craft are now on their way to the red planet, presumably to make soft landings. The first probe, named Mars 4, began its 6-month, 290-million-mi. trip on July 21, 1973, and was followed 4 days later by Mars 5. Finally, the third spacecraft, Mars 6, was launched on August 5, and Tass news agency reports it is functioning normally. The three of them should reach Mars' vicinity in February and March 1974.

Writing in *Spaceflight* for July 1973, Finnish astronomer Heikki Oja of Helsinki Observatory discusses the probable characteristics of the new Soviet Mars landers. The braking system, he notes, is based on the use of a conical heat shield, parachute, and, during the last part of the descent, a short burst from a solid-propellant retrorocket.

He mentions the following equipment aboard the lander: instruments for measuring air pressure and density, a mass spectrometer to determine the martian atmosphere's chemical composition, wind-velocity meter, apparatus for determining soil properties, and a TV camera to take panoramic pictures. There may be other experiments. If the instru-

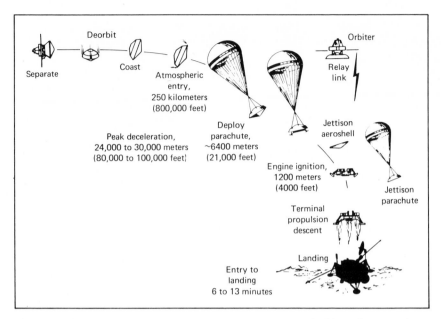

Deorbit
Separate
Coast
Atmospheric entry,
250 kilometers
(800,000 feet)

Peak deceleration,
24,000 to 30,000 meters
(80,000 to 100,000 feet)

Deploy parachute,
~6400 meters
(21,000 feet)

Orbiter
Relay link

Jettison aeroshell

Engine ignition,
1200 meters
(4000 feet)

Jettison parachute

Terminal propulsion descent

Landing

Entry to landing
6 to 13 minutes

FIG. 91. Flight sequence of a Viking lander starting (left) with separation from the orbiter and ending (lower right) with soft landing. (JPL-NASA diagram.)

FIG. 92. The 1-ton Viking Mars lander stands on three pads cushioned by shock absorbers, its basic structure a hexagonal box, on which are mounted a high-gain S-band antenna (back, top), two RTG power supplies, two TV cameras in turrets (one with U. S. flag, front), wind meter and atmosphere sensor (front), and a long extendible soil sampler (left) with claw and magnets on the end. Three retrorockets face downward between the legs and their fuel tank extends to the right. (JPL-NASA photo.)

mented landing capsules of Mars 4 to 6 are similar to that of Mars 3, each is about 9 ft in diameter and will weigh nearly 800 lbs on the planet's surface.

Mars 5 made it but, like Mars 3, failed to transmit after landing. Less is known about the other two, except that the retrorockets failed on Mars 4 (so it swept past Mars in solar orbit), and that argon caused a pump to fail on Mars 6 (p. 228).

NASA launched Viking 1 on Aug. 20, and Viking 2 on Sept. 9, 1975, both from KSC. Each Viking consisted of a 5200-lb orbiter similar to Mariner 9, and a 2300-lb lander equipped with two TV cameras, a seismometer, an atmosphere analyzer, a 10-ft claw, and three small "laboratories" for chemical, magnetic, and biological analyses of martian soil, all parts heat-sterilized before launch. Following Soviet procedure (p. 209), the lander is to spiral down from the orbiter and use a heat-shield cone ("aeroshell") to decelerate in Mars' upper atmosphere to about the speed of sound (800 mi/hr—see Fig. 91). Then a "disk-gap-band" parachute will be released—a parachute with a hole in the middle and an open gap at about three-quarters of the canopy's 55-ft diameter—which can take the shock when it opens. (In 1972 it was tested successfully at 25-mi. altitude in the Earth's atmosphere.) Finally, three small retrorockets are to slow the lander to a few feet per second and it drops 8-10 ft to land on three cushioned legs (Fig. 92).

Before the launches, Carl Sagan wrote about—and suggested changes to—the plans for this dramatic and significant mission.—TLP

Viking to Mars:
The Mission Strategy

CARL SAGAN

(*Sky and Telescope*, July 1975)

This is the summer of the Viking launch to Mars. If all goes well, two landers and two orbiters will be sent there in an event unique in human history—the first extended and closeup reconnaissance of the surface of another planet.

The Viking spacecraft includes orbital experiments for imaging, infrared thermal mapping, infrared water-vapor detection, and radio science; atmospheric-entry experiments for analysis of the neutral and ionized components of the martian upper atmosphere; and lander experiments for imaging of the surface, atmosphere, and astronomical objects; for inorganic chemistry, meteorology, seismometry, magnetic properties of sand grains, organic chemistry of surface samples and analysis of the lower atmosphere, plus three compact biological experiments designed to search for any martian microorganisms.

Viking is an expensive mission, costing almost a billion dollars; it has occupied hundreds of scientists and engineers for many years, some of us for more than a decade; and, if it works, it promises to revolutionize the planetary sciences in general and the study of Mars in particular.

While the orbital television system will obtain photographs of Mars with a resolution two to three times better than the best images from Mariner 9, the prime function of the orbiter is to act as a radio relay link to Earth for the lander and to help certify preselected landing sites. The orbiters do not have, for example, the infrared and ultraviolet spectrometers that were so successful on Mariner 9.

The primary objective of the mission is to search for life on Mars, and this is not something easily performed remotely, as demonstrated by the long and sad history of "canals," "green" coloration, seasonal changes attributed to biology, and "hydrocarbon" absorption features in the martian dark areas. To find life, we must land. But where?

The Soviet Union has made four unsuccessful attempts to place scientific payloads on the martian surface. Mars 3 entered the martian atmosphere and softly landed on its surface but, after 20 sec of featureless television transmission, mysteriously ceased signaling. Mars 5 also successfully entered the atmosphere, but enigmatically failed within seconds of touchdown. Mars 4 suffered a rocket-motor failure, and Mars 6 a pump failure.

Mars 3 entered during the great 1971 global dust storm. Winds above 100 m/sec—more than half the speed of sound—raged during the storm, and it is entirely possible that the spacecraft was blown over. Mars-5's entry in 1973 was not during a storm; its failure must be ascribed to another cause. Whatever the reasons for the losses of Mars 3 and 5, these pioneering efforts have put us on notice: it may be dangerous to land on Mars.

Unlike the Soviet landers, the Viking landers will be carried into orbit around Mars and will descend only after the preselected sites have been "certified" as safe—as well as we can. We do not wish to land in regions that have lately undergone extensive albedo changes (p. 227) or which exhibit other signs of high winds; nor where the surface is rough—pocked with small craters or boulders. The lander then might be unable to establish a radio link with the orbiter or direct contact with the Earth, or its sample arm might be swinging helplessly up in the air instead of being firmly extended parallel to the surface.

We do not wish to land in locales that are too hard, since the sample arm is designed to collect powdery material; nor in places that are too soft, where the lander would sink into a dry martian equivalent of quicksand. Finally, we do not wish to land in a spot that is too high. For the descent parachute to work in the thin martian air, the lander must alight in a region of high atmospheric pressure (6-7 millibars).

These are a great many "do nots," each excluding a significant portion of Mars' surface. It might have been that after all these constraints were applied there were no areas left to land on at all. But luckily it appears that there are several unexcluded regions of more than passing scientific interest.

Landing Sites A and B

The orbiter's perimartium (point of orbit nearest the planet) specifies within a few degrees the martian latitude of the landing site. But we have complete freedom in our choice of longitude, needing only to wait for Mars to turn beneath us. The first mission to be launched, A, is now scheduled to land at 21° north latitude, 30° longitude in a region called Chryse (see Figs. 75, 93), the confluence of four great meandering tributaried channels which may have been carved by running water in a previous epoch in martian history (see p. 228). The backup site for the A lander is safer yet, according to the selection criteria, but scientifically much duller.

We have ground-based radar data near the Chryse site suggesting that its radar roughness is not too high and that the bearing strength of its surface is not too low. But 20 percent of the regions of Mars that have been so studied show a very low radar reflectivity. This could be due to roughness, which will scatter radio waves, or to high porosity, which means low bearing strength.

Whichever explanation is correct, something like one-fifth of Mars seems dangerous on this account. Only such radar studies of potential landing sites can determine the safety; TV pictures are not good enough.

However, the actual landing site in Chryse will not be accessible to ground-based radar until June 1976, a few weeks before the nominal landing date, which is, by a peculiar coincidence, July 4, 1976. A successful landing would make an excellent 200th birthday present for the United States, and a failure would not contribute much to that festive occasion! Thus, there is likely to be a great deal of scurrying, analyzing, thinking, and worrying by scientists at the radio telescope in Arecibo, Puerto Rico, and at JPL around the spring equinox of 1976. But if the landing-site situation is too indeterminate, Viking can wait in orbit for many weeks until a decision has been made.

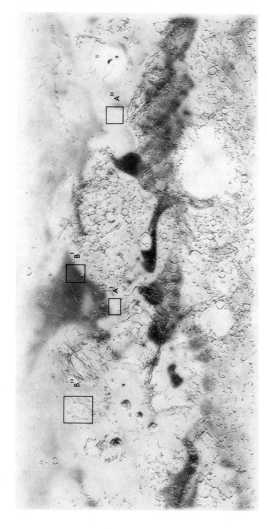

FIG. 93. Prime (A', B') and back-up (A", B") sites for Viking landers, plotted on a map of Mars that shows both topographic features like Figure 84 and albedo features like Figures 75 and 85. The top is 65° N, bottom 65° S latitude, and 0° longitude is centered. (NASA/Lowell Observatory map.)

The nominal landing site for the **B** mission is at 44° north latitude, 345° longitude in a region called Cydonia. Some investigators like it because it is at moderately high latitude, although Mars' transient polar snows have never been recorded south of about 55° north. Others like it because it is a very low region, so low that the high atmosphere pressure (7.6 millibars) increases the likelihood of at least transient quantities of liquid water in the soil. (Two of the three Viking microbiology instruments are designed to search for aqueous life.) But because of the relative geometries of the two planets, we can never obtain from Earth radar data for so high a martian latitude; we have no way of knowing whether a spacecraft can land successfully in Cydonia.

Thus, in the present mission plan, we are committed to a latitude determined by the perimartium of the **B** orbiter before the first biology experiment on the **A** lander can be completed. This is, in my view, a risky strategy for so expensive and significant a mission, and I have been urging substitution of an alternative site at a more equatorial latitude. The matter is under active discussion, and options still remain open. There is no question that the first successful landing on Mars requires extremely difficult decisions.

The Onboard Computers

The key elements in both orbiter and lander are their onboard computers. Each lander possesses—like a human being with his *corpus callosum* cut—two identical brains. Each brain has an 18,000-word memory, with 24 bits per word, stored on two coupled magnetic wires. (A memory of 18,000 words is quite large. Basic English, which is alleged to be serviceable in many layers of American society, consists of fewer than 1000 words.)

A scientific and engineering protocol is already in the Viking memory banks, and a revised protocol will be inserted by radio command while the spacecraft are in interplanetary space during January 1976. Each lander's computers will then be checked by a data dump to see how well they remember and understand the new program. In each case, the computer that has forgotten the least will be put in charge. A third computer on Earth selects the winner. The loser will be put to sleep, but will wait in readiness; in case of an accident or senility in the winner it may be called upon later.

While it is certainly large, an 18,000-word vocabulary is inadequate for everything the Viking lander may be asked to do. (The sample-arm motions alone require several hundred words.) Only 5000 of its words are for functions to be performed exclusively after landing. The remainder are for executive matters, involving the structure of the entire computer program; checkout of engineering functions; and functions performed during descent to the martian surface.

For this reason, the mission controllers have devised a "primary design," according to which every 6 days a set of new commands, comprising hundreds to thousands of words, will be radioed uplink from Earth to Mars. But this corresponds to a very sluggish response to what may be astounding discoveries made on the planet's surface, and an update every 3 days seems to be necessary if we are to perform appropriately responsive experiments on the surface.

One clear lesson from past spacecraft is that enormous scientific payoffs follow from the ability to do new experiments on the basis of what we have just learned. Many of the most famous discoveries made by Mariner 9—the great volcanoes, the surfaces of Phobos and Deimos, variable features, great sand-dune fields, and details of the large sinuous channels—required changes in the TV plan. An entirely preprogrammed Viking mission would be relatively feeble scientifically. It could not even select with the lander cameras the place where the sample arm is to dig. But if we can perform an experiment tomorrow on the basis of what we learned the day before yesterday—that is a scientific capability of stunning potentialities.

The Mission Profiles

Both Viking launches will take place from Pad 41 at KSC. A 10-day minimum interval is required between them. The nominal launch window for the **A** mission is Aug. 11-22, for the **B** mission, Aug. 21-Sept. 9, 1975. Because of holds during countdown, or possibly even the necessity for replacement of components (a very serious matter for a sterilized spacecraft), it is not entirely clear that the windows are wide enough.

If the launch dates slip a little, the price will be less science on Mars. Any launch later than about September 20 will seriously degrade the mission, and might require a 23-month "hold" until the next opportunity in 1977. NASA has a very good launch record, but there has never been as complex a scientific mission as Viking. The launch vehicle is the Titan 3E, a Titan booster with a Centaur second stage, a configuration that has been tried only twice. Its batting average is 0.500.

In the nominal mission, the first (**A**) orbiter-lander combination is injected into orbit around Mars on June 13, 1976, giving it 21 days for site certification before the July 4 landing. After certification, several propulsion maneuvers are required to make the orbit Mars-synchronous, with a 24.6-hr period and a perimartium altitude of about 1500 km over the landing site. The orbit is highly elliptical. The **B** mission arrives in the vicinity of Mars on July 28, 45 days after the **A** configuration. It has about 30 days in orbit for **B**-site certification and scientific investigations before its lander separates and spirals down.

The **A** lander has a working lifetime on Mars' surface of 58 days; the **B** lander, 62 days. The budget for the Viking mission will apparently permit almost no simultaneous operation of major scientific experiments on the two landers. This is a great pity—despite its high cost, Viking is severely hampered by lack of funds.

The descent maneuver begins when a mechanical spring separates the orbiter and lander, giving a relative velocity of 1-2 m/sec. The two follow essentially the same orbit for 2 hrs, during which the lander orients itself for entry and examines the thin upper atmosphere of the red planet. It radios its findings to the nearby orbiter, which relays them directly to Earth as well as recording them on tape recorders for future playback. The entire descent sequence is under control of the active lander computer; the ground "controllers" will be able only to bite their fingernails.

After an initial rocket burn, the lander enters the denser atmosphere, ablation shield first (Fig. 91). After this burns off, the parachute is deployed and then, under control of the accelerometers with backup timing devices, is jettisoned. Finally, the terminal-descent rockets burn, to be turned off only about 10 ft above the martian surface. The entire delicate landing maneuver is actively controlled by the lander, relying on its descent radar and other instruments and maintaining a careful attitude control. It is an intricate servomechanism, making decisions on the basis of its sensory information, as we do. The lander free-falls the final few feet and—many of us sincerely hope—safely lands on its three spring-loaded footpads.

The Lander Program

Immediately upon setting down, at about 4:30 in the afternoon Chryse standard time, the spacecraft initiates a range of engineering and house-keeping functions. It asks itself if it is feeling well. The lander will relay data to Earth via the orbiter, when that is above the lander's horizon, at 16,000 bits/sec; and at other times directly to Earth at the much slower rate of 500 bits/sec. During the next 3 days, it takes seismometric and meteorological data, as well as the first closeup pictures of the martian terrain.

On the third day after landing, the first set of uplink commands arrives from Earth. By the sixth day, a decision will have been made on where to obtain the first soil sample with the sample arm. On the eighth day, after the onboard computer has demonstrated that it truly understands its newly arrived instructions on where to dig a hole, the sample arm gingerly extends itself toward the surface. It can reach a soil sample (or a more interesting object) as much as 10 ft away. With a nervous jittery backhoe motion, it lifts its sample into the air and gradually retracts, telescoping itself until it is only a few inches from the lander's main body. Photographs are taken before, during, and after the sampling operation.

The arm then positions itself over one or more of the three entry bays. One is for the x-ray fluorescence experiment, to examine the inorganic chemistry of molecules with atoms heavier than about mass number 20; another is for the gas-chromatograph mass-spectrometer, to examine the organic chemistry of the samples; and the other is for the three different microbiology experiments. The arm opens its little claw, shakes itself, and deposits the sample into a funnel which is covered by a wire-mesh screen. The experimenters back on Earth will have decided whether they want the same sample for each experiment.

The experiments then do their stuff, which may take some days. The biology experiments, for example, require an incubation period before the results can be radioed to Earth. In all, three samples will be examined by each of the three biology experiments, four by the spectrometer, and five by x-ray fluorescence, all during the nominal life of each lander.

It is hardly necessary to remark that the results of these experiments and those of the imaging experiment may be of the greatest interest for the age-old question of life on other planets. But our ability to exploit the Viking experimental tools depends crucially on the uplink radio capability.

After the **B** mission lands, high-data-rate transmission from the **A** lander is turned off for the economy reasons mentioned above. Only low-data-rate experiments on the direct link to Earth can be performed, chiefly seismometry, meteorology, and single-line video scans. The **A** orbiter, freed of its relay responsibility for the **A** lander, now exuberantly explores Mars from orbit. Many questions posed by Mariner 9 may be answered at this stage.

This is also the first time in the mission when radio occultation experiments will be performed, as the atmosphere and the planet intercept the transmission from orbiter to Earth. Eventually, the **A** orbiter may be called back to service the **B** lander (a function possible only because the longitudes of the nominal **A** and **B** landing sites are very nearly the same). If this occurs, the inclination of **B**'s orbital plane to the equatorial plane of Mars can be increased to 75°, converting the **B** orbiter into a martian polar observatory. Many other elegant mission strategies are possible.

When Mars is in solar conjunction on Nov. 25, 1976, the sun will be between the planet and the Earth, and communication with the Viking spacecraft will be interrupted from November 8 until about Christmas Day. The generosity of Nature is evident here: Viking scientists will finally be able to take time out from data gathering to ponder what the data mean. Many space missions never provide such an opportunity.

If all is still working well, the Viking Extended Mission may begin early in 1977. While the nominal lifetime of a Viking spacecraft is only 90 days, Mariner 9, with a similar life expectancy, performed for a full year, failing only because it ran out of consumables. The Viking landers need no sunlight; they are powered and heated by two Radio-isotope Thermoelectric Generators (RTG). If it is as well engineered as Mariner 9 was, the really interesting part of the Viking mission may begin in January 1977. In any case, if Viking works even moderately well, planetary astronomy will never be the same again.

The Viking orbiters should not be ignored; they each have a general-purpose computer with 4096-word memory, and a tape recorder capable of storing 55 TV frames. They are powered by solar panels backed up with Ni-Cd batteries. Orientation is maintained by eight attitude-control jets so that a sun-sensor is locked onto the sun, and a star-sensor onto the bright star Canopus. Of course, the attitude must be changed for mid-course corrections and for orbit insertion near Mars, as well as the later orbit changes.

Each orbiter also carries an infrared water detector—a grating spectrometer fixed on the 13800A absorption band of water vapor. This downward-looking instrument will map water vapor as low as 1 micron of precipitable water in areas as small as 2 x 15 mi. on Mars' surface. Similarly, an infrared thermal mapper will map the surface temperature to 0°.3C, using four IR pass bands between 6 and 24 microns' wavelength, and covering a 5 x 5-mi. area on Mars' surface. The detectors are antimony-bismuth thermopiles.

Two downward-looking Vidicon TV telescopes (5.5-in., f/5.6) with selectable color filters will photograph the surface at 5-sec intervals, with resolution about 45 m. After site selection, these telescopes can be used for other astronomical observations. A 20-watt transmitter with 58-in. dish will telemeter pictures and other data (including relay messages from the landers) to Earth at 500 bits/min (240,000 bits/min when one of the 210-ft antennas, described on p. 279, of the Deep Space Network can "see" them). Two other orbiter antennas can also be used, one designed for measures of Mars' atmosphere and ionosphere as the orbiter passes behind Mars.

More detailed pictures and diagrams of the Viking landers were published in Sky and Telescope for September 1975. They show the compactness and complexity hidden under the smooth exterior shown in Figure 92. After jettisoning its aeroshell and parachute (Fig. 91), its landing weight is only 1320 lb. Its frame is a hexagonal prism, 18 in. high and 56 in. across, thoroughly insulated with fiberglass and dacron to maintain warm temperatures inside during the martian night. External tanks and legs extend the largest dimension to almost 10 ft. Electric power comes from two RTG's (Fig. 54)—each containing banks of small thermocouples heated by plutonium-238, similar to the Pioneer SNAP-19 (Fig. 106)— which also help to heat the interior.

To save weight, power, telemetry, and cost, the two cameras are facsimile type, scanning slowly (line by line) from 40° below horizontal to 40° above at any azimuth, with resolution 0°.04 (2.4 arc-min.) so as to show 1-mm sand grains near the lander and golf-ball-size rocks 100 ft away. There is also a seismometer inside, and an anemometer and thermometer on a 40-in. boom to measure wind and outside temperature.

The sample scoop can dig with a force of 30 lb, 10 ft from the lander, lift 5 lb, sift it, and dump the sand-dust into any one of three funnels on the top of the lander, each with a screen sifter and petal-like cover (closed when the wind is blowing). The funnels lead to the three soil-analysis experiments. One is an x-ray fluorescent spectrometer to measure proportions of the chemical elements in the soil, as was done remotely on the Moon (p. 125). It can analyze five samples dumped into its funnel. The second funnel leads to a gas-chromatograph mass-spectrometer that will test soil samples for organic traces of former life. Its operation may not go as planned because its titanium-ion pump may be inactivated by argon, now suspected to be a major constituent of the martian atmosphere (p. 228). For this reason the Gas-Exchange Experiment has recently been scheduled last. The third funnel leads to the other experiments, as described in the following article.—TLP

The Viking Lander's Soil-
Analysis Experiments

(*Sky and Telescope*, September 1975)

Above all, attention will be given to the three miniature biology laboratories packed into one cubic foot within the lander body. Each approaches the problem of martian biology from a different point of view.

Pyrolytic Release Experiment. Seeking to detect organisms that exist on Mars and function in a normal martian environment, this experiment exposes a soil sample to a Marslike atmosphere containing carbon dioxide and carbon monoxide labeled with radioactive carbon-14. The sample is incubated for 5 days under simulated martian sunlight less the ultraviolet radiation; then the sample and its chamber are flushed with inert gas (to remove the unused portions of the labeled atmosphere) and immediately heated to 625°C. At this temperature, any organism's cells break down and are vaporized. Then the hot gases are passed through detectors that can identify the radioactive carbon assimilated by the organisms. This procedure can be repeated up to four times with different samples.

Labeled-Release Experiment. Supplied with nutrients any organism takes in certain compounds for its metabolic processes. These almost always involve the consumption and production of common gases, such as hydrogen, oxygen, nitrogen, and carbon dioxide. The Labeled-Release Experiment attempts to detect these processes by moistening a portion of soil with a carbon-14-labeled nutrient solution and incubating it for as long as 11 days. Some gaseous by-products of life processes (such as carbon dioxide, carbon monoxide, and methane) will contain the radioactive label and can then be identified by carbon-14 detectors located in the test chamber. Here again, the experiment can be performed on 4 different samples.

Gas-Exchange Experiment. This device operates on the simple principle that living organisms alter their environment as they live, breathe, eat, and reproduce. Thus, gaseous changes within the experiment's airtight canister will indicate that life processes are going on. Unlike the first two experiments, the gas-exchange device uses no radioactive trace elements, but analyzes the enclosure's atmosphere with a gas chromatograph. The sample to be tested is partially submerged in a nutrient solution and incubated for up to 12 days in an atmosphere of helium, krypton, and carbon dioxide. Gases within the experiment cup are sampled by the chromatograph every few days, checking for traces of molecular hydrogen, nitrogen, oxygen, methane, and carbon dioxide. Although this cycle can be repeated four times, only one sample can be tested.

The detection of life on Mars will not be a yes-or-no proposition during the course of these experiments—someone's monitor will not light up suddenly to signal that life has been discovered! Instead, each scientist will have to look carefully at his experiment's data and compare his results with those of other biology-team members. The team will then decide whether or not the tests indicate the presence of life.

It is entirely possible that no life (as we know it) exists on Mars at this time, but did in the past. If this is the case, remanent organic material may still be present in the soil. The lander's gas-chromatograph mass-spectrometer (GCMS) is designed to identify such components, distinguishing between them by their molecular weights.

Small soil samples are first pulverized and heated until the organic portions are vaporized. These are forced into the gas chromatograph's column, where the temperature is raised $6\frac{1}{2}°C/min$. Each level of temperature releases different organic molecules into the mass spectrometer; they are ionized and focused into a concentrated beam by electronic and magnetic devices. This beam of ions is then measured with an electron multiplier tube.

6

Close-Up Views
of Jupiter

Jupiter, at closest approach, is over six times as far from Earth as Mars is. Getting there with a space probe takes four times as long—21 months. The ringed planet, Saturn, is about twice as far again—see Table 8. Although these flight times may be reduced by nuclear rockets, they show the enormity of space and the value of the light that brings the astronomer data from distant objects. Nevertheless, NASA can fire small space probes at five or six times the speed of the large Apollo spacecraft, and is considering a 1979 shot at Uranus, some 19 AU (1760 million mi.) from Earth. (It would take almost 5 years to get there, and many astronomers who design instruments for the probe may have retired before their data come back!)

Jupiter is an obvious first among the giant planets; it is the largest and closest, and has several puzzling features of scientific interest. Two of its four large moons (Ganymede and Callisto) are as large as the planet Mercury, and the 14 jovian satellites form a miniature solar system whose origin and history probably parallel those of the sun-planets system. Jupiter broadcasts high-frequency (30-3000 MHz) radio waves in an irregular manner. Its atmospheric bands and huge Red Spot are evidence of a complex atmosphere, and its internal constitution is not fully understood (see Vol. 2, Neighbors of the Earth).—TLP

Plans for Pioneer
Flights to Jupiter

RAYMOND N. WATTS, JR.

(*Sky and Telescope*, August 1970)

In late February or early March 1972, the first of a pair of spacecraft (Pioneer 10) will be launched on an ambitious mission to the planet Jupiter. The second (Pioneer 11), nearly identical to the first, is scheduled to fly some 13 months later.

The direct-ascent flights (no parking orbit) will be powered by Atlas-Centaur vehicles with a special third stage . . . giving the 550-lb Pioneer an initial velocity of about 32,400 mi/hr, the fastest for any space probe to date.

During a long voyage of 600 to 900 days, each probe will spend from 6 months to a year passing through the hazardous asteroid belt, where it is estimated there are tens of thousands of minor planets (asteroids— see Fig. 58) at least 1 mi. in diameter, and unknown numbers of smaller ones.

The uncertainty in total travel time to Jupiter is due to a new type of mission planning. Instead of selecting a date for the encounter and striving to get the spacecraft there on time, the Pioneers will be launched whenever they are ready. Each trajectory and encounter date will be decided later and set by remote control, providing more flexibility and greater chance of success, according to NASA's mission planners.

Each Pioneer (Fig. 94) will be equipped for 13 scientific experiments, powered by four radioisotope thermoelectric generators (p. 445) producing a total of 120 watts. A nine-ft radio antenna will point earthward for telemetry, with the craft spin-stabilized at 5 revolutions per minute. The scientific instruments will weigh 60 lbs, and a number of them will sense the environment through which the probe passes.

One experiment will use penetration cells to monitor encounters with small meteoroids (see p. 45). In the asteroid belt four optical telescopes will detect asteroids that are too small to be observed from Earth. Other experiments include plasma detectors and a cosmic-ray-particle detector.

Each Pioneer will spend about a week swinging around the planet, during the period of close approach (within 100,000 mi.—see Fig. 95). A Geiger-tube telescope will search for radiation belts surrounding Jupiter, and if there is a magnetic field, a helium-vapor magnetometer will detect it.

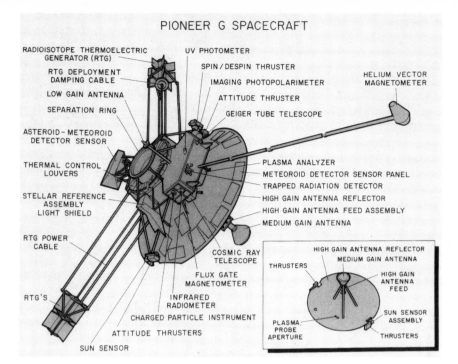

FIG. 94. The Pioneer spacecraft for Jupiter viewed from behind the 9-ft high-gain radio antenna, which is to be pointed at Earth. Simplified front view shown in inset, lower right. (JPL-NASA diagram.)

FIG. 95. A planned trajectory for Pioneer 10 to pass Jupiter, marked with hours before (minus) and after closest approach. Experiments are listed at left and bottom and the relative sizes of Jupiter's images are shown at right with distance (R_j) in Jupiter radii (71,500 km) and camera resolution (RES) in km. (NASA diagram.)

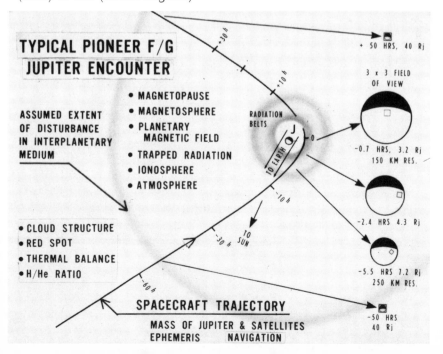

Jupiter's radiation in ultraviolet and infrared light will be measured by a photometer and a radiometer. The planet's visible surface will be photographed in red and blue light by an imaging photopolarimeter, using the spin of the spacecraft to scan the planet in narrow strips, a full picture being completed every 25-50 min.

Swinging behind Jupiter, each Pioneer will be occulted for a short period. At ingress and egress, radio signals coming through the jovian atmosphere will be monitored (see p. 195). The gravitational field of the giant planet will be calculated by the manner in which it alters the spacecraft's fly-by trajectory (see p. 97).

Although most of the asteroids are in orbits between those of Mars and Jupiter, a few come much closer to the sun. Ones that pass close to the Earth (such as Eros) were used by astronomers early on to establish the scale of the solar system—in particular, the number of miles in 1 AU. In early 1976, a 2-mi. asteroid very close to the Earth was discovered at the Palomar Observatory. Its orbit is a little smaller than the Earth's (r = 0.96 AU; see Table 8), with inclination 19°. E. M. Shoemaker (CalTech) estimates the chance of its colliding with Earth at 1/24,000,000 per year, but a close miss will probably throw this 10-million-ton rock into a completely different orbit, eliminating the hazard.—TLP

Pioneer-10 Mission to Jupiter ·

RAYMOND N. WATTS, JR.

(*Sky and Telescope*, May 1972)

The first of two long-range unmanned flights to the outer parts of the solar system got off to a good start on March 2, 1972, when an Atlas-Centaur booster hurled a 570-lb spacecraft aloft from KSC. Pioneer 10 began the journey that will carry it past Jupiter on Dec. 3, 1973, some 87,000 mi. above the planet's cloud tops.

On March 7 and again on March 23-24, Pioneer's engines were fired for a few minutes, the combined effect of these trajectory adjustments being to shorten the flight time to Jupiter by 6½ hrs. Enough fuel remains for several other such maneuvers, if desired, but the present path appears accurate enough for almost all the fly-by objectives.

One purpose of these course corrections was to time the fly-by for several close looks at Jupiter's Great Red Spot. If all goes well, the craft will pass behind Io, the innermost of the four large jovian satellites, allowing a sensitive radio-occultation test (see p. 195) for an atmosphere surrounding Io. However, the arrival time must be controlled to within 8 min. if the occultation by Io is to occur.

A critical part of the mission will begin this summer when Pioneer 10 enters the asteroid belt. A collision with a sizable body could seriously damage the craft, but the probability of such a collision is very slight. As of March 24, Pioneer's detectors had recorded 10 micrometeorite impacts, and its asteroid-meteoroid telescopes had seen one passing meteoroid.

Once past Jupiter, Pioneer 10 will continue on beyond the outer planets and eventually leave the solar system. About 1980, it will be near the orbit of Uranus and at the limit of radio detectability, if its transmitter is still functioning. In about 80,000 yrs, Pioneer 10 will have traveled 1 parsec (3.26 light years), slightly less than the distance to the nearest star.

Because of the infinitesimal chance that this spacecraft might someday encounter intelligent beings elsewhere in our Galaxy, it carries a small gold-anodized aluminum plaque etched with pictures and diagrams designed to convey as much information about our civilization as possible in such a limited space (Fig. 96). The plaque has been etched deeply enough so that it should withstand erosion by interstellar particles during a voyage of at least 10 parsecs and probably more than 100. Thus, Carl and Linda Sagan and Frank Drake of Cornell University, who are largely responsible for the plaque, suggest that it is likely to survive for a longer period of time than any of the works of man on Earth.

Carl Sagan has since admitted that his plaque (Fig. 96) is unlikely to be viewed or read by living beings elsewhere in our Galaxy. It was "written" more for the beings on Earth, to stress our isolated place in the universe and the difficulties of telling someone else where we are and what we do.

As on earlier missions to the Moon (Chapters 2 and 3), the meteoroid hazard proved to be minor. Seven large meteoroids (10 cm) and 24 medium-size ones (1 cm) were detected in the asteroid belt, and many micrometeoroids, probably from old comets, all along the route. The main problem during the long flight stemmed from prolonged periods of inaction and the need for highly accurate course corrections. On Sept. 19, 1972, thrusters were fired briefly, speeding Pioneer 10 by 9 in/sec so that it would pass 330,000 mi. behind Io for the radio-occultation experiment at 9:23 p.m. EST on Dec. 3, 1973.

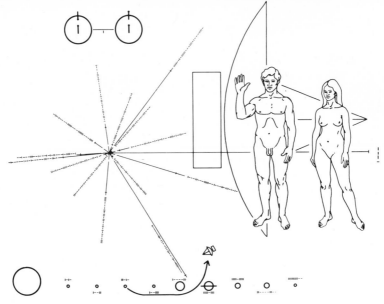

FIG. 96. Plaque attached to Pioneer 10, describing for distant residents of the Milky Way where we are and what we look like. The figures of man and woman on the right are compared in size to an outline of the spacecraft. The bottom line shows schematically the sun and planets, and the route of Pioneer 10. Binary code is used at left to show directions and distances from sun to pulsars ("landmarks") in space. The symbol, top left, shows the unit of distance used from the hydrogen molecule. (Courtesy of Carl Sagan, Linda Sagan, and Frank Drake.)

On April 5, 1973, Pioneer 11 was launched from KSC without difficulty, programmed to reach Jupiter in 1975. Its trajectory will later be modified on the basis of what Pioneer 10 learns.—TLP

Pioneer 10 Passes Jupiter

RAYMOND N. WATTS, JR.

(*Sky and Telescope*, January 1974)

After a 620-million-mi. journey that began on March 2, 1972, the 570-lb spacecraft Pioneer 10 reached the vicinity of Jupiter on December 3, 1973, passing 81,000 mi. from the planet's surface. At this climax of its highly successful mission, Pioneer was so distant that the signals from its 8-watt radio transmitter took 46 minutes to reach Earth.

During closest approach the unmanned probe was moving 23 mi/sec as it transmitted a closeup view in color of the planet's Great Red Spot.

Fourteen minutes later, as planned, Pioneer passed behind Jupiter's satellite Io, causing a 91-sec interruption in communications.

Early results of the fly-by were released by scientists assembled at NASA's Ames Research Center at Mountain View, California. The first observational evidence for helium in Jupiter's atmosphere was obtained by Pioneer's ultraviolet photometer. Although the readings from the magnetometer were not easy to interpret, they did confirm that the polarity of Jupiter's magnetic field is opposite to that of the Earth's field (south magnetic pole near Jupiter's north pole).

Scientists reported that the asteroid-meteoroid detector aboard Pioneer was seriously damaged by Jupiter's radiation belts. The spacecraft began to encounter the trapped radiation as far as 4.2 million mi. from the planet. The belts appear to be enormously more intense and flatter than the Earth's Van Allen belts.

Transmission of pictures of Jupiter and the four Galilean satellites began on November 4 and was to go on until January 3. The planned 336 pictures come from an imaging photopolarimeter that measures red and blue light from the planet as the spacecraft (spinning 5 times/ min) scans a strip of Jupiter. Even before the closest approach, the pictures surpassed in resolution the best taken by telescopes on Earth.

After leaving the vicinity of Jupiter, Pioneer 10 will continue to monitor the interplanetary medium as the craft heads outward through the solar system. The value of such measurements is enhanced by similar ones being made by Pioneer 11. For example, it is possible to find the speed of a solar-wind gust or magnetic disturbance traveling out from the sun if it has been detected by both probes.

It is hoped that Pioneer 10 can continue to measure the solar wind out to the orbit of Uranus, which it will reach in 1979. Eight years later, the spaceship will pass Pluto's orbit as it heads toward stars in the constellation Taurus.

*The intensity of Jupiter's radiation belts (see Fig. 7 for the Earth's), in which electrons, protons, and other ions move at high speed back and forth between the magnetic poles, is an indication of the planet's strong magnetic field (20,000 times the Earth's magnetic moment, with north magnetic pole 15° from Jupiter's south pole) and is related to Jupiter's 30-3000 MHz radio emission. Luckily, none of the Pioneer electronics was damaged by the radiation belts and the photographs radioed back to Earth were spectacular. Jupiter has never been seen in crescent phase before; it is always seen near full phase from Earth.—*TLP

Pioneer Observes Jupiter

RAYMOND N. WATTS, JR.

(*Sky and Telescope*, February 1974)

On Nov. 26, 1973, after 634 days of relatively uneventful spaceflight, Pioneer 10 plunged through the shock wave caused by the interaction of solar-wind particles with the magnetic field of the planet Jupiter. The space probe's mission had taken it farther from the Earth (and faster) than any man-made object had gone before.

As Pioneer approached Jupiter, its instruments were turned on and tested. The first images of the planet were received on November 4, when the craft began daily observation periods lasting 3-8 hrs. On November 25, intensive round-the-clock data-gathering commenced.

The next day, Pioneer's helium-vector magnetometer sensed an abrupt tripling in the intensity of the surrounding magnetic field, to a new level of 1.5 gammas. Accompanying this change was more than a hundredfold increase in the temperature of the solar wind, and a drop in its velocity to less than 150 mi/sec (about half what it had been). In all, 6 of the spacecraft's 11 instruments recorded some kind of change associated with the passage through the bow shock where the solar wind bounces off Jupiter's magnetosphere.

The probe was about 4,790,000 mi. from Jupiter at that time.

November 26 was also when scientists at Ames Research Center adjusted Pioneer-10's attitude in space for the last time prior to the encounter with Jupiter. Further use of the thrusters (rockets) was to be avoided, so that the probe would behave like a body in free fall and permit a measurement of the planet's gravitational field. Ten bursts from the hydrazine thrusters tilted the spin axis of the craft to the proper angle for the coming closeup imaging and polarimetry of the planet. The photopolarimeter (see pp. 258 and 264) began working 23 hrs a day.

On November 27 the magnetometer registered another jump in the field strength, to about 5 gammas, after which the field became steady. Jupiter's magnetopause (Fig. 99) had been crossed, somewhat closer to the shock front than had been predicted. Pioneer had entered the more orderly field of the planet itself—the jovian magnetosphere. The plasma analyzer noted the disappearance of the solar wind, and the energetic-particle counters showed substantial increases in both protons and electrons trapped in the radiation belt.

On November 30, the University of Southern California's ultraviolet photometer began a 38-hr observing session, taking data on Jupiter's atmospheric composition and searching for auroral activity (like "northern lights" on Earth). This instrument detected helium in the atmosphere (as expected) and an ultraviolet glow in the atmosphere of satellite Io.

Later, on the 30th, the magnetic field abruptly dropped back to its interplanetary level, and the solar wind was detected once more by the plasma analyzer. This surprising situation continued for 11 hrs. Evidently an increase in the solar wind was temporarily pushing the weak magnetosphere inward. In fact, 8 days earlier, Pioneer 11 (also on its way to Jupiter and now crossing the asteroid belt) had detected a substantial increase in the solar wind, and a 100-km/sec increase in its velocity—a "gust" that briefly dented the envelope around Jupiter.

FIG. 97. Jupiter viewed by Pioneer 10 at 1.6 million mi. on Dec. 1, 1973, at 19:02 UT, showing the Great Red Spot near the limb at lower right. The south pole is slightly right of top center. (NASA photo.)

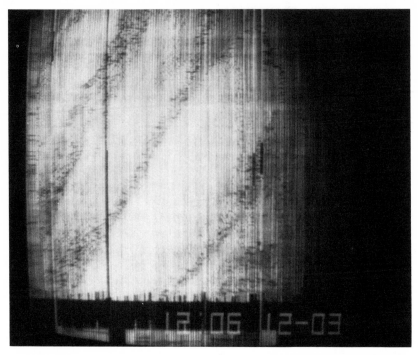

FIG. 98. Raw image of Jupiter as transmitted from Mariner-10's imaging photopolarimeter. After computer processing the image looks like Figure 97. (NASA photo.)

High-resolution views of Jupiter's clouds came on December 1, when Pioneer was close enough to return 100 or more picture elements in scans across the planet's disk, compared with the 80 for Earth-based photographs of high resolution (see Glossary).

In Figure 97, the planet's equatorial and polar regions seem less turbulent than the middle latitudes. Note the four small bright spots, each about 2500 mi. in diameter and ringed with darker material, in the southern (upper) hemisphere. The north tropical latitudes have several bright swirls, at scales down to the limit of resolution. Also remarkable is the waviness in the low-latitude boundaries of the bright north tropical zone and south temperate zone. The Red Spot seems to have somewhat pointed ends; the rim of the oval is sharply defined (to the resolution limit) and darker than the interior of the spot.

Although this view of Jupiter was the best available early in January, about 80 images (now being processed) were obtained closer to the planet, and their resolution may be up to six times better.

How were these full-color pictures made? If the spacecraft's red and blue-light images were combined by themselves, the result would be an unrealistic purple disk. So, during the computer processing, the relative amount of red and blue light at each point on the disk was used to predict the green intensity at the same point, using statistical data from three-color photometry with Earth-based telescopes. This gave an artificial green image that, when added to the red and blue ones, made a lifelike picture of the planet in color. This procedure happens to work especially well for Jupiter, because red and blue are its principal hues to begin with.

By December 2, Pioneer 10 had crossed into the inner region, characterized by a strongly dipolar magnetic field, where the direction of polarity was found to be in the sense opposite of that of the Earth's. Jupiter appeared 5° in diameter to Pioneer, showing a 3-in. disk on the 8-in.-square TV viewing screens at Ames. The spacecraft was still 22 radii away, but perijove, closest approach to Jupiter, was due in just 24 hrs.

These last hours were the most hectic. As soon as the craft passed the 19-radius mark, a sharp rise in electron flux was detected. Early on December 3, Pioneer passed Callisto at 880,000 mi.; 90 min. later it came within 280,000 mi. of Ganymede. The CalTech infrared radiometer made heat measurements of both moons, finding temperatures of —261° and —193°F, respectively.

Then, at about noon, when Pioneer was within 200,000 mi. of Europa and 7½ radii from Jupiter, the proton flux increased sharply. The spacecraft was then so close to the planet that its whole disk could not be seen at one time on the viewing screen.

By the time Pioneer reached the 5-radius mark, James Van Allen noted that the University of Iowa's Geiger-tube experiment was recording a tenfold increase in counts every 1½ hours. Fortunately, the flux level peaked about 75 min. before Pioneer's closest approach to Jupiter, and although two cosmic-ray detectors became saturated, none of the craft's instruments was damaged.

In these last few hours, one disappointment was the loss of a planned high-resolution image of Io (which was within 220,000 mi.), and of part of a close-up of Jupiter. Evidently, spurious commands had placed the photopolarimeter in one of its nonimaging modes (not scanning a picture). Jupiter's great distance from the Earth meant that 92 min. elapsed between the sending of each signal and its return, so there was no chance to correct the erroneous commands in time.

At 6:25 p.m. PST December 3, Pioneer swept only 81,000 mi. above the jovian cloud tops. Ten minutes later, the probe crossed the equatorial plane, and then passed behind Io. At 7:42 p.m. the signals stopped again, as Pioneer 10 passed behind Jupiter, and 34 min. later it entered Jupiter's shadow. The occultation of radio signals ended at 8:46 p.m., and in another 20 min. Pioneer emerged into the sunlight once more. Its trajectory had been bent through nearly a right angle, and the probe proceeded off in a direction roughly tangent to Jupiter's orbit around the sun.

Near Jupiter, the density of dust particles was about 300 times greater than in interplanetary space. Meteoroid hits, which had averaged one every 600 hrs on the outbound journey, picked up to 1 every 2 hrs during the fly-by. But Pioneer's instruments survived well, and scientists are now hard at work analyzing and interpreting the data collected by this eminently successful deep-space probe.

The Fly-By of Jupiter

TOM GEHRELS

(*Sky and Telescope*, February 1974)

Pioneers 10 and 11, prepared at a cost of about $100 million, are spin-stabilized craft (p. 270), unlike the larger Mariners (Chapters 4 and 5), which are stabilized by means of gas jets. Each Jupiter Pioneer has a main communications dish antenna 2.7 m. in diameter, which is always kept pointing earthward while the spacecraft spins around the dish axis at about 5 revolutions/min.

We were worried about Jupiter's radiation, since predictions of its intensity were revised upward after launch, when our instruments were already on their way and we could not redesign them or add shielding.

On the back of Pioneer's antenna are 13 panels used for detecting impacts of interplanetary particles of mass about 10^{-8} to 10^{-9} gm.

Instrumentation for other experiments is located in the instrument bay. . . .

When the spacecraft was still 108 Jupiter radii (each 71,600 km) from the planet, it first encountered the bow shock produced by Jupiter as it plows through the solar wind. This was unexpectedly far from the planet and increased our concern about strong radiation. A few days later, there was excitement at Ames because a storm of solar wind had pushed the magnetosphere back, and turbulent conditions were observed until, at about 50 radii, the magnetosphere was crossed again.

On the most pessimistic appraisal, we expected our optical telescope to begin to show radiation damage (p. 44) about 6 hrs before the closest approach to Jupiter and, indeed, at that time there were some uncommanded telescope pointings and operations. But soon things were normal again, and the other instruments and spacecraft components survived, too. Our photomultiplier tubes were responding to extraneous energetic protons coming right through the spacecraft, actuating an automatic protection device that decreased the sensitivity of detectors for some of the observations near Jupiter. But that was all!

When Pioneer passed behind the planet, the cutting off of the sun's radiation caused a slight shrinkage of the booms extending from the craft. The situation resembled that of an ice skater tucking in his arms to spin faster: Pioneer came out of eclipse with a spin period of 12.6216 sec instead of its normal 12.6227. This curious effect was, however, harmless to the scientific studies.

The observations during the fly-by enabled several investigators to find a 15° tilt between Jupiter's magnetic axis and its axis of rotation. The center of the magnetic field appears displaced slightly north of the planet's equatorial plane, possibly an explanation of the differences between the visible features of the northern and southern hemispheres.

The polarity of the magnetic field is opposite to that of the Earth and the field strength at the surface of the planet was inferred to be 4 gauss (see Fig. 99 and App. V).

The most striking finding is the exceedingly strong concentration of the trapped radiation toward the plane of the magnetic equator. Because of the 15° tilt mentioned above, the equatorial intensity peak swept past the spacecraft in approximate synchronism with the rotation period of Jupiter, which is about 10 hrs. On future missions, a probe could get closer to the planet without radiation damage by avoiding the equatorial plane near the time of closest approach.

Fields and particles share Jupiter's rotation rate out to about 35 radii, and the energies of the trapped particles are particularly high within 10 radii. Most investigators were deeply relieved that Pioneer 10 did not go in closer than 2.85 radii from the planet's center; even at 3 radii, the flux of electrons with energies greater than 3,000,000 electron volts (see App. V) was found to be 5×10^8 particles/cm²sec. For protons with energies greater than 30 million electron volts, the intensity was 4×10^8. Pioneer 10 took quite a bombardment!

The infrared radiometer (40 microns wavelength) observed the detailed temperature structure over the planet's disk. The total heat leaving Jupiter is 2½ times that which the planet receives from the sun.

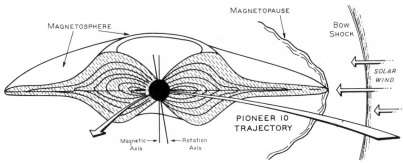

FIG. 99. Diagram of Jupiter's magnetosphere based on Pioneer-10 measurements. The north pole of rotation is slightly left of up (rotation axis), inclined 15° to Jupiter's south magnetic pole. Solar wind (and sunlight) come from the right, and deform the magnetosphere. The shaded area is filled with high-speed ions, protons, and electrons, of higher intensity (shown by contours) near the planet. (NASA diagram.)

Presumably, Jupiter is still contracting gravitationally, converting potential energy to heat.

From the radio-occultation experiment it was found that Jupiter has a layered ionosphere and that Io has an ionosphere. The radio tracking of Pioneer 10 showed gravitational attraction by the four large satellites.

The co-investigators associated with me in the imaging photopolarimeter project are David Coffeen, Charles KenKnight, William Swindell, and Martin Tomasko at the University of Arizona; Martha Hanner and Jerry Weinberg at the State University of New York, Albany; and Robert Hummer at the Santa Barbara Research Center. Also, a dedicated NASA crew supports this complicated threefold experiment, consisting of the imaging of Jupiter and its satellites, photometry and polarimetry (measuring polarization—see Glossary) of the same objects, and mapping the interplanetary zodiacal light (p. 334).

Of the 11 instruments on board, our 2.5-cm telescope-polarimeter is the most complicated to operate because the telescope is pointable (see Fig. 100). During the few weeks of the encounter, about 15,000 commands were transmitted to it. Pointing is essential to keep observing the planet during a fly-by in which the phase angle (sun-Jupiter-spacecraft angle) changes by many degrees. Our main goal is to study jovian molecules and aerosols by their light scattering (changes in intensity, color, and polarization) as a function of phase angle.

About 80 images of Jupiter were produced with resolution better than that of Earth-based telescopes. Our resolution was not as fine as would be possible with television, but we got pictures and photopolarimetry over a far wider range of phase angle than is observable from Earth.

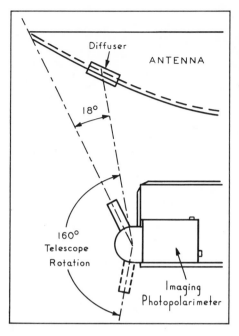

Diffuser

ANTENNA

18°

160°
Telescope
Rotation

Imaging
Photopolarimeter

FIG. 100. The 1-in. telescope of Pioneer-10's imaging photopolarimeter can be turned from 10° off the forward rotation axis (down) to 10° off the backward (Earth-directed) axis for scanning with the spacecraft spin, but the last 18° are blocked by the high-gain antenna dish. In the full backward position, the instrument views the sun through a diffuser in the antenna dish for calibration. (University of Arizona diagram.)

Our immediate task is to rectify the images received from Pioneer 10, producing pictures for a preliminary study of the forms of Jupiter's clouds. The intensities, colors, and polarizations of selected regions will then be compared with theoretical models.

Radiation Dosage
Near Jupiter

(*Sky and Telescope*, May 1975)

The word "lethal" has often been loosely used to describe the roughly doughnut-shaped belt of trapped high-energy particles that surrounds Jupiter. But now three biophysicists at the University of Rochester's school of medicine have calculated the radiation dose that would have been encountered by any organisms on board Pioneer 10 when it flew by Jupiter in December 1973.

M. W. Miller, G. E. Kaufman, and H. D. Maillie summed the fluxes measured by seven electron detectors and eight proton detectors on Pioneer 10 while the craft was within about 100 radii of Jupiter. The effect can be expressed in *rads*, one rad being the absorption of 100 ergs of energy per gram of irradiated biological material.

The radiation dose on the outer skin of the spacecraft was at least 4.9×10^5 rads from electrons and 2.9×10^5 rads from protons. This would have been enough to kill more than 99.9 percent of any spore-forming bacteria on the surface and practically all of the nonspore formers.

Inside the spacecraft, the radiation dose was between 2.8×10^5 and 5×10^5 rads. This would have resulted in a spore survival rate of about 0.05-0.01. For almost all "higher" forms of life—such as seeds, plants, algae, worms, insects, and others—the radiation dose inside Pioneer 10 was supralethal. For man and other mammals the interior dose far exceeded the lethal level.

Manned flights to Jupiter are unlikely, with astronauts confined to the spacecraft for over three years, lethal radiation near the planet, and no rocky terrain from which samples can be selected. The disklike shape of Jupiter's magnetosphere shown in Figure 99 was not entirely confirmed by Pioneer 11 (p. 266). In fact, two magnetometers (one helium-vector, one fluxgate) on Pioneer 11 gave discordant results on Jupiter's magnetic field, raising the suspicion that the strong particle radiation somehow affected the magnetometer readings. However, the helium-vector magnetometer readings seem to be supported by other observations. Still, experts conclude that Jupiter's magnetosphere is a large, blunt-nosed region at least 80 Jupiter radii (5.7 million km) in radius on the sunward side, possibly deformed at times by "gusts" in the solar wind.

The Galilean satellites, Jupiter's four large moons, Io, Europa, Ganymede, and Callisto, first observed by Galileo in 1610, are somewhat more suitable for landing but still in Jupiter's radiation belts. Io has a high-albedo (white) surface, probably dominated by salt (NaCl), and an unusual atmosphere containing sodium. Ganymede and Callisto have low density (1.8 gm/cm^3) and are probably mostly water with an icy crust (see Table 8), peppered with meteorites. According to current geochemical theories, their formation, in a rotating gas-dust nebula around the evolving Jupiter, was similar to planet formation in the solar nebula around the proto-sun (see p. 155). In its early stages, Jupiter probably played a role similar* to the sun's—it reached a fairly high temperature as gas and dust fell into it, releasing gravitational energy. (Even today, Jupiter radiates over twice the energy it receives from the sun—it is almost big enough to be a star.) Thus, the theory goes, Io and Europa, close to Jupiter, have higher density (about 3.5 gm/cm^3), like the terrestrial planets near the sun. Data obtained by Pioneers 10 and 11 require

* Note one major difference: Jupiter rotates rapidly, while the sun rotates very slowly.

many more months of study before we understand these Galilean satellites.—TLP

Ganymede from
Pioneer 10

(*Sky and Telescope*, January 1975)

When Pioneer 11 flew past Jupiter in December 1974 and Pioneer 10 went by just a year earlier, an important task for both was to obtain high-resolution photographs of the giant planet's four Galilean satellites. Thus, Pioneer 10's imaging photopolarimeter—a Maksutov telescope of 1-in. aperture, described on p. 258—made use of the spacecraft's 5 revolutions/min. spin to scan a 14°-by-14° field that included Ganymede, Jupiter's third, large satellite.

About 21½ hrs before its closest approach to Jupiter, the spacecraft transmitted the images in Figure 101 during a 6½-min. interval.

Revolving around Jupiter every 7.2 days at a distance of 665,000 mi., Ganymede has a diameter of about 3300 mi., making it larger than the planet Mercury. The Pioneer-10 pictures show dark- and light-mottlings, the resolution being about 240 mi. The close correspondence between the blue and red pictures indicates both the reality of the main features and the absence of large-scale color contrasts. Some of the white areas might be frozen water. (Radar observations from Earth are interpreted as reflections from rocky and metallic meteorites embedded in ice.)

FIG. 101. Pioneer-10's views of Ganymede from 485,000 mi., in blue light (3900-5000A) left, and red light (5950-7200A) right. Computer improvements of these photos (*Science 191*, 1237, March 1976) show features very like lunar maria and craters. (NASA photos.)

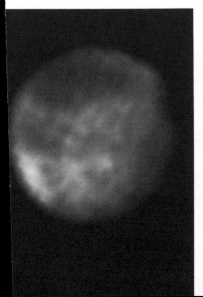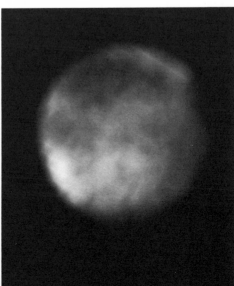

Pioneer 11 Retargeted

(*Sky and Telescope*, May 1974)

An important decision has been made by NASA to alter the course of Pioneer 11, so that it will travel to Saturn after it passes Jupiter. Pioneer 11 was launched in April 1973, and would have continued on out of the solar system after passing Jupiter in December 1974.

The revised plan calls for firing the spacecraft's thrusters in late March or mid-April to slow it, so it will come to within 42,000 km of Jupiter's cloud tops on December 5, on a trajectory inclined 55° to the planet's equator, going from south to north. The close approach will accelerate the spacecraft to a speed of 175,000 km/hr relative to Jupiter. This high speed, combined with the steep path through the intense jovian radiation belts, should reduce radiation damage to acceptable limits and provide a greatly improved estimate of the belts' thickness.

Under the new plan, Pioneer 11 will fly in front of Jupiter (to the left as seen from Earth) as it moves along in its orbit. The spacecraft will then pass behind the planet, emerging on its right side. During this encounter, Jupiter's gravitational attraction will sling the spacecraft across the solar system to reach the vicinity of Saturn in September 1979.

That will be 6½ yrs after launch, well beyond the spacecraft's design lifetime, but there is a fair possibility that it will be able to go on transmitting scientific data.

Such an encounter with Saturn, the first such, could provide valuable information about the planet's radiation belts (if any), the nature of its rings (see Vol. 2), and its temperature. And if the Pioneer's imaging system continues to function, it could return the first closeup views of a planet not yet observed from nearer than about eight AU.

On its new course, Pioneer 11 will go by Jupiter at only one third the distance of Pioneer-10's closest approach and traverse a much wider range of Jovian latitudes than its predecessor. Hence, the course change will provide investigators with much better and more varied data on Jupiter than would have been possible from a repetition of the Pioneer-10 mission.

Pioneer 11: Through the Dragon's Mouth

(*Sky and Telescope*, February 1975)

Pioneer 11 passed Jupiter in early December 1974 after 606 days of travel outward through the solar system.

It came three times closer to Jupiter than Pioneer 10, but along a S-N trajectory that quickly traversed the flattened radiation region, instead of crossing through its full extent. This strategy greatly reduced Pioneer-11's total exposure to the high-energy particles, leaving the instruments undamaged to observe Jupiter at close range. In the words of Robert S. Kraemer, director of NASA's planetary programs, Pioneer 10 tickled the dragon's tail, but Pioneer 11 flew into its jaws!

The path that Pioneer 11 followed carried it to a minimum distance of about 26,600 mi. above Jupiter's cloud tops on Dec. 3, 1974. This is only 0.60 of the planet's equatorial radius. Approaching Jupiter from below its south pole, the craft crossed the planet's magnetic equator (the central plane of the radiation belts) at an inclination of 54° (see Fig. 102).

FIG. 102. Pioneer-11's encounter with Jupiter in a perspective view toward the sun, Earth, and Mars, as of the fly-by date, Dec. 3, 1974. Jupiter's gravitational attraction flips Pioneer 11 across the solar system to meet with Saturn in September 1979. (Diagram by J. K. Beatty.)

Safety was only one factor in choosing this encounter track; scientists wanted to obtain the first closeup photographs and measurements of the jovian polar regions. Pioneer 11 flew counter to the planet's rotation, so that it could observe the magnetic field and radiation belts around the whole circumference of Jupiter; Pioneer 10, moving in the direction of the rotation, could not do this. The encounter also works as a slingshot to carry Pioneer 11 toward Saturn, as shown in Figure 102.

The photographs of Jupiter were taken with the imaging photopolarimeter, one of 12 instruments mounted on the spacecraft, as shown in Figure 103 (which omits the flux-gate magnetometer next to the cosmic ray telescope). It is a 1-in. telescope of 3.4-in focal length, with a photoelectric light detector (Fig. 100). Incoming light is split by a prism, according to the polarization (see Glossary) into two separate beams. Each beam is further divided into a red and a blue image, by means of filters transmitting 5940-7200A and 3900-5000A, respectively.

The telescope scanned the disk of Jupiter in strips only 0°.03 wide as the spacecraft spun about 5 times/min. From 25 min. to 110 min. were needed to complete the scans for a picture, depending on the distance from the planet. Color pictures were formed from a red and a blue image, with an admixture of green chosen to give hues best matching those seen by terrestrial observers (see p. 255).

Elaborate computer rectification of the images was needed to remove the changing curvature of the scan paths, and to compensate for smear

FIG. 103. Experiments mounted on Pioneer 11. The helium-vector magnetometer (1) is on a 15-ft boom. Two shorter booms (not shown here) hold two RTG power sources each, and there are also 12 small RTG heaters to keep the spacecraft warm. See also Fig. 94. (NASA diagram.)

PIONEER/JUPITER EXPERIMENTS

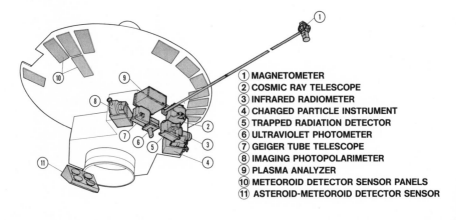

1. MAGNETOMETER
2. COSMIC RAY TELESCOPE
3. INFRARED RADIOMETER
4. CHARGED PARTICLE INSTRUMENT
5. TRAPPED RADIATION DETECTOR
6. ULTRAVIOLET PHOTOMETER
7. GEIGER TUBE TELESCOPE
8. IMAGING PHOTOPOLARIMETER
9. PLASMA ANALYZER
10. METEOROID DETECTOR SENSOR PANELS
11. ASTEROID-METEOROID DETECTOR SENSOR

FIG. 104. Pioneer-11 view of Jupiter's northern hemisphere from 750,000 mi. above 50° N latitude, showing the north pole on the terminator at top, convection cells in northern latitudes, and the northern atmospheric bands. (NASA photo.)

caused by the planet's rotation and the spacecraft's motion. Even so, Pioneer 11 was traveling so fast near closest approach that no pictures could be taken for 6 hrs. During the 52 hrs centered on perijove, 25 pictures of Jupiter were obtained, and one each of its satellites Io, Ganymede, and Callisto. Figure 104 is a later view of Jupiter's northern hemisphere.

Because Jupiter radiates more heat into space than it receives from the sun there are strong currents in the atmosphere.

The bright zones are rising currents of gas, whereas the dark belts are descending currents. Pioneer-10's infrared experiment showed that the tops of the bright zones are cooler than the tops of the belts by about 9°C, which implies that the zones reach about 12 mi. higher than the belts.

The famous Great Red Spot in Jupiter's south tropical zone (Fig. 105)—a 20,000-mi.-wide feature—is now generally believed to be the vortex of a violent, long-lasting cyclone, as first proposed by the late

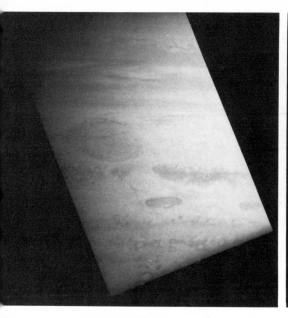

FIG. 105. Pioneer-11 view of the 20,000-mi. Great Red Spot in the bright south-tropical zone, looking northward, in red light (left) and blue light (right). (NASA photo.)

G. P. Kuiper. The spot extends about 5 mi. higher than the surrounding cloud tops. [Figure 105 was released in June 1975, after four months' "computer processing" at the University of Arizona, greatly improving the detail, and eliminating defects in the illustrations published in February.—TLP]

Pioneer 11 was still some 4,000,000 mi. from Jupiter on November 27, when it passed for the first time through the bow shock-wave that the planet's magnetic field creates in the solar wind. In the course of a day or two, there were three crossings of the bow shock and three crossings of the magnetosphere—the boundary of the huge volume around Jupiter that the solar wind cannot enter.

The intense inner magnetic field extends about 850,000 mi. out from the planet's center, and its weak disklike extension much farther (Fig. 99).

At closest approach to Jupiter on December 2 the fluxgate magnetometer indicated a field strength of about 1.2 gauss, and the helium vector magnetometer gave a similar value. (Pioneer 10, at three times the distance, registered 0.2 gauss.) It can be concluded that the magnetic field at the visible surface of Jupiter is about 4 gauss, or some 10 times as strong as the field at the Earth's surface.

The Geiger counters and other charged-particle (ion) detectors aboard Pioneer 11 made highly successful observations of Jupiter's radiation belts (like the Earth's Van Allen belts, Fig. 7). The high-energy electrons and protons which form the inner radiation belt are trapped inside the planet's inner magnetic field, whereas particles gradually escape from the weak disklike outer field. For protons with energies greater than 3.5 million electron volts, the peak counting rate was some 150 million particles/cm²/sec, about 40 times the peak rate recorded by Pioneer 10.

It now appears that the inner radiation belt contains well-marked structural features, resulting from its continual stirring by the four Galilean satellites. This phenomenon is most marked for Io, according to J. A. Van Allen. At the distance of Io from Jupiter, protons having energies near one million electron volts are depleted by a factor of 70.

Electrons with energies below 1 million electron volts are absent within the orbit of Io. Similar but less marked electron depletion occurs along the orbits of Europa, Ganymede, and Callisto.

The ultraviolet spectrometer reobserved the remarkable hydrogen cloud associated with Io. Discovered by Pioneer 10, this cloud of hydrogen atoms stretches 120° along Io's orbit, and fades out in Jupiter's shadow. Ganymede and Callisto seem not to have analogous clouds. Since Io's gravity is too weak to retain hydrogen atoms, the hydrogen cloud implies a source in Io—possibly outgassing of the satellite.

Radio Doppler data show how the trajectory of Pioneer 11 was affected by the gravitational field of Jupiter. John D. Anderson (JPL) says that there was no indication of any anomalies such as would be caused by mass concentrations inside the planet. Instead, Jupiter is in hydrostatic equilibrium, probably fluid throughout.

After leaving the vicinity of Jupiter, Pioneer 11 will need almost 5 yrs to reach Saturn. The craft is now in an elliptical orbit around the sun, inclined 15°.6 to the plane of the ecliptic.

Current plans call for sending Pioneer 11 on an inclined trajectory between the innermost ring and the globe of Saturn, within 2300 mi. of the planet's cloud tops, some 12,000 mi. from Titan.

This maneuver was considered preferable to a second slingshot boost to Uranus because of Pioneer-11's ebbing power supply.

Because solar radiation at Jupiter's distance from the sun is only 1/27 as intense as at the Earth, Pioneer 11 is powered not with solar cells but with four small SNAP-19 thermoelectric nuclear generators (Fig. 106). Their fuel is radioactive plutonium-238. At Jupiter, this system provided about 130 watts, but it is losing power at the rate of 8-9 watts per year.

FIG. 106. SNAP-19 nuclear 28-volt DC power source (radioactive thermal generator). Each Pioneer spacecraft carried four of these (Fig. 103), each weighing 30 lbs, with 8.3 lbs of plutonium providing the heat. The cylinder is 11 in. in diameter. (NASA diagram.)

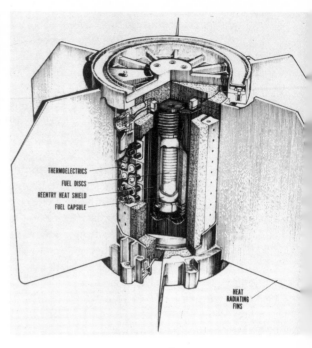

THERMOELECTRICS
FUEL DISCS
REENTRY HEAT SHIELD
FUEL CAPSULE

HEAT RADIATING FINS

Minimum operation of all onboard equipment requires 83 watts, the scientific experiments using 26. Therefore, Pioneer will have just about enough power to operate all equipment aboard during its fly-by of Saturn. There would not be enough for a useful encounter with Uranus.

During the journey from Jupiter to Saturn, Pioneer-11's solar-wind, charged-particle, and magnetometer experiments will continue to explore the interplanetary medium.

This high-risk pass close to Jupiter, with its slingshot continuation on to Saturn, exemplifies several aspects of space science as well as the technical capabilities developed in the last 10-15 years. First, there is the precise computerized application of celestial mechanics—the accurate timing and aiming of a space probe to a target 500 million mi. away, then utilizing Jupiter's gravitational force to flip the probe across the solar system to meet with Saturn over 4 yrs later. Second, there is the complex magnetosphere with its intense radiation belts, detected by Pioneer 10, assessed, and then further explored by the retargeted Pioneer 11 without lethal damage to electronic equipment aboard the spacecraft. Third, there is the wide variety of significant data obtained: (a) the ring current of electrons extending in an equatorial sheet out to about 7 million mi.; (b) the double shell of high-energy (35-MeV) protons at 80,000 mi. and 160,000 mi. from Jupiter's magnetic axis; (c) the huge magnetosphere pulsating in phase with the planet's rotation about 2 million mi. above and below the equatorial plane (see Fig. 99) and "stirred" by the four Galilean satellites in their orbital motion; and (d) 10-hr bursts of very high-energy electrons and protons (cosmic rays) shot out of the magnetosphere on either side of the equator by processes not yet understood. (In March 1976, Pioneer 10 passed through and detected the tail of Jupiter's magnetosphere beyond the orbit of Saturn, 9.5 AU from the sun, and 430,000,000 miles from Jupiter.)

The photographs, only a few of which are reproduced here, show that the surface of the atmosphere we see is in a convective layer, pocked by many rising (warm—up to 245°K) and falling (cool—down to 100°K) currents, and banded by jet streams moving at relative speeds of 100-250 mi/hr. Over the poles, the clouds are of a different nature—possibly showing more aerosols (liquid droplets). Accurate measures of the Pioneer orbits (see p. 267) indicate that Jupiter is fluid throughout, with a density of 0.23 gm/cm³ at 68,000 km from the center (3000-km deep), where the temperature must be 13,000°K (twice the temperature of the sun's surface). The brownish color of Jupiter's atmosphere (and Saturn's) is interpreted by Carl Sagan as an indication of organic molecules formed there, as in the Earth's primitive atmosphere (where they gave rise to living organisms).—TLP

7

Spacecraft Design and Workshops in Space

Aerospace design is a well-developed branch of engineering, and we can't possibly do it justice in this chapter. Nevertheless, several threads of spacecraft design can be traced through previous chapters of this volume (particularly Chap. 2) and in Volume 1 (Sputnik, Vanguard, Explorers, OSO, Telstar, and Mariner 1), Volume 2 (Mariner 2, power sources), and Volume 3 (TIROS, IMP, and Ranger 7). Some of these threads are obvious design requirements:

1) An adequate launch rocket of two or three stages (see p. 17).

2) A power supply for controls, instruments, and radio-TV. Solar cells use sunlight; the RTG (p. 242) uses nuclear power, and batteries are often used, sometimes to smooth out peaks in the power requirements.

3) Radio-receiving antennae for up-link radio commands from Earth, and directional (high-gain) transmitting antennae for telemetering to Earth data on spacecraft orientation, fuel supply, power supply, environment (temperature, etc.), scientific observations, and other information.

4) Stabilization by gyros and/or gas jets, and orientation control. Unmanned spacecraft are often spin-stabilized (p. 256), but need jets to change the spin-axis direction.

5) Thermal coatings to control the spacecraft temperature.

6) Meteoroid "bumpers" (p. 5) and radiation shielding (p. 44), especially for photographic materials, astronauts, and electronic gear.

7) Tanks of oxygen, water, etc. for manned flights; also dump tanks and airlocks (enclosed vents) for waste disposal.

8) A vacuum-tight pressure hull and access hatchway for manned spacecraft.

9) Thruster (rocket motor), thrust control, and fuel for changing orbit or soft landing. (Parachutes are sometimes used for the latter.)

10) Nose-cone heat shield for re-entry into Earth's atmosphere.

11) Light-weight, reliable observing instruments with low power requirements.

12) Cushioning to protect fragile instruments and astronauts from acceleration and vibration during launch and re-entry.

Other threads lead to ground installations, many of them strongly linked with spacecraft design:

a) A launch pad and gantry (supporting tower) matching the size of booster rocket.

b) A network of sensitive, directional radio receivers spaced in longitude so that the spacecraft generally has a radio line-of-sight to at least one receiver. On many space missions, when down-link radio is interrupted, transmissions are tape-recorded, then broadcast from the tape when radio contact is reestablished.

c) A control center, where down-link messages are received and recorded for later use, and where commands are written by a "controller" (engineer or scientist) for up-link transmission.

d) A recovery ship and/or helicopters and other aircraft for recovery of the spacecraft after re-entry. Many early spacecraft were left in low orbit around the Earth from which they slowly descended and were burned up in the atmosphere. Others were abandoned in solar orbit (p. 48) or escaped from the solar system (Pioneer 10, p. 249).

On the rocket requirement, early ideas were big, but it took the enormous Saturn booster to start NASA's major space missions.—TLP

Rocket Propellants—The Key to Space Travel

FREDERICK I. ORDWAY III

(*Sky and Telescope*, February and March 1959)

Despite impressive recent advances in the rocket art, the task of getting a manned space vehicle onto the Moon remains titanic. To place the small 3-lb Vanguard test satellite into orbit in March 1958 required a 27,000-lb thrust from the main-stage rocket, plus second and third stages generating an additional thrust of over 10,000 lbs. The Explorer satellites relied on the 83,000-lb thrust of a Jupiter C, plus additional upper staging, to attain orbital velocity (see Table 10).

TABLE 10. WEIGHTS AND THRUSTS OF EARLY LAUNCH ROCKETS

Rocket	Approx. weight in lbs	Approx. thrust in lbs
Vanguard	22,600	27,000
Jupiter C	65,000	83,000
Pioneer/Thor	110,000	150,000
Sputnik-III launcher	185,000	260,000
Atlas	200,000	360,000
Titan	220,000	360,000

But none of these rockets comes close to meeting the requirements for getting a payload of 600 lbs to the Moon. For this purpose a rocket weighing about a million lbs and having huge booster engines is needed (Fig. 107), and the U.S. Army and Air Force are developing liquid-propellant rocket engines to provide 1,000,000 and 1,500,000 lbs of thrust. The Russians have apparently solved this problem (see Table 2).

When the propellants—oxidizer and fuel—are burned in a combustion chamber, large quantities of hot gases are formed at high pressure. The heat energy of the chemical reaction is the power source. The hot gases rush out through the nozzle-exit, and the reaction to the force that accelerates the combustion products yields the thrust to propel the missile.

There are two basic ways of improving performance: (1) increase the discharge velocity of the gaseous exhaust and (2) increase the mass of the material expelled in the exhaust—the *momentum thrust* of the engine is the product of the exhaust velocity and the rate of propellant mass-flow. In the long run, increasing the exhaust velocities will be the more important of the two.

Consider a single-stage rocket, carrying a quantity of fuel that is completely used at the time of burnout. During this combustion period the missile is constantly accelerated, reaching its peak velocity at the moment of burnout. It is evident that the total mass of the exhaust propellant is of primary importance—the bigger the propellant mass, and the smaller the vehicle mass, the better.

This characteristic of a propulsion system is its *mass ratio*—the ratio of the fully fueled vehicle weight to its weight after all propellants have been expended.

The maximum speed, v, that a single-stage rocket can attain is related to the mass ratio, M, and the exhaust velocity, c, by the formula:

$$v = c \log_e M,$$

where e is the base of natural logarithms, the number 2.718.

FIG. 107. Size of a million-lb thrust booster compared with the USAF Atlas and Army Jupiter missiles. (NASA-Rocketdyne drawing.)

1,000,000-pound-thrust vehicle   **Atlas**   **Jupiter C**

For a missile to move with the speed of its own exhaust ($v = c$), $\log_e M$ must be 1, that is, the initial loaded weight has to be 2.718 times the empty weight. For larger M, it also follows from the relation above that a missile can move faster than its own exhaust.

Engineering advances during the past decade have made possible larger and larger mass ratios. The first Viking rocket, for example, had $M = 3.5$, whereas Viking 12 had $M = 4.5$ or more. If we could get an exhaust velocity of more than 10,000 mi/hr and design a single-stage vehicle with a mass ratio of 7.4, it would go at twice its exhaust velocity.

However, even this model would not escape permanently from the Earth. The capabilities of single-stage rockets are definitely limited.

The solution to this problem is multistaging, used successfully in placing artificial satellites in orbit and in the recent lunar-probe firings. The over-all mass ratio of a staged rocket is equal to the product of the mass ratios of the individual stages. Thus a three-stage missile, each component having a mass ratio of four, would have an effective ratio of $4 \times 4 \times 4 = 64$.

Further improvement depends on the energy available in propellants, which are compared by the useful concept *specific impulse*. This is the impulse (thrust $\times$ duration) per unit mass of a propellant, expressed in units of pound-seconds per pound. It is an important number—the final height reached by a missile is proportional to the square of the specific impulse.

While specific impulse is determined principally by the composition of the fuel, it also depends on the operating pressure in the combustion chamber and on the ratio of the nozzle exit area to throat area.

To achieve spaceflight with chemical propellants, we need those that give the most energy per unit weight. Therefore, let us note some of the recent research in high-energy fuels, both solid and liquid.

Solid Fuels give upwards of 200 lbs of thrust for each pound of fuel consumed per second but values above 240 are still rare (Table 11). Experts believe that there is little hope of getting above 300, for stable solid chemical propellants contain rather limited energy.

Some solid fuels can provide exhaust velocities of 8000 mi/hr but are generally unstable, with undesirable physical or ballistic properties. In Table 11 the burning rate refers to the speed at which the burning surface advances toward the head of the combustion chamber.

One disadvantage of a solid rocket is that the propellant must be carried in the combustion chamber, the size of which limits the amount of fuel.

JPL at CalTech has carried out valuable research on "internally burning charges," offering nearly constant pressure and thrust. Combustion takes place on the surface of a longitudinal hole through the charge, gradually progressing outward toward the chamber wall. This permits light, thin-walled construction in place of the heavy walls used earlier. Since the burning takes place on the inside, the combustion chamber is protected until this process has terminated.

Liquid Fuels have many advantages. They can be carried in separate, detachable light-weight tanks under low pressure, and the combustion chamber into which they are fed can be much smaller and more efficient than for solid propellants. Specific impulses range from about

TABLE 11. CHARACTERISTICS OF SOLID PROPELLANTS

Combination	Density	Flame tempera- ture	Specific impulse	Burn- ing rate	Pressure range	Exhaust charac- teristics
Amino ethanes (composite)	—	—	200	0.3-0.6	300-2000	—
Ballistite (double base)	1.2-1.7	4500-5400	210	1.4	1000-3000	Black smoke, high flash
Black powder (composite)	1.2-2.1	3600-5400	70	0.1-0.5	100-1000	Gray smoke
Buna and sulfo rubbers (composite)	—	—	210	0.4	100-800	—
Cordite (double base)	—	—	180	—	1000-3000	Black smoke, high flash
Galcit 161 (composite)	1.8	3600-4500	190	1.6	1300-3700	White smoke
Lox-rubber (liquid-solid)	—	—	225	—	100-500	Smokeless
NDRC-EJA (composite)	1.8	3600-4500	180	0.2-1.0	600-1000	Gray smoke
Polymethane (composite)	—	—	215	—	500-2000	—
WASAG DEGN (homogeneous)	—	—	182	0.2-0.8	700-4000	Black smoke, high flash

Units: *Density*, water = 1; *Flame Temperature*, °F; *Specific Impulse*, lb-sec/lb; *Burning Rate*, in/sec; *Pressure Range*, lbs/in².

150 to perhaps 400 lb-sec/lb, depending on the pressure in the combustion chamber (see Table 12).

Whatever combination of rocket motor and liquid fuel we use, it is never possible to exploit 100 percent of the available energy, because of incomplete combustion, heat losses to the chamber walls, and kinetic energy carried away in the exhaust gases themselves. As a rule, only about 60 percent of the heat content of the combustion materials is actually converted into directed kinetic energy that drives the rocket forward.

The desirable properties to be sought in a liquid propellant are high heat of combustion, high reaction rate, and low molecular weight of the combustion particles. In practice, it is necessary to balance low molecular weight against high temperature to reach the highest practicable specific impulse. It is unlikely that temperatures much above 7000° will be used.

We can anticipate the use of oxidizers with a greater content of

TABLE 12. CHARACTERISTICS OF LIQUID PROPELLANTS

Combination	Specific impulse	Chamber temperature	Combusion products
Red fuming nitric acid-aniline	210	5065	25
Nitric acid-ammonia	237	4220	21
Red fuming nitric acid-unsymmetrical dimethylhydrazine	249	5200	22
Liquid oxygen-alcohol	259	5560	22
Hydrogen peroxide-hydrazine	262	4690	19
Liquid oxygen-gasoline	263	5770	22
Nitrogen tetroxide-hydrazine	263	4950	19
Liquid oxygen-hydrazine	280	5370	18
Oxygen difluoride-ammonia	295	6380	18
Liquid oxygen-hydrogen	364	4500	9
Fluorine-hydrogen	373	5100	9

Units: *Specific impulse*, lb-sec/lb at a chamber pressure of 500 lb/in^2; *Chamber temperature*, °F; *Combustion products*, average molecular weight.

fluorine—the most powerful oxidizer known—and there will probably be more use of the light elements lithium, beryllium, and boron in such compounds as beryllium borohydride, . . . methyl hydrides and organo-metallic compounds offer higher exhaust velocities than do straight hydrocarbon fuels.

Probably the upper limit of specific impulse to be obtained from liquids will be around 375 lb-sec/lb, with a fluorine-hydrogen system.

Ion Propulsion, in which electrified atomic particles are accelerated by either electric or magnetic fields to obtain very high exhaust velocities, offers interesting possibilities. The greatest thrust is obtained with charged particles of relatively high mass. The theoretical limit of velocity is the speed of light itself!

Ernst Stuhlinger of the Army Ballistic Missile Agency proposes to form the ions of alkali metals on hot tungsten surfaces. Of the alkali metals, cesium appears the most attractive, because it has a high atomic weight (133), low ionization potential for its first electron, and low melting and boiling points. But some 1200 kilowatts of electrical energy would be needed to heat the cesium and provide the power for the accelerating electrostatic field. The rocket would have a total initial mass of 730 tons (45 for the electric-thrust chambers).

It is evident from the nature of the ion-propulsion system that an ion-drive vehicle will accelerate very slowly. Thus, it would be impractical to use ion propulsion where we require short-period, high-acceleration—as from boosters at launch.

Nuclear Power for rockets may be considerably further from realization than is an ion-propulsion system. Any practical type of atomic rocket engine that can be visualized today would require a working fluid heated by a nuclear reactor. The nuclear process generates enormous heat, which must be supplied to the working fluid or propellant. If hydrogen could be used at a temperature of 8000°F, a specific impulse of 1000 lb-sec/lb would result. Exhaust velocities would be up to 13,000 mi/hr.

Ion propulsion has been tested successfully on eight NASA flights between 1962 and 1970, using mercury ions as well as cesium, with the purpose of developing a high-specific-impulse (6000 lb-sec/lb), low-thrust (0.03-lb) unit for station-keeping and attitude control after launch by chemical boosters. A mercury-ion thruster on SERT 2, launched on Feb. 3, 1970, operated at 60 percent of its maximum thrust over 4 yrs later, and a cesium-ion thruster was used for station-keeping on ATS 5 in geosynchronous orbit. Although these ion-propulsion units require almost 1000 watts' power supply, they would be used intermittently and may last forever. It may be that their very high specific impulse will be useful in carrying payloads from Shuttle to higher orbits (see Chapter 10), although their small thrust will require 35 days to lift 3000 lbs from Shuttle's 90-min. orbit to geosynchronous (24-hr) orbit.

*The latest development is an ion thruster using solid teflon as a fuel. The teflon is vaporized and ionized periodically by a very high voltage, which then ejects the ionized plasma in a series of pulses. NASA has tentative plans to use one of these ion-propulsion thrusters in a Solar-Electric Propulsion Stage in the 1980 fly-by of Comet Encke (see Chapter 10).—*TLP

Another Successful Saturn

RAYMOND N. WATTS, JR.

(*Sky and Telescope*, March 1964)

The heaviest satellite to date was boosted into orbit on Jan. 29, 1964, by a 16-story-tall Saturn rocket. It was the first operational use of Saturn's hydrogen-fueled second stage. The satellite's weight of 19 tons included nearly 6 tons of sand ballast. The previous over-all weight record (14,293 lbs) was set by Sputnik 7 in early 1961.

With some 1½ million lbs of thrust, the powerful Saturn is intended to carry experimental two-man capsules into orbit soon, a step toward manned lunar exploration (see Chapter 3).

This fifth Saturn firing was the first with a live second stage. More than 1000 performance measurements were telemetered to Earth.

Although some engineers had high hopes for nuclear rockets to power long spaceflights, they have never been used. The Los Alamos Scientific Laboratory built and tested three "Kiwi" reactors in 1959, 1961, and 1962, each pumping hydrogen through U^{235} (uranium)-graphite fuel rods with a carbide coat to prevent hydrogen-carbon reactions, but the results were disappointing. Westinghouse tested a more powerful "Nerva" rocket in 1968 that was more successful, but NASA had decided by then that chemical rockets were cheaper and developing adequately for its planned space missions.

* * *

Another early development was the huge "radio telescope"—a pointable dish-shaped antenna with very high gain (sensitivity) along its axis. The disk acts like a telescope mirror for high-frequency radio waves, concentrating them on a "feed horn" at the focus. With such a receiver, the

FIG. 108. Goldstone 210-ft, 2500-ton radio dish, used as a receiving antenna (radio telescope) or as a transmitter, pointable (by altitude-and-azimuth controls) in any direction above the horizon. The four-leg structure in the middle holds a Cassegrain secondary mirror that reflects radio waves collected by the 0.8-acre dish down to the feed cone at the center where selected frequencies (wavelengths) are amplified and recorded on magnetic tape. (JPL-NASA photo.)

faint radio transmissions from spacecraft several hundred million kilometers distant can be detected, amplified, and transmitted to mission
control, as noted in preceding chapters. The British built the first large
(250-ft) radio dish at Jodrell Bank for radio astronomy (see Vol. 2, Telescopes), and the Australians had a 210-ft dish at Parkes, New South
Wales (now moved to Aurora Valley).

In 1965 NASA added a 210-ft dish (Fig. 108) at Goldstone, California,
at a cost of $14 million. The aluminum dish is a paraboloid, accurate to
0.5 cm, built to withstand winds up to 120 mi/hr. The 210-ft Parkes dish,
an 85-ft dish (later replaced with a 210-ft) at Madrid, Spain, and the
Goldstone 210-ft—spaced round the world to allow reception at any time
—make up the Deep Space Network used for Apollo flights and interplanetary space probes. They generally work in the S-band of radio frequencies—2100-2300 megahertz (MHz = million cycles/sec), and the
Goldstone receiver has a ruby-maser that amplifies the S-band signals
by 40,000 times, so that it can receive data at a much higher rate. For
Earth orbiters, the NASA Space Tracking Network is used; it consists
of some 10 or 12, 30-ft dishes in Florida, Bermuda, Chili, Ascension
Islands, Canary Islands, Canton Island, Guam, and Hawaii, and on ships
of the U.S. Navy placed to provide nearly continuous coverage. Data received is tape recorded, and, in most cases, relayed to the control center
by microwave.

* * *

Starting in 1964, the Soviet Union and later the United States began
developing workshops in space where three cosmonauts or astronauts
in a large vehicle could work on several scientific or engineering projects.
The idea at first was to understake several scientific experiments, but as
time went on it appeared that production processes might be done more
effectively in an orbiter than in factories on the ground. Two features of
an orbiting spacecraft could be thus exploited: (1) zero gravity (o-g) and
(2) the hard vacuum of space above the Earth's atmosphere.

In o-g many chemical reactions go much faster because the reactants
don't separate in the solution as they do in 1-g on Earth (gases up, solids
down). Also, molten metals and glass can be cast in accurately spherical
shapes (ball bearings, for example). If a concave mirror is used to focus
sunlight on minerals or partly purified metals, extremely pure materials
can be produced in space vacuum as the impurities boil off without contaminants from heaters or containers.

NASA's developments of the 1960's led to Skylab in 1973, and 1975
plans call for many routine flights of the Space Shuttle, starting in 1980.
With the chance for regular work in space, the space business came of
age.—TLP

Soviet Three-Man Spaceship

RAYMOND N. WATTS, JR.

(Sky and Telescope, December 1964)

Three men dressed in lightweight gray suits and blue jackets walked across the broad concrete runway and climbed the ladder to an elevator. They waved farewell. . . . Then the trio was hurled into space to spend a day in orbit. This was the scene described by a Tass reporter covering a launch at the Tyuratam Cosmodrome in Kazakhstan on Oct. 12, 1964.

The three-man team was composed of a pilot, Col. Vladimir M. Komarov; a physician, Boris B. Yegorov; and an experienced designer of spacecraft, Konstantin P. Feoktistov. The interior of the spacecraft, named Voskhod (Sunrise), is lined with a soft, spongelike white synthetic fabric. The crew's seats were close together in a row, facing the instrument panel. Among the instruments named were a clock, a globe indicating the craft's position, a radio, telegraph key, and many buttons and switches. The pilot had a black nylon handle for controlling the ship.

There were stores of food, warm clothing, flashlights, and inflatable suits to provide comfort in space and for survival in case the craft were forced to land outside the planned area in Kazakhstan. Voskhod was designed to float should it come down at sea. The cabin also carried a television camera, permitting Soviet audiences to view the cosmonauts in flight each time they passed over the U.S.S.R.

Voskhod traveled in an orbit inclined about 65° with apogee and perigee heights of 254 mi. and 111 mi., respectively, quite similar to that of Cosmos 47, launched on October 6, evidently as a rehearsal for Voskhod.

After circling the globe 16 times, Voskhod landed on October 13, on a state farm 350 mi. north of Baikonur.

Both scientific experiments and engineering tests were made during Voskhod's flight. To orient the capsule, an ion engine was used, as well as the normal gas jets. An externally mounted television camera monitored the heat shield during re-entry.

The scientific work ranged from navigation tests to psychological investigations. The cosmonauts made observations with an ordinary sextant, demonstrating the feasibility of simple stellar navigation without radio contact with the ground.

Yegorov's main task was to study the vestibular reactions of the crew, including, for example, how sense of balance was affected. He watched for changes in muscular capacity, and experimented with voluntary and involuntary eye movements. Visual acuity, color sensitivity, and muscle tone did not appear to change during the flight.

Other experiments involved electrical stimulation of the inner ear. In one test, the three men shook their heads sharply, and both Feoktistov and Yegorov felt vague dizziness and general discomfort. Periodically, blood samples were drawn from all crew members and stored for later analysis of sugar content, cholesterol, proteins, chlorides, and for counts of red and white cells.

Soyuz 9 Sets
Endurance Record

RAYMOND N. WATTS, JR.

(*Sky and Telescope*, August 1970)

The longest manned space flight to date ended successfully on June 19, 1970, when Soyuz 9 landed in Soviet Kazakhstan. In a shirt-sleeve environment, cosmonauts Andrian Nikolayev and Vitaly Sevastyanov had spent 17 days, 16 hrs, 59 min. in space, breaking the 13-day record set in 1965 by Frank Borman and James Lovell in Gemini 7.

Colonel Nikolayev, the Soyuz-9 commander, was making his second flight, having flown in Vostok 3 in 1962. (His wife, Valentina Tereshkova, was the first woman to venture into space.) Flight engineer Sevastyanov was a civilian, on his first trip.

Two groups of Soviet ships, one in the Atlantic and the other in the Indian Ocean, served as communication relays during the long voyage. The specially equipped flagship, *Cosmonaut Vladimir Komarov*, has large parabolic antennas housed in radomes for protection at sea.

The astronauts lived in relative comfort in their mahogany-paneled craft . . . and conducted a number of scientific experiments during the flight. In addition to an extensive biomedical program, their work ranged from observations of the sun's radiant energy to photographing the Earth in color and black and white, recording different types of geological features.

Shortly after landing the cosmonauts were pronounced fit and healthy by Soviet doctors. However, some medical experts voiced concern about the apparent deterioration of the cosmonauts' eyesight.

The Russians were not able to use a giant new booster on this launch (nor on any launch up to 1976), reportedly because it exploded or other-

wise failed in launch tests. Roughly equivalent to NASA's Saturn 5-B, it burns liquid fuel (kerosene) and should develop more than twice the thrust of the old Soyuz booster.

Somewhat before this time, NASA assembled several scientific experiments for a manned mission devoted to astronomy. It was called the Apollo Telescope Mount (ATM) and was eventually combined with Skylab for two endurance records set in 1973–74. The astronomers who proposed these experiments began to worry about contaminants in the space around ATM—from waste disposal, stabilizing jets, and all sorts of leaks. Astronauts had reported snowflakes around Gemini and early Apollo spacecraft—when urine dumps suddenly evaporated, some of the water was cooled below freezing. These flakes, and gases escaping the spacecraft, coasted along with it in orbit, forming a cloud that might degrade space observations.—TLP

Atmospheres Surrounding Manned Spacecraft

NATALIE S. KOVAR AND ROBERT P. KOVAR

(*Sky and Telescope*, March 1968)

Astronomers have long looked forward to the time when they would be able to observe for extended periods from above the Earth's atmosphere. Balloons and rockets permit using both photographic and photoelectric methods of data collection. Their experimental packages are recoverable, but their data-gathering times are short.

During the Gemini program, wide-field ultraviolet stellar spectra were obtained for Karl G. Henize aboard Gemini 10 and Gemini 12. Also, the faint zodiacal light and gegenschein (p. 334) were photographed from above the Earth's airglow (light from high in the atmosphere) for Edward P. Ney on Gemini missions 5, 8, 9, and 10. Although the Gemini vehicle provided a rather primitive observing platform, the quality of the returned data showed strikingly the promise of manned vehicles as astronomical tools.

An objective of the Apollo program is to provide a platform from which numerous experiments can be performed. The Apollo Applications Program (AAP) will involve no extra-vehicular activity (EVA); all astronomical equipment is enclosed in a special canister secured to the cabin wall directly behind an airlock (Fig. 109). Depending on the program, data may be recorded photographically from inside the spacecraft (through a window in the canister) or by other means from outside the spacecraft. At the end of an experiment, the astronaut will close the

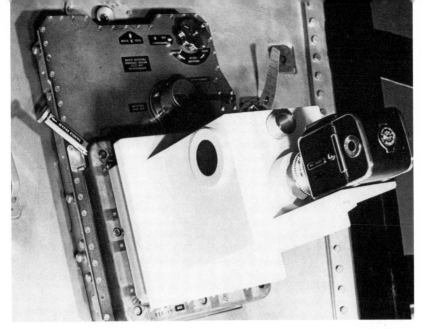

FIG. 109. An experiment canister attached to the airlock in the Apollo cabin wall. When the airlock is opened, the vacuum-tight canister is evacuated, and instruments inside exposed directly to space. (NASA photo.)

airlock, remove the canister, and replace it with another experimental package. All equipment must fit within approximately 8 × 8 × 16 in., the dimensions of the canister.

Larger observing instruments will be carried on the ATM configuration. This is made up of the Service Module, Command Module, Docking Adapter, Airlock Module, and Saturn-4B Workshop. As Figure 111 shows, extending outward from the docking adapter is the telescope mount, to which the large instruments will be attached. Since they will be serviced periodically by astronauts on EVA, photographic observations can be carried out, with the exposed plates and film being returned to Earth for analysis. The astronomical programs planned are listed in Table 13.

Astronomical observing in space is not without difficulties; several astronomers have pointed out that a permanent "debris atmosphere" may surround a manned spacecraft—it could seriously interfere with measurements of faint sources.

First evidence of debris around manned spacecraft came in early 1962 with the flight of John Glenn in Mercury 6. He reported a cloud of particles that at times looked like snowflakes accompanying his orbiting vehicle. In Mercury 7, to demonstrate that these particles were definitely associated with the capsule, M. Scott Carpenter rapped against the cabin wall several times, each time seeing a new group appear. Cos-

TABLE 13. ASTRONOMICAL EXPERIMENTS FOR THE APOLLO PROGRAMS

AIRGLOW HORIZON PHOTOGRAPHY   *Naval Research Laboratory*
Photography of the night and twilight horizon airglow in the visible and ultraviolet to obtain altitudes and intensities of various airglow layers. Ultraviolet photography of the sunlit Earth to record its appearance in the ultraviolet and to examine the atmosphere's ozone distribution.

SOLAR SPECTRA IN THE X-RAY REGION   *Naval Research Laboratory*
Photography of the solar spectrum from 10A to 100A, using an instrument of high spectral resolution.

EXTREME ULTRAVIOLET SOLAR SPECTRUM (ATM)
                                              Naval Research Laboratory
Photography of the sun at discrete wavelengths between 160A and 650A. Line spectra of selected areas of the sun in the ultraviolet from 800A to 3000A.

SOLAR X-RAY SPECTRA (ATM)   *American Science and Engineering, Inc.*
Study of solar flares at wavelengths 2-10A, recording images and spectra of the emitting regions. During nonactive periods, solar images in the 10-60A range will be obtained.

SOLAR ULTRAVIOLET SPECTROMETER   *Harvard College Observatory*
Photoelectric spectra of selected areas of the sun, in the 300-2250A range, with a resolution of about 5 arc-sec.

SOLAR X-RAY TELESCOPE (ATM)   *Goddard Space Flight Center*
Recording of data on coronal emission in the interval 3-100A, using selected wavelength bands. The total solar x-ray flux between 1A and 20A will also be measured.

WHITE-LIGHT CORONAGRAPH (ATM)   *High Altitude Observatory*
Coronal photography at distances from the sun's center of 1½-6 solar radii, with a resolution of about 5 arc-sec in the visible spectrum from 4650A to 6150A. The purpose is to obtain detailed observations of various coronal features and to record their development with time.

ULTRAVIOLET STELLAR SPECTRA   *Northwestern University*
Stellar spectra in the 1300-3000A range obtained with a wide-angle objective prism. This will yield line spectra and energy distributions in the ultraviolet.

X-RAY TELESCOPE   *American Science and Engineering, Inc.*
Measurement of positions and dimensions of cosmic x-ray sources to within several seconds of arc; also, obtaining spectra of these sources. Selected objects will be observed for possible x-ray emission and a search will be made for new x-ray sources.

---

monauts Komarov, Feoktistov, and Yegorov aboard Voskhod 1 also saw glowing particles through the vehicle's view ports. It seems probable that this debris came from water dumps. Astronauts have had difficulty in seeing stars in the daytime through a window that suffers from smudging and launch contamination. Nevertheless, a few star sightings have been made during spacecraft "daytime" by a number of astronauts. The best observations were made when no direct sunlight or earthlight illuminated the windows. L. Gordon Cooper, on Mercury 9 in May 1963, was able to see 4th-magnitude stars in the daytime.

Ney, evaluating his zodiacal-light experiment on Gemini 5, concluded that the background surface brightness of the sun-illuminated spacecraft atmosphere was about a billionth (10^{-9}) of the sun's mean surface brightness. The faintest star visible through this "background" would have been about magnitude $+2.5$.

D. C. Evans and L. Dunkelman at GSFC interpret the astronauts' reports from Gemini 6 to conclude that crewmen could see stars as faint as $+4.5$ in the daylight. They blame star-sighting difficulties on the general background level of light in the cockpit, the physiology of the eye itself, and on sunlight and earthlight scattered by the windows.

Thus, there are two points of view, one that star observations are hampered by an atmosphere surrounding the vehicle, the other that the causes are strictly within the spacecraft. Before considering experiments to resolve this question, let us examine possible sources of a spacecraft atmosphere.

There are both continuous and occasional contributions to the spacecraft environment. Controlled events, such as waste dumping and thruster firing, add heavily to the spacecraft atmosphere and may cause serious contamination of windows and exposed optical surfaces. Optical samples flown on Gemini 12 showed a significant amount of material deposited on them by engine exhaust and staging events. C. L. Hemenway and Elizabeth L. Hallgren analyzed micrometeoroid detectors flown on Gemini missions 10 and 12, also detecting deposits from weak atmospheres about the two spacecraft.

Continuous sources of contamination include outgassing through small leaks, and diffusion through the cabin walls. Materials consisting mainly of propellants (N_2O_4, Aerozine 50, monomethylhydrazine), hydrogen from batteries, and oxygen and water vapor from the cabin atmosphere, are added to the vehicle's environment.

Leakage rates for the Gemini cabins were experimentally measured prior to launch. The minimum continuous leakage was (for Gemini 3) 0.0042 gm/sec, and the maximum (for Gemini 11) 0.014. The crew-systems division of Houston's MSC estimates the leakage from the Apollo vehicle as 0.03 gm/sec and from the ATM cluster as 0.1. This last figure is equivalent to almost 20 lbs in a day!

Of course, during periodic events such as hydrogen purge of the fuel cells (an 80-sec discharge every 7 hrs), these rates will be substantially increased. Properly timed, these events will not be so serious to astronomical observations. Of great concern is whether the contaminants move away from the spacecraft's vicinity or whether they remain with it in orbit. For altitudes of 160 km to 500 km, atmospheric drag is the main dispersive mechanism. However, with the very slight atmospheric drag at these altitudes, a debris atmosphere will accompany the craft for substantial lengths of time.

Spectral analyses of the contaminants found on spacecraft windows indicate that the contaminant was a form of silicone. When heated above 420°K (about 300°F), the silicone of the nose-cone fairing produces a vapor of silicone oil. During launch, the spacecraft surfaces become covered by this oil, which then acts as a glue that traps particles during the orbital period and the 130 m.^{2} of silicones on the cluster will give a steady loss rate of 0.03 gm/sec (about 10^{20} molecules/sec) at a temperature of about 120°F. The silicone molecules probably coat the optical surfaces open to space, and also produce a kind of silicone "fog" about the entire ATM configuration. Thorough studies are required before definite conclusions can be made of the effects of this fog on astronomical observations. The AAP experiments through the airlock will in general require only brief exposure to space and can be cleaned between uses.

As for the debris atmosphere, we note that the radiance will result entirely from scattering by particles, and the column mass-density along a line of sight will depend on the velocity of particles and the rate at which mass is leaving the spacecraft.

Actually, water vapor is our chief concern, since it forms ice particles that scatter light. The cabin atmosphere will contain about 3 percent by mass of water vapor. The minimum velocity of the leaking material outside the spacecraft is the speed of sound in the cabin atmosphere— 330 m/sec.

Experiments with carbon dioxide and water support our assumption that all water vapor leaking through the cabin walls will form ice particles upon reaching space. These will scatter incident sunlight, producing a general background glow. The amount of sunlight scattered by the ice particles has been calculated and the resulting values are graphed in Figure 110 for directions from near zero to 180° from the sun (solar-elongation angles). We assumed spherical particles of refractive index 1.30 and a particle-size range from 0.2 to 5.0 microns.

When earthlight is included, the background brightness for Gemini 11 is about a 10-billionth (10^{-10}) of the sun's brightness, as shown by the dashed curve in Figure 110. It is evident that daytime observations of the zodiacal light could not have been made from Gemini 11. Even

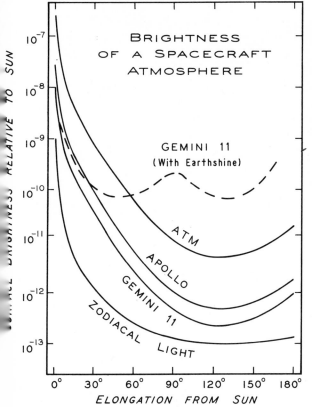

FIG. 110. Plot of spacecraft-atmosphere brightness *vs* angle from sun (elongation) for several spacecraft, compared with the zodiacal light. The dashed curve shows the added brightness from the half-illuminated Earth when Gemini 11 was over the terminator. (NASA diagram.)

Apollo and ATM background would swamp the solar corona and zodiacal light.

Definite conclusions must await results from coming experiments, including two planned for the AAP series. One of these is a triple-disk externally-occulting coronagraph (p. ooo), designed by G. P. Bonner of MSC.

In another experiment, J. A. Muscari and others at the Martin-Marietta Corporation have proposed that a photometer be employed to measure sky brightness as a function of solar elongation, both when the debris atmosphere is illuminated by the sun and when it is in darkness.

Despite the space-contamination problems, NASA firmed up plans for ATM, AAP, and a variety of space-engineering projects in a 100-ton cluster of "modules" (Fig. 111) renamed Skylab, to be visited during a 10-month period by three separate crews of three astronauts each. The

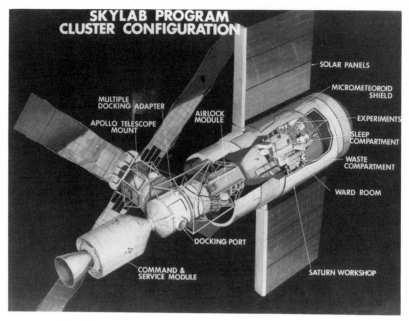

FIG. 111. The Skylab cluster, seen from the down-sun side, with CSM (left) attached to one docking port in the MDA. The ATM has four solar panels in addition to the two large panels on the Workshop. (NASA diagram.)

FIG. 112. Floor plan of Skylab's crew quarters, right-hand end of the Workshop in Figure 111. The access door was only in the full-scale models at MSC and MSFC—the Workshop in orbit is vacuum tight, with 5 lbs/in² cabin atmosphere inside. (NASA diagram.)

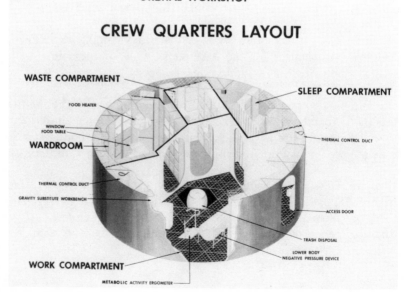

10,000-cu-ft, 65-ton Workshop was made by remodeling an S-4B third-stage rocket and outfitting it with air conditioning, a wardroom, food-heating units (stoves), teletype, sleep compartments (bedrooms), shower bath, toilet, freezers, vacuum cleaners, and storage cabinets for everything the crews would need (partly shown in Fig. 112). In 1970, Skylab was the largest spacecraft yet designed, with two docking ports for Apollo CSMs and six large panels of solar cells—four on the ATM and two on the workshop—that would provide 21 kilowatts of solar power. It would be launched without crew, the ATM folded up in front of the Multiple Docking Adaptor (MDA), as in Figure 113. Each crew would be launched separately in an Apollo CSM to dock on the MDA after the ATM is swung out to its proper position on the side by remote control.

Experiments for Skylab included those listed in Table 13, photography of the Earth for studies of resources (see p. 296), and ATM studies of the sun.—TLP

The Apollo
Telescope Mount

RAYMOND N. WATTS, JR.

(*Sky and Telescope*, October 1970)

The ATM will be the first manned solar observatory in space. It will have eight instruments to be used in the visible region of the spectrum and in the ultraviolet and x-ray regions.

The ATM's octagonal external frame will surround a large cylindrical package nearly 7 ft in diameter and 11¼ ft long. The cylinder's interior is divided into four equal parts by a 10-ft-long cross that serves as an optical bench to support the telescopes (Fig. 114). To maintain a stable inside temperature, the cylinder will be shielded from the sun's direct rays, and liquid coolant will circulate through its skin to keep the interior surface at 50°F.

Closed-circuit television and TV data transmission to Earth are to play a big part in the astronauts' operation of the ATM. Two display monitors on the control panel in the Multiple Docking Adapter will be hooked to the five TV cameras in the experiments package that show what view of the sun each astronomical instrument is recording.

White-Light Coronagraph. High Altitude Observatory (HAO) is here continuing its long-term program to observe the outermost atmosphere of the sun, using external occulting disks to block out the bright photosphere (see p. 321).

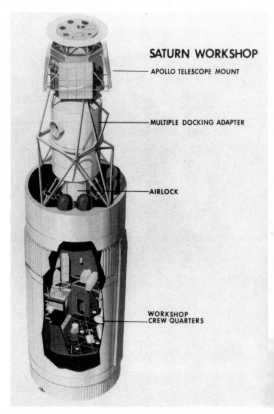

APOLLO TELESCOPE MOUNT

MULTIPLE DOCKING ADAPTER

AIRLOCK

WORKSHOP
CREW QUARTERS

FIG. 113. ATM folded up for Skylab launch on top of the MDA, below which is the Airlock Module (see Fig. 111) partly inside the Saturn cylinder (bottom) that houses the Workshop. The tubular trusses holding the ATM are so made that it can be swung to the right and attached to the MDA, with its four solar panels unfolded, as in Figure 111. The two large Workshop solar panels are folded down close to the Saturn cylinder. (NASA drawing.)

Gordon Newkirk, Jr., is principal investigator for the HAO coronagraph. His group wants to determine the three-dimensional structure of coronal streamers, the relation between streamers and plages (bright patches in the chromosphere), the solar wind's relation to the corona and surface magnetic fields, and whether solar radio bursts have optical counterparts.

Far-Ultraviolet Solar Scanner. Harvard Observatory, with Leo Goldberg in charge, will observe the evolution of solar flares simultaneously in many spectral lines. Operating in the wavelength region 300-1310A, this instrument will map the sun with a spatial resolution of about 5 arcsec and a spectral resolution of 1.3A. Each picture will be 5 arc-min. square, built up of 60 scan lines. A hydrogen-alpha telescope is included for pointing control, and also to provide red-light photographs at a wavelength of 6563A for comparison purposes.

Two Ultraviolet Instruments. At the U.S. Naval Research Laboratory (NRL), J. D. Purcell and R. Tousey are preparing a coronal spectroheliograph for the extreme ultraviolet and a chromospheric spectrograph for the far- and near-ultraviolet. The slitless spectrohelio-

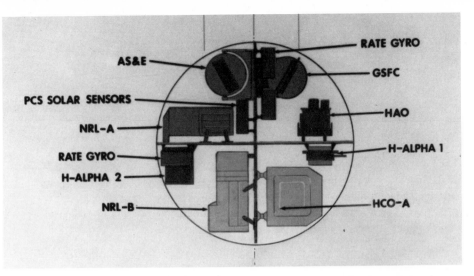

FIG. 114. ATM instruments mounted in the 82-in. cylinder. (NASA diagram.)

graph will record large images at many different wavelengths in the range 150-650A, where lines of highly ionized iron (Fe IX-Fe XVI) are quite intense, as well as lines of many other elements found in the corona. A concave grating forms the images on UV film contained in four cameras, each holding 200 strips of 35×258 mm size. When the 200 exposures have been completed, an astronaut will have to go on EVA to change cameras (and to retrieve the last one before returning to Earth). Exposures may be as short as 1/10 sec or as long as 5 min., but most will be 5-60 sec.

The NRL slit spectrograph is a large double-dispersion, concave-grating instrument. It works at a resolution of 0.08A in the wavelength range 970-1970A, and at 0.16A in the 1940-3940A range. One use of this spectrograph will be to record the solar spectrum from 12 arc-sec inside the limb to 20 arc-sec above it. Such observations will provide information on the physical conditions in the sun's atmosphere through the chromosphere out to the corona.

Other studies to be made with this high-dispersion spectrograph involve quiet regions on the sun, dark filaments, the chromospheric network, supergranulation of the photosphere, and the spectra of flares. Film changing and camera retrieval will again be done by an astronaut on EVA.

Inside the ATM module there will also be an extreme-ultraviolet monitor telescope, connected to the astronauts' television screen, and perhaps to observing stations on Earth. It will allow direct inspection of the sun in the range 170-550A to permit identifying solar features and events that would otherwise be invisible.

X-Ray Telescope. This project of J. Milligan and his collaborators at NASA's Goddard Space Flight Center (GSFC) is to obtain pictures of the sun in soft x-rays (wavelengths 5-60A), showing the disk and the inner corona out to 1.5 radii from the solar center. The grazing-incidence telescope (see Chapter 9) contains nearly cylindrical paraboloid and hyperboloid surfaces placed in tandem to focus x-rays on the film. There are six different filters for as many spectral intervals. The Goddard scientists hope to obtain observations giving better understanding of the sun's x-ray emission from the quiet corona at times of low solar activity, from active coronal regions above sunspots and plages, and from solar flares, eruptive prominences, or radio bursts.

X-Ray Spectrographic Telescope. At American Science and Engineering (AS&E), R. Giacconi and W. Reidy are principal investigators for equipment to study x-rays from solar flares. Their main instrument is similar in operation to the GSFC x-ray telescope described above, but it has two coaxial sets of paraboloid-hyperboloid surfaces, the outer one 30.5 cm in diameter, the inner about 23. The total energy-collecting area is 42 cm^2 and the focal length is 213.4 cm. Spatial resolution of 2 arc-sec will be achieved over 48 arc-min. Although it can be used for direct photography, this telescope will operate chiefly with an objective transmission grating to obtain an x-ray spectrum, with the images recorded on 70-mm film supplied in 1000-ft rolls. The film magazine will have to be changed by the astronauts on EVA every 14 days.

Since the astronauts cannot see the sun directly at x-ray wavelengths, a photomultiplier detector will give a signal when solar x-rays rise above a critical level, and will also be used to determine exposure times for subsequent x-ray pictures with the main instrument. A small auxiliary x-ray "finder" telescope will present the sun's image on a cathode-ray tube (TV screen) where an intense spot at the location of a growing flare will permit the astronaut to point the main instrument at that place.

Skylab was a giant in several ways: it was the largest (15,000 cu ft), heaviest (100 tons) spacecraft yet put in orbit; despite a serious launch accident (see p. 71), it was repaired; and it went on with three visiting crews to get down-looking Earth observations of sea conditions, water pollution, crop failures, forest conditions, geologic formations, and urban spread, observations of the atmosphere, and astronomical observations of a comet (p. 305), the sun (p. 323) planets, stars, nebulae, and galaxies, at a total cost of about $2.5 billion. There was so much space inside the Workshop that the astronauts could do o-g ballet, and a good deal was learned about the effects on humans of living in o-g for long periods.

Day-to-day supervision was in the hands of a NASA executive at the Manned Spacecraft Center in Houston, Kenneth (Kenny) Kleinknecht, to whom a great deal of credit must go for the mission's success. He was backed by a team of several hundred engineers and specialists in Houston, the flight controllers in Mission Control with two of the largest computers in existence, and NASA's Space Tracking Network around the world. He had to handle some 20 or 30 scientists who demanded priority for their experiments and three teams of three astronauts each who felt (with reason) that they were overworked in Skylab. In the Daily Skylab Conferences, engineers sometimes said something couldn't be done. Kenny's rejoinder was "Why the hell not?" and he usually proceeded to do it.

However, Skylab was delayed, and the Soviet Union continued the Space Race with an orbiting workshop in 1971. It later became clear that the Soviet Intercosmos Council is emphasizing the design of long-endurance spacecraft with closed systems recycling oxygen and water. In 1974, they announced the successful test of chlorella-cells nutrient solution illuminated by sunlight. Eight gallons of concentrated chlorella illuminated by ultraviolet lamps to simulate sunlight above the atmosphere were able to recycle oxygen (from CO_2) and water (from urine) for one man in a 5-cu-m. sealed cabin filled with air.

Academician Boris Petrov, head of the Council, announced plans in 1974 to follow up the Salyut program. The idea is to build a large station in orbit by assembling easily launched "blocks," using "space tugs" (p. 401) to fit them together. A space station suitable for a crew of 10 men was mentioned, and a duration of 10 years, with frequent changes of crew.—TLP

Soviet Soyuz 10 Links
with an Orbiting
Space Station

RAYMOND N. WATTS, JR.

(Sky and Telescope, June 1971)

Although the technological details of the Soviet Union's two major space flights in April 1971 have not been released, it is clear that progress has been made toward an orbiting workshop in which cosmonauts can conduct research for long periods. (The American Skylab launch is now delayed to April 1973.)

At present a gigantic unmanned vehicle named Salyut ("Salute") is in a nearly circular orbit 165 mi. above the Earth. Since it can stay aloft at least through mid-June, additional attempts to place men aboard may be made before atmospheric drag brings Salyut down.

Apparently a first attempt was made on April 22, when three men were sent aloft in Soyuz 10 from the Baikonur Cosmodrome. Rigorous preflight training presaged a long mission, yet after linking up with Salyut for 5½ hrs, they returned to Earth in an unprecedented night landing in Kazakhstan.

The work space aboard Salyut is estimated at 1400 cu ft. It is equipped with solar cells and a large array of antennas for data transmission to Earth. On April 22 Soyuz 10 was launched under the command of Col. V. A. Shatalov, whose crew consisted of A. S. Yeliseyev and N. N. Rukavishnikov. After reaching a rather high orbit, they fired a retrorocket to permit a rendezvous with Salyut, and 15¼ hrs later docking was completed. The cosmonauts first viewed Salyut through a telescope while still 9 mi. away.

According to Shatalov, the mission's main objective was to check a new rendezvous system and new locking mechanism. It is believed that cosmonauts arriving at Salyut in Soyuz should be able to enter it directly through the docking connection, instead of by EVA. But no such visit took place while Soyuz 10 and Salyut were linked. Instead, "onboard systems were checked out and dynamic characteristics were evaluated."

Note that the 20-ton Salyut space station was two years ahead of NASA's Skylab workshop. The concept of a space station visited by several crews was the same, although the Russians' limited booster thrust kept Salyut much smaller than Skylab. Soyuz 10 docked successfully, but the crew could not open the hatch on Salyut 1. Soyuz 11, launched 42 days later, docked with Salyut on June 7, 1971, and this crew succeeded in opening the hatch. The three cosmonauts worked 22 days on experiments in the space station. All this, and the tragic deaths of Debrovolsky, Volkov, and Patseyev on their return to Earth is covered in Chapter 2, p. 70.

Three later Salyuts were flown successfully, all of them in rather low orbits. The re-startable booster rocket was used to "kick" each one to a higher orbit after atmospheric drag had lowered the orbit. In 1975 Salyut 4 carried a large solar spectrograph of 270-mm aperture and about 7-m. focal length, recording high-resolution spectra on film later returned to Earth by Soyuz. There were problems in pointing it accurately to specified places on the solar disk (see p. 363), and no data have been released.

The NASA space station, Skylab, almost suffered catastrophe during launch (p. 71), but before this happened the plans were finalized.—TLP

Progress Report on Skylab

RAYMOND N. WATTS, JR.

(*Sky and Telescope*, January 1973)

On May Day 1973 at Cape Kennedy, three astronauts are expected to blast off from Launch Complex 39B for a rendezvous in space with the first Skylab space station, placed in a 270-mi.-high orbit from Pad A the day before.

Of the all-Navy team, Capt. Charles Conrad, Jr., is a veteran of the Apollo-12 Moon landing but Cdrs. Joseph P. Kerwin (a medical doctor) and Paul J. Weitz (an aeronautical engineer) have never flown in space.

Once in orbit, these men will maneuver their modified Apollo Command Module to dock with Skylab. At that moment, the combined spacecraft (Fig. 111) will be 118 ft long and some 90 ft wide with its solar panels deployed.

Skylab will be the crewmen's home and workshop for 28 days, after which they will re-enter the Command Module and return to Earth.

ATM solar observing will require a small part of the Skylab astronauts' time—100 hrs are scheduled during the first mission. This, and some of the other experiments planned, are made possible only through stabilization of the entire 95-ton spaceship by a system of three massive gyroscopes with mutually perpendicular axes. Each 142-lb flywheel rotates 9100 times/min., with so little friction that in the absence of external braking it would take 36 hrs to come to rest. Constructed by Bendix Corp., this system will hold the orientation of the spacecraft to within 3 arc-min. of the desired direction at all times, despite venting of gases or motion of the crew.

Since the astronauts must measure the mass of the food they eat, a special spring balance has been supplied for use in zero gravity. An object to be weighed is attached as a load to a coil spring, whose period of oscillation is measured to give the mass in kg. Similarly, each astronaut will measure his own body weight while strapped in a spring-mounted chair.

In a group of engineering experiments, the astronauts will use an electric furnace for studies of solidification and crystal growth of metals and alloys. For example, it is expected that doped (purposely contaminated) semiconductor crystals of high chemical uniformity can be produced in space, because of the absence of gravity-induced thermal convection. The crew will also study flammability, since aircraft experiments have shown that during short periods of weightlessness flames tend to go out.

No fewer than 132 experiments directed by scientists from 24 countries involve Earth resources. Skylab's 270-mi.-high orbit, inclined 50° to the equator, will carry it over 75 percent of the Earth's surface, its ground track repeating itself every 5 days. Cameras and sensors will observe our planet at visible, infrared, and microwave wavelengths. By comparing photographs made in several spectral regions, such terrain features as vegetation type, vegetation health, and soil moisture can be determined.

Experiments suggested by 19 U.S. high-school students have been approved for Skylab, selected by NASA and the National Science Teachers Association from more than 3400 proposals.

In one case, astronauts will grow radishes under weightless conditions, to find whether the presence of light is enough to guide the plants to send their roots down and their sprouts up. In another experiment, a spider will be released to ascertain if the web it spins differs from the ones it produces on Earth. A third student proposal involves infrared monitoring of volcanoes as a possible way to predict eruptions.

First Skylab Mission
Proves Successful

RAYMOND N. WATTS, JR.

(*Sky and Telescope*, August 1973)

All the dramatic misadventures and problems of Skylab 1 are recounted in Chapter 2 (p. 71). However, the Skylab's first inhabitants spent much of their time fruitfully in housekeeping, making medical records, and carrying out their scheduled astronomical observations. From the viewpoint of the ground-based experimenters, the first visit to Skylab was a great success.

Despite the power problems that plagued much of the mission, ATM was operated for 88 hrs, and the Earth Resources Experiment Package (EREP) obtained highly refined photography of the Earth's surface (Fig. 116).

On June 7, 1973, when the astronauts successfully cut the little metal strap that had kept the remaining solar wing of Skylab from deploying, the array of solar cells moved outward very slowly, because the hinges had all but frozen in the frigid shade of space. However, when the spacecraft was maneuvered to permit sunlight to warm the hinges, the wing extended well enough to boost Skylab's power by as much as 3000 watts.

FIG. 115. Launch damage to Skylab viewed from the CSM off the bottom of the Workshop. The undamaged ATM is in the background with the shadow of one of its solar panels falling diagonally across the foreground Workshop, whose meteor-bumper and thermal shield are missing. At the left are broken wires and tubes where one large solar panel was torn off. At lower right is part of the other large solar panel only partially folded out. (NASA photo.)

The astronauts then settled into a more nearly normal routine, being active some 16 hrs a day. Conrad reported, "We've adapted very well. If you are just reading you float free and wind up anywhere, on the ceiling, on the floor, even in the corner, sometimes ricocheting, and it doesn't seem to bother you." He also noted the very quiet environment.

On EVA June 19, Conrad and Weitz retrieved ATM film cassettes and replenished film for instruments that would operate automatically until the second Skylab crew arrived. They also repaired an ATM battery-power regulator and removed a small threadlike contaminant from the occulting disk of the white-light coronagraph. Solar astronomers were greatly pleased by an ultraviolet observation of a hole in the corona.

Finally, on June 22, and after cleaning house, the astronauts entered the Command Module and separated from Skylab for the return journey to Earth, after photographing the now-empty Workshop-ATM (Fig. 117). Everything went well on the re-entry, and the module parachuted

FIG. 116. EREP photograph of western Lake Erie from Skylab, showing Lake St. Clair and the Detroit-Windsor area at upper left, Toledo and the mouth of the Maumee River at lower left, and Cleveland at lower right. (NASA photo.)

FIG. 117. Skylab, photographed from the CSM after the first crew left on June 22, 1973, showing the "umbrella" shading the sunward side of the Workshop, and the fully opened large solar panel. (NASA photo.)

into the Pacific Ocean 840 mi. southwest of San Diego and only 6½ mi. from the recovery ship USS *Ticonderoga*.

Although somewhat unsteady after being weightless for 28 days, the three men stepped onto the deck of the carrier without too much difficulty and appeared to be in good condition. They suffered a bit of land sickness later, but recovery was quick. The many medical tests performed in orbit and the preliminary checkouts after the return to Earth give no reason why man should not be able to spend 56 days in space.

Hence plans are rapidly being advanced for the next crew's mission, Skylab 2, scheduled to begin July 28.

Another Skylab Success

RAYMOND N. WATTS, JR.

(*Sky and Telescope*, November 1973)

After 59 days in space, the second team of astronauts to occupy the huge Skylab satellite landed safely in the Pacific Ocean on September 25, 1973. Despite several crises, the mission of Alan L. Bean, Jack R. Lousma, and Owen K. Garriott was precisely conducted and highly successful.

A Saturn-1B rocket carried the three astronauts into space at 7:11 a.m. on July 28, after one of the smoothest countdowns ever. (In fact, Lousma fell asleep in the Command Module about 45 min. before liftoff.)

Later, a major problem developed when the crew had observed "snow" near the spacecraft, which turned out to be leaking oxidizer (nitrogen tetroxide) from a cluster of attitude-control thrusters on the CSM. The flow was shut off, and the crew used the other three clusters to dock with the Multiple Docking Adapter on Skylab (see Fig. 111). Six days later, the astronauts and ground controllers discovered a second oxidizer leak, on the opposite side of the CSM, dangerous because the caustic and explosive oxidizer was accumulating inside the spacecraft. The first leak lost about 50 lbs of nitrogen tetroxide and the second 12 lbs by the time it was turned off.

Since the oxidizer would be needed to manuever the Command Module when it came time to re-enter the Earth's atmosphere, plans were begun for what might have been the first space rescue. The plan was to send two fresh astronauts up to the Skylab, where they would rendezvous and enter through a second docking port to collect Bean, Lousma, and Garriott. Then all five would return to Earth in a specially prepared Command Module, leaving the disabled CSM attached to the Skylab.

Meanwhile, the three men in Skylab all got motion sickness, and their first EVA was delayed three times. Finally, on August 6 Lousma and Garriott climbed out of the Skylab to replace film packs in the ATM, and to rig a new sunshade over the parasol on the damaged sunward side of the Workshop (see p. 72). In all, they spent 6 hrs, 31 min. outside the Skylab, 2½ hrs longer than anticipated, but they got the job done.

Once over the motion sickness, the astronauts worked with increasing efficiency, and by mid-August they did not have enough assigned work to keep busy, so more solar observations were made. On the second EVA, Garriott and Lousma worked outside Skylab for 4½ hrs on August 24. They disconnected six of the nine Skylab stabilization gyros and replaced them with new ones. Garriott changed film for the ATM experiments and the pair returned to the Skylab cabin.

By the mission's end, the astronauts had completed 305 hrs of solar observations, taken some 77,000 photographs of the sun, studied more than 100 solar flares, and observed a major solar disturbance on August 21.

They also took 16,800 pictures of the Earth for use in crop surveys, population-growth studies, land-use planning, sea-surface monitoring, and searching for natural resources. Of particular interest will be their surveys of drought-stricken Mali and Mauritania. Their pictures of Alpine snows may be of value in predicting avalanches.

In tests of space-welding techniques, the Skylab crew demonstrated that strong new alloys can be formed under weightless conditions. The absence of thermal convection in molten metal permits making castings of exceptional uniformity.

One of the astronauts' final chores, occupying most of their last day before heading for home, was to transfer their films and 18 mi. of magnetic tapes to the Apollo Command Module. Analysis and interpretation of this vast collection of data will keep Earth-bound scientists busy for months.

The third and final mission to the Skylab workshop is planned for about November 11. While in space, the new crew will observe Comet Kohoutek; NASA is considering extending the duration of the flight to permit observations in January after the comet's perihelion passage.

Skylab Revisited

RAYMOND N. WATTS, JR.

(*Sky and Telescope*, January 1974)

The third and final Skylab mission began on November 16, 1973, as three astronauts were launched from KSC; it was the first spaceflight for all three: Lt. Col. Gerald P. Carr of the Marine Corps, Lt. Col. William R. Pogue of the Air Force, and Edward G. Gibson, a solar physicist.

With them inside the Command and Service Module was nearly a ton of supplies, including a new far-UV camera, film, recording tape, a new treadmill exerciser, tools, and several hundred pounds of food. If all goes well, their stay in space will last up to 84 days.

Eight hours after launch and on the third attempt, the crew in the Command and Service Module docked with Skylab, as it orbited 270 mi. above the Earth.

For the first few days, weightlessness troubled the astronauts. Gibson said: "It's really an effect of body orientation. In o-g I can move into a room sideways or upside down and not recognize it. Or I recognize it, but don't feel at home in it." However, Gibson found that when he turned his body "upright," as he had worked in training, "all of a sudden my mind would flash and say, 'Yes, I know where I am.'"

One of the first tasks facing the new crew aboard Skylab was to replace fluid in leaking cooling lines that draw heat from electronic equipment. Pogue accomplished this by punching a hole in the plumbing line and pumping in some of the 40 lbs of liquid brought along for the purpose.

On November 22, Pogue and Gibson spent 6½ hrs outside Skylab on EVA. They moved half way around the space station to repair the drive mechanism of the EREP microwave altimeter, and change four canisters of film in the ATM.

On the eighth day of the mission, one of three large gyroscopes (see p. 295) that control Skylab's maneuvers failed, but the observing instruments could still work effectively.

The astronauts were scheduled to go on EVA again on Christmas Day to photograph Comet Kohoutek as it passed the sun, and to change film on the ATM.

By the end of November, the Skylab astronauts had observed Comet Kohoutek with binoculars. To take far-UV photographs of it, they have a special camera designed and built at NRL by Thornton Page and George Carruthers. It is a new version of the camera carried to the Moon by Apollo 16 (p. 331) and is to photograph the hydrogen halo of the comet.

The microwave altimeter, repaired by Pogue and Gibson, recorded irregularities as large as 15 m. in the open sea surface over which Skylab passed. These were later explained (Science, Dec. 27, 1974) by mounts and trenches in the sea bottom.—TLP

Skylab Mission Completed

RAYMOND N. WATTS, JR.

(*Sky and Telescope*, April 1974)

The third crew of Skylab astronauts came down safely in the Pacific Ocean on Feb. 8, 1974, after circling the Earth 1214 times in the Skylab spacecraft, which they had boarded on November 16, after riding up to it from KSC. They set a record of 85 days in space.

Skylab carried three control-moment gyros, one of which failed early in the mission. A second was giving intermittent distress signals throughout, causing concern that this mission might have to be ended prematurely, but it was made to continue functioning by keeping its bearings as cool as possible.

The crew's health appeared excellent. Preliminary medical examinations after landing indicate that the deleterious effects of weightlessness are significant only for short flights. The body seems to become acclimated after the first few weeks in space, and the hitherto troublesome losses in body weight and bone calcium level off.

On February 1, as their time in space drew to a close, Carr, Gibson, and Pogue completed their last photographic sweep of the Earth, bringing the total for all three Skylab missions to 89 sweeps. Figure 118 shows an interesting photo of Hawaii. On February 3, Carr and Gibson spent 5 hrs, 19 min. on EVA, retrieving film and bringing in a piece of the spacecraft's metal skin for analysis of the effects of exposure. This brought the total EVA time on Skylab to 4½ hrs. Gibson continued work on the ATM up to the last possible moment (Fig. 119).

Then, on February 6, the astronauts fired the steering rockets for 3 min., raising the orbit enough so that Skylab will not succumb to atmospheric drag for 9-10 yrs. Into the Command Module they loaded 1700 lbs of film and material samples that were to come back to Earth with them. They also packed a special bag of food, film, filters, tele-

FIG. 118. EREP photograph of Hawaii from Skylab in February 1974, showing Hilo Bay at upper right, snow on Mauna Kea (above center) and on Mauna Loa (below center). (NASA photo.)

FIG. 119. Edward G. Gibson practicing at the ATM control console in the full-scale Skylab replica at JSC before the Skylab-3 launch. Scientist-astronaut Gibson used these controls in the MDA overtime to get the many observations of the sun reported in Chapter 8. (NASA photo.)

printer paper, and electrical cables, to be left on Skylab for possible future study. Perhaps Skylab will be visited by some future astronaut team to pick up the bag of left-overs.

At 10:34 UT on February 8 the astronauts unlatched their CSM and prepared for re-entry. There were tense moments when the electrical controls that fire the small maneuvering rockets would not work at first try, but the re-entry went well and at 15:17 parachutes dropped them into the Pacific Ocean 170 mi. west of San Diego, California, where the helicopter carrier USS *New Orleans* was standing by.

Skylab Is Stabilized in Orbit

(Sky and Telescope, June 1974)

Empty since Feb. 8, 1974, the huge Skylab spacecraft continues to circle the Earth every 93 min.

It is hoped that sometime before Skylab re-enters the atmosphere, astronauts may revisit it to examine its condition and retrieve the "time capsule" of samples left in it. To facilitate such a visit, the spacecraft has been stabilized so that its longest axis points vertically downward, with its docking port facing away from Earth. Tests show that this orientation should minimize drag forces that could cause Skylab to tumble.

A few hours after the astronauts left it, the spacecraft was deactivated by ground command, and it is unable to transmit data on its attitude. Hence, the vehicle is checked by radar from time to time by the Air Force North American Defense Command (NORAD). In April such a check showed that the space station was still vertical and not tumbling.

Since I was involved (with the far-ultraviolet camera) on Skylab 3, I cannot refrain from describing the day-to-day operation in Mission Control, where I had desk space in the "Corollary Science Support Room" along with five other Principal Investigators (PI's), mostly concerned with Comet Kohoutek, which passed close to the sun on Dec. 28, 1973. There were other SSR's for Earth Observations, Solar Observations, and Medical Research, each supported by a team of flight controllers—two men on duty for each 8-hr. shift.

Starting about a week in advance, the Mission Control organization discussed plans for each day's work, and finalized them one day in advance so that a "pad message" could be radioed up-link to the teleprinter in *Skylab* instructing each astronaut as to what he must do from 8 a.m. to 5 or 6 p.m., Houston time. In exceptional cases, pads included extra work during the crew rest period, but this was discouraged. All the PI's met twice a week from 9 p.m. to midnight with the "mission scientist," Bob Parker, to iron out differences as to which experiment should get what time. Observations had been roughly allotted in a pre-flight "Mission Requirements Document" that became our bible in these sessions. Bob·Parker, a scientist-astronaut, had to decide which experiments could get a little more time when conditions warranted it, and who would have to take a little less.

During the day in the SSR, PI's heard by down-link radio what observations had been made, and whether anything went wrong. It was worth being on hand because if an instrument went wrong, another experiment was substituted and gained an extra hour or so of observing time. A PI could not talk directly with the astronauts, but he could submit messages to the flight controller giving extra information or asking questions about his experiment, and these might be included in the evening radio discussion with the Skylab crew. Some flexibility was allowed by an "alternate pad" transmitted up-link with the regular pad message, telling the astronaut that under certain circumstances he should make alternative observations.

The Mission Control system was well thought out and worked well. Although the days were long (8 a.m. to midnight), they were great fun, and most of the SSR gang felt a letdown when Skylab 3 ended on Feb. 8, 1974, after 85 days. Soon after this, the data analysis began—a year's work by each PI on his photographs, or spectra, or instrument readings, following a Data Analysis Plan that had been prepared before the start of the mission to show the results expected from the experiment. Most of the Skylab-3 cameras and spectrographs were used on Comet Kohoutek, which received more attention than any previous comet. It was expected to become very bright in mid-December 1973, but disappointed the experts, and we could scarcely see it on the ground. Scientist-astronaut Ed Gibson was able to provide ten fairly detailed drawings, only two of which can be reproduced here. His estimates of colors in the comet, verified by astronauts Carr and Pogue, and the shape of the sunward spike, were later used to estimate the size and nature of dust particles exuding from the comet's nucleus.—TLP

Comet Kohoutek Drawings from Skylab

EDWARD G. GIBSON

(*Sky and Telescope*, October 1974)

We, the crew of Skylab 3, were very fortunate to have been in Earth orbit and able to observe Comet Kohoutek during its perihelion passage (closest to the sun) at the end of 1973. It was a very impressive sight. We made sketches of our observations on each of the 9 days following perihelion, when the comet could be observed from inside Skylab. These sketches, first sent to Houston by down-link TV, were drawn when we could find free time. Ten-power binoculars were used for most of our views. After I drew the sketches in black and white, they were reviewed by the other two crew members to assure that my renditions agreed with what they saw.

Comet Kohoutek was considerably brighter at perihelion than its pre-perihelion appearance would suggest. The comet went on a splurge as it passed the sun, shedding more matter than expected. But the splurge was brief and was soon compensated by a fading.

On Dec. 29, 1973, one day after perihelion, the comet was first seen (Fig. 120) at sunset while Carr and I were outside on EVA and looking through sun visors. It was extremely bright and had an unusual sunward spike, which was faint relative to the coma and to the tail immediately behind the coma (the hazy region around the nucleus). A very faint diffuse feature could be seen in the acute angle between spike and tail.

The over-all color of the comet was bright yellow. The colors of the two relatively faint parts, that is, the spike and the adjacent feature, were difficult to perceive. However, the spike was judged to be a lighter yellow than the coma and tail, perhaps only as a result of its faintness, and the diffuse feature was not perceptibly different from light yellow.

On each sketch I wrote the visible length of the comet in degrees, and the equivalent in Earth-Moon (EM) angular separation as we saw it. The top sketch is a somewhat subjective picture, and the sketch in the middle is an attempt to show contours of equal brightness. The third sketch, which was made postflight, is an attempt to illustrate the observed color, texture, intensity, and form. White is used to imply a higher intensity even though the hue itself may have been uniform.

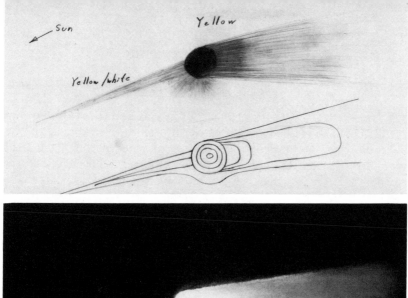

Sun

Yellow

Yellow/white

DEC 29, 1973 PERIHELION +1 DAY

FIG. 120. NASA drawings by E. G. Gibson.

During days perihelion +1 through +9, the comet changed appreciably in form, color, texture, and length. On +2 the sunward spike was fainter, the over-all length had increased to 6°-7°, and a very definite orange cast was apparent. The next day, +3, the spike continued to fade, the length reached its maximum of 7°-8°, and the color returned to light yellow.

The apparent increase in the tail length from +1 to +3 occurred most likely because the axis of the tail became more nearly normal to our line of sight and the angular distance of the comet from the sun increased, so the eye could become better adapted to faint light levels. Since Skylab was moving around the Earth at approximately 4°/min., the time between our sunset and the comet's setting was short during this period.

By +4 days, it was questionable that the sunward spike could be seen at all. The over-all color was a very light yellow and the apparent length of the main tail had decreased to 5°-6°. On +5 the spike was not observed, and the comet was white with no perceivable yellow or orange cast. The apparent length was only 4°-5° (Fig. 121).

From +6 to +9 days the color shifted from white to violet-white and the texture of the tail changed from nearly smooth to very mottled. The mottling in the drawings indicates its general appearance, not the exact geometry. The transition from a uniformly white-yellow texture to a mottled white-violet is perhaps due to a transition from our observing the dust to our observing the gas in the tail. Two separate and distinct tails, dust and gas, were looked for but could not be distinguished.

From +5 to +7 days the tail again lengthened, reaching 6°-7°. This probably happened because the airglow from the sun was no longer in front of or immediately adjacent to the comet. After +7, however, the apparent length and brightness decreased, as the comet continued to move away from the sun. On our last sketch, perihelion +9 days, the tail was only 3°-4° long.

To have viewed Comet Kohoutek as we did from Skylab was a rare opportunity, and our only regret is that everyone on the ground could not see it equally well while it was so close to perihelion.

[The pictures with this article were presented and described by the author at the Comet Kohoutek Workshop, held on June 13-14, 1974, at Marshall Space Flight Center in Alabama. At that conference, more than 40 papers reporting studies of this comet were given by researchers from six countries.—Ed.]

Skylab provided detailed data on bodily changes in nine men who spent a total of 400 man-days in o-g. All exercise, sleep, food intake, and excrement were accurately recorded—in fact, most of the work and activities between 8 a.m. and 5 p.m. each day were tape recorded and are on file. An interesting report (Science, May 30, 1975) is based on Bill Pogue's records of "light flashes" that he saw during an hour in the dark on two separate days when he recorded the time and nature of each flash. The flashes had been reported earlier by Apollo astronauts and interpreted as cosmic rays passing through the vitreous humor of the eyeball. Pogue's flashes were well correlated with the known flux of cosmic rays that changed as Skylab moved around its orbit, but there were numerous additional flashes as Skylab passed through the South Atlantic Anomaly (p. 44) where protons trapped in the Van Allen belt bombarded Skylab. Pogue's flashes match well the measured flux of protons, showing that the human eye can be used as a particle detector, and confirming the radiation hazard in spacecraft.

*The Soviet Union released less information but was pursuing similar goals in 1974.—*TLP

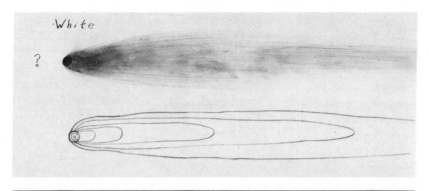

JAN 2, 1974 PERIHELION +5 DAYS

FIG. 121. NASA drawings by E. G. Gibson.

Soyuz-13 Manned Mission

RAYMOND N. WATTS, JR.

(*Sky and Telescope*, February 1974)

Two cosmonauts traveled 128 times around the world in an 8-day mission during December 1973, making the first Soviet manned space flight since Soyuz 12 in September.

Major Pyotr Klimuk and Flight Engineer Valentin Lebedev, both 31 years old, were launched from the Baikonur Cosmodrome on December 18. Their Soyuz-13 spacecraft was placed into an 89.6-min. orbit, inclined 51.6° to the Earth's equator, at a height ranging from 140 mi. to 169 mi.

Stated purposes of the mission included testing several navigational systems, taking ultraviolet solar spectrograms, and trying a procedure for growing food during future protracted space flights.

One in-flight investigation was a study of blood circulation in the cosmonauts' brains under conditions of weightlessness. In the food-production experiment, each of two connected cylinders held a different kind of bacterium that could live symbiotically. The expected result in one cylinder was an edible protein mass.

On December 26, the Soyuz-13 spacecraft landed safely about 125 mi. southwest of Karaganda, in Soviet Central Asia.

Two Soviet Cosmonauts
Spend 15 Days in Space

(Sky and Telescope, September 1974)

On July 5, 1974, two Russian spacemen successfully docked their Soyuz-14 craft with the orbiting scientific station Salyut 3, thus completing the first Soviet space rendezvous in three years. The cosmonauts, Col. Pavel Popovich and Lt. Col. Yuri Artyukhin, had been launched on the 3rd from the Soviet space center at Baikonur. The 3-room, 25-ton Salyut was boosted into orbit on June 25.

The two spacecraft remained joined for 2 weeks as they traveled around the Earth once each 89.7 min. in a nearly circular orbit inclined 51.6° to the terrestrial equator. During this interval, Popovich and Artyukhin performed various experiments, some to test the feasibility of manufacturing new products in space. Others included surveying for possible mineral deposits in Central Asia, and tests on the recently modified Soyuz.

By carrying only two men instead of three, the Soyuz can accommodate astronauts wearing pressure suits—a feature not employed on the ill-fated Soyuz 11 (see p. 68). After 15 days aloft, the crew was returned safely, landing near Karaganda in Central Asia on July 19.

The Soyuz-14 mission was particularly significant to United States space officials, since it employed equipment and techniques that will be used in the joint American-Soviet flight scheduled to begin July 15, 1975. During that mission, a three-man Apollo spacecraft will dock with an orbiting Soyuz carrying two men, and the combined crews are expected to conduct five joint experiments.

In 1975 the Apollo-Soyuz Test Project (ASTP) was carried out jointly by NASA and the Soviet Union, as already mentioned (p. 36). Further plans for workshops in space—NASA's Shuttle, its Spacelab, and Large Space Telescope (LST) are covered in Chapter 10.—TLP

8

Optical Observations of the Sun, Earth, and Stars

Although Skylab produced more astronomical observations than any previous space mission, many other data have been obtained from rockets shot briefly above the atmosphere, and from unmanned orbiters. Some of the many sounding rockets used, including the highly successful Aerobees, are shown in Figure 122. They are fired mostly from two NASA launch sites: White Sands, New Mexico, and Wallops Island, off the Virginia coast. They must be launched when the wind is low; their instruments are returned to ground or sea by parachute. Early rockets were more primitive, but by 1970 an Aerobee 350 could carry 450 lbs of instruments above 160 mi. for 2-3 min., observing in a stabilized orientation above most of the Earth's atmosphere (see Fig. 123). Such flights have several advantages over orbiters; they are cheap, film and instruments are recoverable, and they require much less preparation time. Also, they were available well before orbiters were.

While above the ozone layer (p. 331), ultraviolet (UV) observations could be made at wavelengths shorter than 3000A, and the goal of the first scientific rocket flights organized by Richard Tousey of NRL was the UV spectrum of the sun. In 1966, Tousey's group got a spectrum

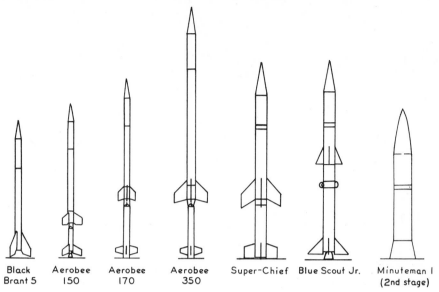

| Black Brant 5 | Aerobee 150 | Aerobee 170 | Aerobee 350 | Super-Chief | Blue Scout Jr. | Minuteman I (2nd stage) |

FIG. 122. Sounding rockets used for astronomical experiments in the 1960's and 1970's. The Aerobee 350 is 65 ft long. Two sets of fins are on the multiple-stage vehicles. (NRL diagram by G. R. Carruthers.)

from 2000A down to 1500A showing, among other things, the strong absorption by neutral silicon (Si) below 1680A. At shorter wavelengths, the solar spectrum appeared to be only emission lines from the chromosphere, an indication that Si absorption is strong, and silicon more abundant than previously thought.

Background information on the sun is given in Volume 3, Origin of the Solar System, and Volume 5, Starlight. In a way, it is surprising that the nearest star retained several features that remained mysterious for so long. Its bulk composition and internal composition were fairly well understood by 1965—largely due to electronic computers used to simulate conditions throughout the interior of many "model stars"—larger, smaller, older, and younger than the sun—but the abundances of several of the 92 elements in the sun are still uncertain. The sun is mostly hydrogen and helium, with smaller amounts of oxygen, neon, carbon, nitrogen, silicon, magnesium, sulphur, argon, iron, and nickel, and minute amounts of other elements. This list of elemental abundances is important for theories of the origin of the solar system (p. 155). Nuclear reactions in the sun's core should produce beryllium and boron, but the neutrinos produced by these reactions have not been detected—another solar mystery.

There was (and is) even greater uncertainty about the sun's atmosphere —the huge shell extending several million mi. above the 865,000-mi. photosphere that we see or photograph as the sun's edge in white light.

FIG. 123. Sequence of events in a rocket-astronomy mission, starting with launch at lower left, and ending with parachute landing and recovery at lower right. Winds cause most of the horizontal motion. Since the rocket is not used after burnout, the payload (instruments) is often separated before despin. (NRL drawing by G. R. Carruthers.)

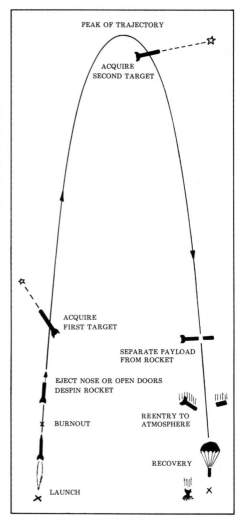

PEAK OF TRAJECTORY

ACQUIRE
SECOND TARGET

ACQUIRE
FIRST TARGET

SEPARATE PAYLOAD
FROM ROCKET

EJECT NOSE OR OPEN DOORS
DESPIN ROCKET

BURNOUT

REENTRY TO
ATMOSPHERE

RECOVERY

LAUNCH

Various layers have been identified: the reversing layer, chromosphere, K-corona, F-corona, and outer corona, each of successively lower density. Farther out, the corona merges with a faint haze called the zodiacal light, and the solar wind (mostly ionized H) is blowing outward at 300 to 600 km/sec as far as the most distant planets.

The layers provide convenient names, but the solar atmosphere is not simply layered, as several space experiments have shown. There are sunspots enduring 1-2 months in the photosphere, brief flares a bit higher, hour-long prominences in the chromosphere, and longer-lived streamers in the corona (p. 290). All these are linked somehow with the 22-yr sunspot cycle (a maximum of sunspots every 11 yrs and a minimum in between, such as in 1973). Theoreticians explain this as due to inter-weaving magnetic fields carried by the ionized gas in the convective layer just below the photosphere. (Historical records show that sunspots failed to appear, for some unknown reason, between 1645 and 1715.)

Temperature in the sun's atmosphere plays an important role in ionizing the gas and causing turbulence (convection). From early NRL rocket spectra, estimates could be made of temperatures above the photosphere because the far-UV light (wavelength less than 1680A) comes from higher regions in the chromosphere. In 1969, Harvard astronomers Gingerich, Parkinson, and Reeves, using data from a rocket carrying a spectrophotometer (see Glossary), were able to draw a curve of temperature vs. height, showing a minimum (4200°K) about 600 km above the photosphere. Above that level, the temperature rises sharply, reaching a million degrees in the corona.

FIG. 124. Quick-look scan of the solar corona from OSO 6 in the 625A line of Mg X. The 4-digit "words" telemetered to GSFC represent brightness in each picture element ("pixel") in the array of 24 columns and 64 lines. The cross-hatching (bright areas), N S E W directions, and dashed circle (sun's outline) were added later by Harvard astronomers studying the corona. (HCO diagram.)

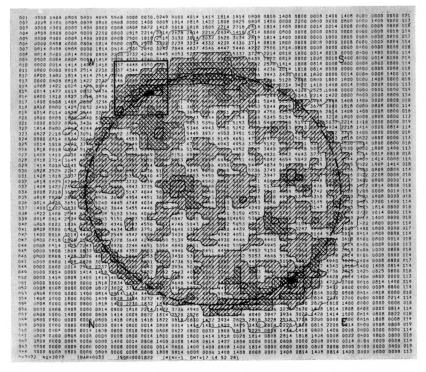

Another NRL Aerobee rocket, launched by NASA in 1968 from White Sands for Richard Tousey's group, obtained visible-light photos of the corona showing long streamers, several million kilometers above the photosphere, that seemed to originate in "active regions" on the surface. The NRL group also got the first extreme-UV (XUV) photos taken in light of 170-500A coming from the very-high-temperature corona and x-ray "hot-spots" (pp. 322, 353).

Turning, now, to orbiters, the sun was extensively studied by NASA's Orbiting Solar Observatories (OSO), a series that started in 1967. By 1969, OSO 6 carried seven instruments, one of them a spectroheliograph that could scan the sun in light of a single (selected) wavelength. Figure 124 is a computer printout of the data telemetered from OSO 6, showing brightnesses in the 625A emission line of Mg X (nine-times ionized magnesium) which are high over active regions. This "mosaic" took less than an hour to prepare. More data came from OSO's 5, 6, and 7.—TLP

What Two Sun-Observing Satellites Tell Us

ROGER J. THOMAS AND STEPHEN P. MARAN

(*Sky and Telescope*, May 1971)

In 1969, two highly successful satellites of the Earth were launched: OSO 5 on January 22 and OSO 6 on August 9. Data from these observatories in space have accumulated rapidly, and on Dec. 2-4, 1970, more than 80 solar astronomers attended the "Second Workshop on Orbiting Solar Observatories" at GSFC.

SOLAR STRUCTURE

Harvard Observatory astronomers E. M. Reeves, E. H. Avrett, A. K. Dupree, G. L. Withbroe, and M. C. E. Huber based their analyses on ultraviolet spectroheliograms and spectral scans obtained by OSO 6 which show the dependence of temperature on height in the solar atmosphere (Fig. 125).

From a value of roughly 6000°K at the photosphere, the temperature decreases to a minimum of about 4300° at a height of several hundred kilometers and then increases through the chromosphere until it reaches the coronal value of more than 1,000,000°. In the thermal "plateau" that extends from 1900 km to 2050 km, the temperature remains nearly constant at about 20,000° in an isothermal layer. The evidence for this

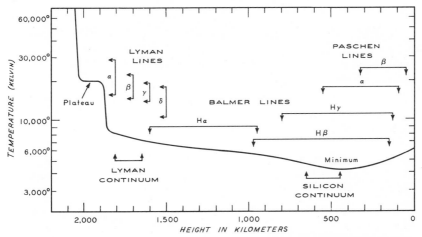

FIG. 125. Temperature *vs* height above the photosphere in the sun's atmosphere as summarized by Harvard astronomers. The Paschen, Balmer, and Lyman spectral lines are produced by atomic hydrogen in the IR, visual, and far-UV parts of the solar spectrum, and are caused by atoms at the height regions indicated by arrows. "Silicon continuum" is a band of wavelength near 1700A in the far-UV absorbed by silicon atoms around 500 km above the photosphere, and "Lyman continuum" near 900A is absorbed by hydrogen atoms around 1700 km. (HCO diagram.)

plateau comes from detailed investigations of the Lyman-beta line of hydrogen at 1026A wavelength and the slight solar-limb brightening in the 1336A resonance line for singly ionized carbon.

Withbroe has used observations of ultraviolet lines taken at various distances from the solar limb by OSO 6 and OSO 4 to make improved abundance determinations for such elements as iron, silicon, and magnesium at different levels in the solar atmosphere. He finds that these abundances do not vary significantly with height, and that they agree with the previously measured coronal values.

New Facts About Flares

Ever since their discovery in 1859, solar flares have intrigued and puzzled astronomers. Their known characteristics are still largely unexplained and new aspects of the flare event are still being found. For example, Kenneth J. Frost of GSFC showed OSO-5 records of hard x-rays (0.049—0.44A) from a flare on March 30, 1969, that had two maxima about 3 min. apart.

R. J. Thomas described a high-resolution flare observation made possible by three unusual circumstances:

1. The total solar eclipse of March 7, 1970, was visible as a partial eclipse from OSO 5.

2. A flare occurred during the eclipse and was the only strong x-ray source on the sun at that time.

3. The part of the Moon's limb that occulted the flare was perpendicular to the part that uncovered it, as shown by the white lines in Figure 126.

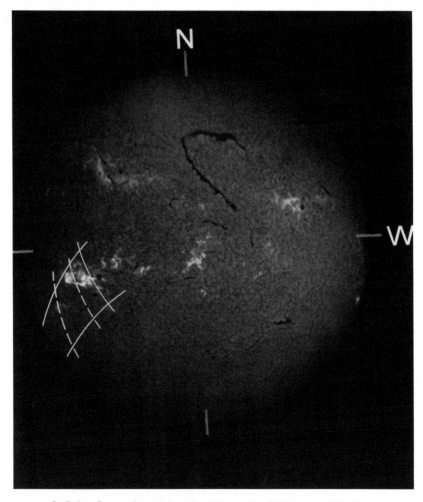

FIG. 126. Solar flare eclipsed by the Moon for OSO 5 on March 7, 1970. The photograph is a spectroheliogram in hydrogen 6563A light (Hα) showing the bright flare (left) and other clouds of hydrogen. The white lines show how the Moon's curved edge moved upward first covering the flare, then (dashed lines) uncovering it. (GSFC diagram.)

Analyzing the x-ray-intensity changes during immersion and emersion, Thomas derived a map of the source with resolution of about 1 arc-sec. This flare's x-rays came from two distinct regions: a "diffuse area" comparable in size to the visible active region (200,000 km across), and a "central core" of hotter and denser gas (about 7000 by 20,000 km). Comparison with optical observations made from the ground shows that the x-ray core was located at the point where the longitudinal magnetic field strength was changing most rapidly with position in the flare. In fact, the core was elongated, apparently so as to bridge two regions of opposite magnetic·polarity.

The OSO-7 Year
of Discovery

STEPHEN P. MARAN AND ROGER J. THOMAS

(*Sky and Telescope*, January 1973)

On Sept. 29, 1971, a launch-vehicle malfunction sent OSO 7 into an unplanned eccentric orbit, spinning rapidly and apparently unable to lock onto the sun. Larger and heavier (1400 lbs) than any previous satellite in the series, OSO 7 was in a flat spin at roughly 60 turns/min. with the sun on the bottom portion of the OSO "wheel," instead of on the sail section that bears the solar cells. Without electrical-power generation, the batteries began to drain. After 8 hrs and 2352 radio commands, the sun was acquired by a sensor with a 90° field of view and bursts of control gas pitched the OSO over until the sail faced sunward—electricity began to flow from the solar cells. The OSO-7 scientific findings reveal hitherto unknown processes in solar flares and some remarkable properties of the corona.

Nuclear Reactions on the Sun's Surface

Gamma-ray emission lines provide evidence of nuclear reactions in solar flares, according to David J. Forrest of the University of New Hampshire, who designed a gamma-ray detector for OSO 7. The data clearly show line emission at energies of about 0.5 MeV (million electron volts) and 2.2 MeV, with weaker enhancements at 4.4 MeV and 6.1 MeV (1 MeV energy = 0.0124A; see p. 356).

The lowest energy is undoubtedly the 0.511-MeV line produced by the mutual annihilation of electrons and positrons. Positrons (positive electrons) are antimatter particles and they cannot long survive in the ordinary-matter environment of the sun. Thus, the positrons involved in the large flares of August 4 and 7, 1972, must have been produced in the flares themselves. Theory suggests that short-lived radioactive isotopes (see Glossary) of carbon, nitrogen, and oxygen were formed by nuclear reactions in the flares, then decayed to stable isotopes by releasing positrons.

The other strong emission can be identified with the 2.22-MeV line of deuterium (heavy hydrogen). This radiation occurs when an excited deuterium nucleus, produced by the collision of a neutron and a proton in the flare, decays to a lower state, emitting a gamma ray.

Some flare-electron streams also produce bursts of hard x-rays. Another OSO-7 instrument, designed by Laurence E. Peterson of the University of California at San Diego, detected a burst of hard x-rays that occurred on Nov. 16, 1971, at 5:19 UT, lasting 1 min., as observed at energies of 30 keV (thousand electron volts) to 44 keV, that is, at wavelengths of 0.41-0.28A. Associated with it was a longer burst at low energies, with the characteristic spectrum of thermal radiation from a plasma at 20,000,000°K. Calculations based on the "bremsstrahlung" (p. 357) thick-target theory show that the energy carried by the fast electron stream was sufficient to heat the surrounding plasma to 20,000,000°K.

HOLES AND POLES IN THE CORONA

Werner M. Neupert of GSFC obtained maps of the corona's radiation at the wavelengths of several emission lines, each arising mainly from ions at different characteristic temperatures, in the range 800,000°-3,000,000°K. The emission was fairly evenly distributed on the lower-temperature maps, but increasingly uneven at the higher temperatures, where it was concentrated toward the lower latitudes.

Typical coronal temperatures out to about 1.5 solar radii above the surface are 3,000,000°-4,000,000° in active regions, about 1,700,000° in the quiet corona, and roughly 1,000,000° near the poles. Occasionally, however, temperatures close to 1,000,000° are found in small regions called *coronal holes* at low latitudes.

George L. Withbroe of Harvard summarized physical conditions in a typical coronal hole as follows. The temperature is roughly 600,000° lower than in the quiet corona, and the gas pressure is roughly one-third of the normal value. The temperature gradient (rate of temperature increase with altitude) near the bottom of the hole, at the chromosphere-corona transition zone, is 10 times smaller than in the transition zone of the quiet corona. This last result implies that the transfer of heat by electron conduction downward from the corona to the chromosphere is also 10 times smaller in a hole. Theory suggests that the solar wind escaping from the corona should be enhanced in the region of a hole. There seem to be no unusual structures in the photosphere beneath coronal holes, and the magnetic field strengths there are not abnormal.

Coronal Streams and the Interplanetary Blast

A white-light coronagraph, with an external occulter and an SEC vidicon imaging detector, was flown on OSO 7. Developed by Naval Research Laboratory scientists under Richard Tousey, it is systematically monitoring the outer corona (see p. 289). For the first time we can study coronal streamers as they form, evolve, and dissipate.

From the first year's observations, Tousey reported that almost all coronal streamers are straight, being directed outward from the sun in the region surveyed by the coronagraph, 3 to 10 solar radii. Arched and loop structures, sometimes seen in the inner corona, are very rare in the outer. On Aug. 2, 1972, when a general expansion of the corona in the northeastern quadrant was observed in association with a large flare, a bright streamer was disrupted and bright plasma clouds were ejected from the sun. The largest clouds, each roughly 30 times the Earth's diameter, sped outward at velocities reaching 1100 km/sec. The clouds originated in the low corona or the high chromosphere, and were identified with four Type-II solar radio bursts that were recorded by the radio telescope at Culgoora, Australia.

It has long been known that such radio bursts signify the outward motion of solar electron clouds; but in addition to the clouds, there was an over-all outward motion of the gas in a large region of the corona—the "interplanetary blast," which involved at least 12 times as many particles as were concentrated in the clouds, and released more than five times as much energy.

(In its elliptical path, the OSO-7 spacecraft experiences significant atmospheric drag at perigee and, as a result, it will re-enter the lower atmosphere and be destroyed in late 1973 or early 1974.)

Note that the OSO orbiters and Skylab could observe the sun and its corona day by day, not only on rare eclipse days or during a brief rocket flight. For ground-based observatories, the faint corona (1 millionth the brightness of the sun) is obscured by air-scattered sunlight (the blue sky) unless the Moon occults (eclipses) the sun. Above the atmosphere it is fairly simple to occult the bright photosphere with a proper-sized disk at a camera's focus. Such an instrument is called a coronagraph. Another instrument used both on the ground and in space is the spectrohelio-graph, which records the sun's image in one wavelength (color) of light.

The Orbiting Solar Observatories (OSO's 5, 6, 7, and 8) and Skylab carried x-ray telescopes that show hot spots where a bubble of very hot (4,000,000°K) gas breaks the surface, sometimes starting a flare, promi-nence, or coronal streamer, and adding a "puff" to the solar wind. (High-energy astrophysics—x-ray and gamma-ray observations—are reserved for Chapter 9, but these x-ray descriptions of the sun cannot be omitted from this chapter.) Most of the x-ray emission in 1-60A wavelengths comes from superhot gas in the solar atmosphere, but other sources much farther out in space are noted in Chapter 9.—TLP

X-Ray Bright Points
on the Sun

(*Sky and Telescope*, June 1974)

Pointlike sources of x-ray emission widely spread over the sun's disk and typically associated with small bipolar magnetic features (regions with well-defined N and S magnetic poles) may be of fundamental importance in understanding the solar cycle, according to L. Golub at AS&E in Cambridge, Massachusetts.

Golub studied photographs taken last year with the x-ray spectro-graphic telescope aboard Skylab (see p. 292), which has the great advantage of permitting continuous observations.

On a typical solar x-ray photograph, there are about 100 bright points. The total number on the entire sun is approximately 500, when due allowance is made for those on the far side and for those masked by overlying coronal features. The mean lifetime is about 8 hrs, so some 1500 bright points are formed each day.

In general, one of these objects is first seen as a diffuse cloud, inside of which a bright core later appears. The cloud grows to a maximum diameter of 20 arc-sec, on the average, or a little more than 1 Earth diameter. Then cloud and core gradually fade to invisibility.

A small fraction of the x-ray bright points show a much more spectacular behavior, brightening by several orders of magnitude within minutes. Just how fast such flaring points brighten and fade is a question that awaits study of later Skylab observations.

Some Results from
Skylab's Solar
Experiments

(Sky and Telescope, July 1974)

The scientific experiments aboard Skylab produced an enormous amount of data, most of which is still being studied and interpreted (see p. 305). Four dramatic photos taken by ATM instruments are shown in Figures 127-130. Others showed the corona during "interplanetary blast" (p. 321), beautiful arch prominences from the chromosphere, and coronal holes changing as the sun rotated.

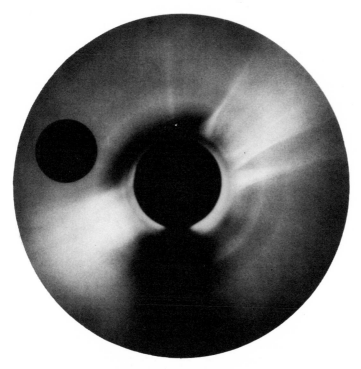

FIG. 127. Moon (left) silhouetted on the solar corona, June 30, 1973, as photographed by the ATM white-light coronagraph. The central black circle is an occulting disk in the instrument, blocking out the very bright sun; the dark ring above it and dark areas below are also instrumental. (NASA photo.)

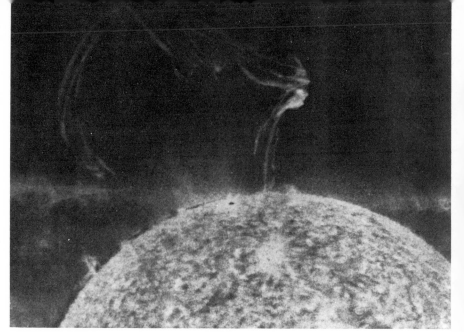

FIG. 128. Spiderlike prominence extending 500,000 mi. above the photo-sphere, photographed in 304A (XUV) light of He II (singly ionized helium) by the NRL spectroheliograph. (NRL photo.)

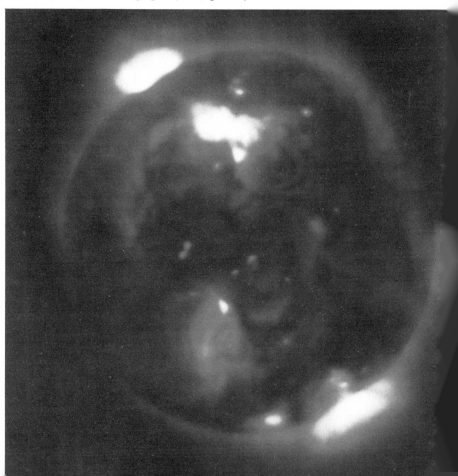

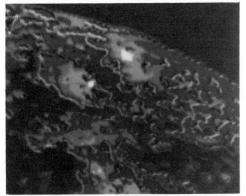

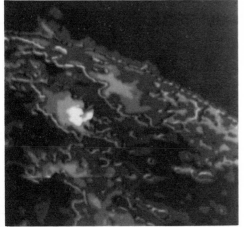

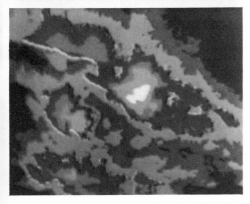

FIG. 130. XUV photos of the solar atmosphere at three levels: in 977A light of C III (top), 1032A light of O VI (center), and 625A light of Mg X (lower), taken simultaneously by the HCO spectroheliometer on Sept. 11, 1973. The top photo shows doubly ionized carbon at 60,000°K (lowest level), center shows 5-times-ionized oxygen at 300,000°K (just below corona), and lower shows 9-times-ionized magnesium at over 1,000,000°K (in the corona). (HCO photos.)

FIG. 129 (LEFT). X-ray photo of the sun taken from Skylab on May 28, 1973, by the AS&E x-ray experiment. The brightest areas are above active regions in the photosphere. There is a dark coronal hole near the sun's north pole (right), and others near the south pole (left) and center of the disk. Note also the many bright points in these "soft" x-rays of 3-60A wavelength (see Chapter 9). (AS&E photo.)

Most of the coronal holes are near the sun's poles. Since they allow stronger solar wind through—like holes in an inflated balloon around the sun—there should be more solar wind perpendicular to the ecliptic plane than in the plane—probably half again as much as we detect on Earth (in the ecliptic plane). In fact, solar cosmic rays are probably different in the polar directions, also, and clearer views of the surface layers at the poles may be of value. For this reason, NASA and ESRO are studying ways of putting satellites in orbits highly inclined to the ecliptic plane.

Observation of the sun from unmanned satellites continues. NASA launched one for West Germany, and replaced the very productive OSO 7, which burned up in the atmosphere in 1974.—TLP

First Helios Successful

(Sky and Telescope, May 1975)

The German Helios-1 spacecraft, launched Dec. 10, 1974, from KSC, passed 0.301 AU (28 million mi.) from the sun on March 15. The 815-lb probe came closer to the sun than any previous spacecraft.

Sensors recorded a temperature on board Helios of 155°C and observed a flux of solar radiation 10.4 times stronger than that at the Earth. Because the temperature was somewhat lower than expected, the second Helios, which is to be launched this December, will be sent even closer to the sun.

The German space-operations center near Munich, in command of the probe, announced that all onboard experiments were operating normally. Helios carries instruments to measure the solar wind, electromagnetic radiation, cosmic rays, micrometeoroids, and the zodiacal light. With an orbital period of 192 days, it will again come to perihelion on September 23.

The Last OSO Satellite

STEPHEN P. MARAN AND ROGER J. THOMAS

(Sky and Telescope, June 1975)

If all goes well, the largest of NASA's Orbiting Solar Observatories will be launched from KSC on June 18, 1975. This satellite, to be named OSO 8 once in orbit, marks the end of a highly successful series of spacecraft (see p. 316); its individual instruments are listed in Table 14.

TABLE 14. INSTRUMENTS CARRIED BY OSO 8

Instrument	Principal Investigator	Range	Res.	Location
Multichannel ultraviolet spectrometer	Roger M. Bonnet Laboratory for Stellar and Planetary Physics	1000-4000A	1"	Sail
High-resolution ultraviolet spectrometer	Elmo C. Bruner University of Colorado	1050-2300A	5"	Sail
X-ray crystal spectrometer and polarimeter	Robert Novick Columbia University	2-8 keV	4°	Wheel
Hard-x-ray telescope	Kenneth J. Frost Goddard Space Flight Center	20-5000 keV	5°	Wheel
Cetestial x-ray spectrometer	Peter J. Serlemitsos Goddard Space Flight Center	2-60 keV	5°	Wheel
Mapping x-ray heliometer	Loren W. Acton Lockheed Missiles and Space Co.	2-30 keV	2'	Wheel
Extra-solar extreme ultraviolet monitor	Charles S. Weller Naval Research Laboratory	150-1230A	9°	Wheel
Soft-x-ray telescope	William L. Kraushaar University of Wisconsin	0.13-35 keV	5°	Wheel

Range denotes either the working range of wavelengths in angstrom units (A), or the range of photon energies in thousands of electron volts (keV). *Res.* is angular resolution.

Two devices, mounted in the "sail" section of the spacecraft, can look steadily at the sun or scan across its face (or selected regions), producing maps of ultraviolet emissions. Both of these ultraviolet spectrometers are attached to Cassegrain telescopes having primary mirrors 6 in. in diameter and spacial resolution about 0.01 AU.

OSO 8 will be launched shortly before the minimum of the present sunspot cycle. A principal aim is to gain a better understanding of the transport of energy from the sun's surface through the chromosphere into the corona, and possibly downward from corona to lower layers. It has been known for some time that even the undisturbed regions of the solar atmosphere are not made up of uniform layers, but consist rather of a variety of three-dimensional structures, such as spicules and granulation cells (probably up-drafts), arranged in a pattern called the solar network. There are still basic questions relating to the formation, energy transfer, and physical properties of these phenomena.

Although solar activity will be at a low level during the mission, some activity is almost always present on the sun. In fact, solar-minimum conditions allow us to examine sunspots, plages, filaments, coronal holes (p. 320), and other features in simpler arrangements and in relative isolation.

Solar astronomers from the Colorado and Paris groups will coordinate their observations at a special solar-observing center located at the University of Colorado, in Boulder. There they will receive and analyze spectroheliograms and magnetograms from ground-based observatories and examine data telemetered from the spacecraft with the aid of specially programmed computers and display terminals. The Boulder facility will be linked to the OSO Control Center at GSFC, where commands are formulated and relayed to NASA tracking stations for transmission to the satellite.

For 12 years Pioneer 6, in solar orbit, has been bringing in data useful in such a solar-observing center. The Soviet Salyut 4, with its large solar spectrograph, might also contribute, after its films are returned to Earth and processed. But the Russians have so far released none of the data.

So much for the sun. Solar observations will undoubtedly continue throughout the space age, and solar power from Earth orbiters (p. 390) may achieve industrial importance.

As noted earlier (p. 296), an Earth orbiter can also look downwards, and photographs of the Earth's surface and atmosphere have achieved increasing importance. Back in 1968, the Gemini astronauts had taken over 1000 photos with hand-held cameras, showing geologic structures which were studied by John O'Keefe at GSFC. On the Gemini pictures he discovered an ancient (Quaternary) volcanic field south of Palomas, Mexico. He also noted broad areas of sand scouring in the Sahara Desert.

Later on, NASA launched several unmanned orbiters specifically designed for Earth observations. The best known of these are the weather satellites from which pictures of cloud patterns are obtained and used by weather forecasters all over the U.S. Starting in the early 1960's there were several NASA series, including Nimbus, ATS, and ESSA, leading to the Synchronous Meteorological Satellites (SMS), 22,590 mi. above the equator, where they can telemeter data for a weather map of almost a full hemisphere every 30 min. for a lifetime of about 5 yrs. In the late 1970's there should be five SMS's spaced equally around the equator for continuous, worldwide weather coverage.

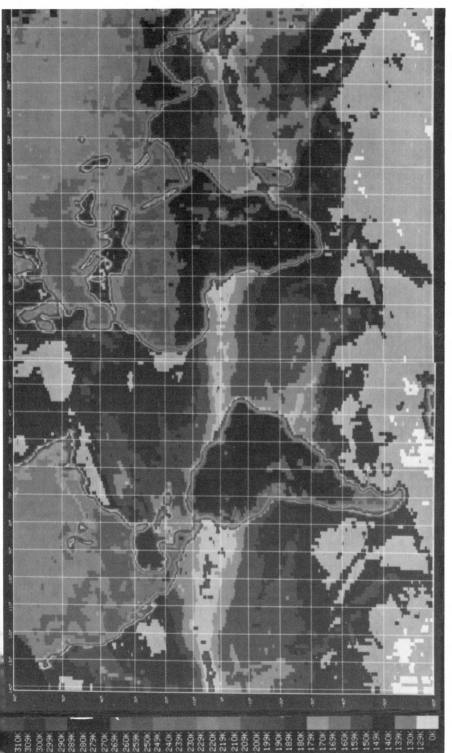

FIG. 131. Brightness-temperature map of the Americas, Africa, and Europe from the microwave radiometer on Nimbus 5. Temperature of the 1.55-cm radio emission is given at left. (NASA chart.)

Another space-technology application is a network of geosynchronous communication satellites for radio-telephone and TV transmission— already described in Chapter 2 (pp. 48 and 50). One downward-looking radio experiment on Nimbus 5, a meteorological satellite in near-polar orbit, gets measures of the Earth's emissivity at 1.55-cm wavelength. A color-coded map of the Earth (Fig. 131 with colors replaced by shading) shows high emissivity over the vast, hot, vegetated areas in Australia, South America, and central Africa, and differences due to soil moisture in the central United States. Ocean surface temperatures are also shown.

Color-coded maps are one way of handling the enormous amounts of data about the Earth's surface and atmosphere. In fact, data handling has become the major problem with the Earth Resources Technology Satellites (ERTS) and Land Satellites (Land Sats), each of which can telemeter 10,000 pictures a week, each picture of an area 115 mi. on a side, with resolution better than ½ mi., showing, among other things (in four colors), the ripeness of crops, any diseases that may be affecting them, water pollution, mineral outcrops, and ocean currents. As NASA-sponsored data analysis becomes more sophisticated[1] and these data are combined in different ways, we can expect frequent maps of crop con-

---

[1] In early 1976 JSC perfected computer mapping of water resources from these photos.

FIG. 132. Layers and "temperature profile" in the Earth's atmosphere at middle latitudes. Names of the layers are given near the bottom, interface names near the top. The temperature maximum ("Stratopause") at about 50-km height is caused by the ozone layer, another name in common use. Higher than 120 km, "temperature" takes on a different meaning (in gas of very low density). (SAO diagram by Luigi Jacchia.)

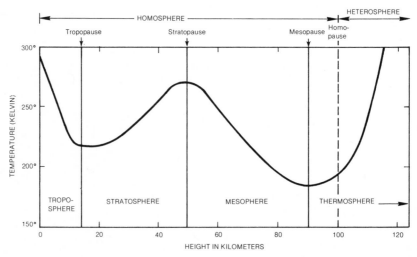

ditions, forest growth, and good mining locations, warnings of drought, and other practical benefits, world wide. (Soviet scientists are reportedly eager to get ERTS data, but are restrained by the politicians.) The most recent analysis of ERTS photos of northern Canada show evidence of glacier advance, possibly indicating the start of a new Ice Age on Earth.

Aside from weather, the Earth's atmosphere has been studied from spacecraft to very high altitudes, revealing things only guessed at before the space age. In a series of three articles (Sky and Telescope, March, April, and May 1975), Luigi Jacchia of SAO describes satellite observations that have determined the temperature (Fig. 132), density, and composition at various altitudes. Our understanding of these upper layers—the ozone layer near the top of the stratosphere, and the thermosphere where hydrogen rises above heavier atoms and molecules to escape from Earth through the geocorona (Fig. 134)—is important in rocket astronomy, radio transmission, and the recent concern about man-made freon gas affecting our exposure to ultraviolet sunlight.

In some ways, our atmosphere is like the sun's—and about as complex. One interesting view was photographed from the Moon.—TLP

More about Apollo 16

RAYMOND N. WATTS, JR.

(Sky and Telescope, July 1972)

Some of the first scientific results from Apollo 16 were announced in mid-May 1972 by MSC in Houston. Of particular interest to astronomers are the far-ultraviolet (far-UV) photographs taken with the NRL camera-spectrograph (Fig. 52).

This camera, designed by George R. Carruthers, was set up in the shadow of the lunar module to photograph the Earth and selected celestial objects in light of wavelengths shorter than 1600A. The semiautomatic device is a combination camera and spectrograph, with an electron intensifier. A Schmidt telescope of 3-in. aperture focuses the picture on a potassium-bromide cathode that emits electrons in proportion to the number of far-UV photons striking it. A 25,000-volt potential accelerates the electrons toward a special photographic film (Kodak NTB-3) for particle detection. The focusing magnet surrounding the camera ensures that the elecron image is a faithful reproduction of the far-UV image. Such an "electrographic camera" is 20 times faster than a similar optical camera with special film for the far-UV.

The plan was for astronaut John Young to place the camera as far from the LM as possible but still inside its shadow. He was to aim at

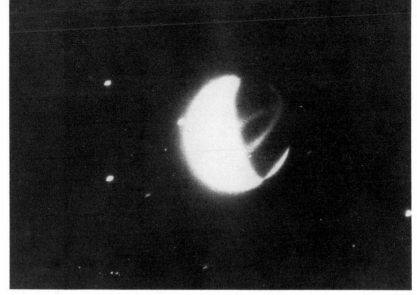

FIG. 133. Far-UV photo of the Earth showing oxygen airglow, equatorial bands, and polar-aurora "caps," a 10-min. exposure taken from the Apollo-16 landing site on April 21, 1972, in light of wavelength 1230-1550A. The north pole is up and sunlight comes from the left. A blue star in Capricorn is very near the Earth's limb at left. (NRL-NASA photo.)

FIG. 134. Far-UV photo of the Earth showing the geocorona, a 15-sec exposure taken a few minutes before Figure 133 in light of wavelength 1050-1550A dominated by hydrogen Lyman-alpha (1216A). On longer exposures the geocorona extends over 40,000 mi. (NRL-NASA photo.)

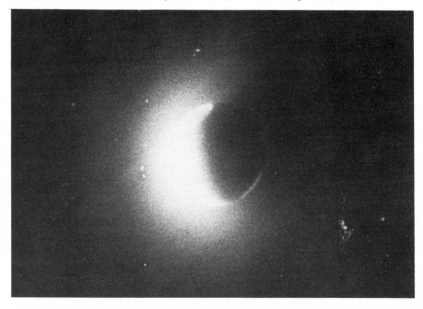

a selected object and then proceed with other duties while the camera took three pictures in the Lyman-alpha light of hydrogen (1216A), three pictures through a filter blocking Lyman-alpha, and two spectrograms. He was to return from time to time to repoint the camera for its next series of exposures.

These operations proceeded as scheduled, but during the delay in the LM landing the sun rose higher in the lunar sky and the landing craft's shadow became shorter. Consequently, some of the objects that Carruthers and his NASA colleague Thornton Page hoped to observe were obscured by the LM when the camera was placed in the shortened shadow.

Especially dramatic is a far-UV photograph of the Earth (Fig. 133), taken through a filter transmitting radiation between 1230A and 1550A but blocking out Lyman-alpha. The light comes mainly from atomic oxygen and molecular nitrogen. As expected, the polar auroral zones show clearly; the two luminous bands, inclined 15° either side of the magnetic equator, had been detected earlier but never photographed. They are believed to be caused by high-level currents of ions and electrons.

To show the glowing geocorona of hydrogen gas that surrounds our planet, photographs were taken in the 1050-1550A interval, which includes Lyman-alpha. Figure 134 reveals the geocorona as a roughly spherical glow, extending well over 40,000 mi. from the Earth, but with a "dimple" on the side away from the sun. The so-called geotail, which was thought to be pointing away from the sun is notably absent.

Spectrograms of the Earth confirm the great extent of the geocorona in hydrogen light and show emission lines of other elements. One is the helium line at 584A, never before observed in the Earth's atmosphere.

The far-UV camera also photographed clusters of galaxies to ascertain if they contain large amounts of intergalactic hydrogen—suggested as one solution to the riddle of cluster masses (see Vol. 8, *Beyond the Milky Way*). Several such photographs of clusters were taken, but inspection of the films does not show any conspicuous excess of Lyman-alpha glow. Accurate measurements of the films are required.

Other deep-sky objects photographed were the Loop nebula in Cygnus, the Large Magellanic Cloud, and Milky Way stars in Sagittarius (Fig. 135). In all, 180 frames were exposed on 11 different targets while the astronauts were on the Moon. [The camera was left on the Moon, but a duplicate has since been used on Skylab. See p. 302.—TLP]

The Earth's oxygen-nitrogen airglow (Fig. 133) has also been measured in visible light from many Earth orbiters, starting with Mercury 7, and

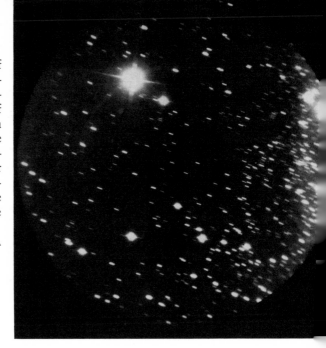

FIG. 135. Far-UV photo of hot blue stars in the constellation Sagittarius, looking toward the center of the Milky Way Galaxy, a 30-min. exposure from the Moon in light of wavelength 1230-1550A. Star images are "trailed" (elongated east-west) by the Moon's rotation. The brightest star (upper left) is σ Sag. (NRL-NASA photo.)

including Gemini flights 5, 8, and 9. A night-side view from a spacecraft at an altitude of 200 km or more shows a greenish band above the horizon, where the line-of-sight passes through several thousand kilometers of airglow in the lower thermosphere (Fig. 132), but even this reduced brightness adds to the background light of the night sky and masks the fainter glow of the zodiacal light.

The zodiacal band can sometimes be seen on a clear, dark night, far from city lights (see Vol. 2, Neighbors of the Earth); it is sunlight scattered from interplanetary dust—micrometeoroids—and fades away from the solar corona (p. 314) to about one-thousandth the coronal brightness at 30° from the sun along the zodiac—the band around the sky where the planets move, marking the plane of the solar system. White-light photographs taken first on Gemini 9 for Edward Ney (University of Minnesota), and on later Apollo flights and Skylab (p. 282), show the zodiacal light fainter than expected. In fact, there is a major discrepancy with Pioneer measurements of meteoroid density between Earth and Jupiter (p. 249). Either the micrometeoroid surfaces are almost completely black, or the zodiacal-light measures are too low. A possible explanation is that micrometeoroids are crystalline and reflect sunlight from separate faces, appearing dark between the bright diamondlike flashes.

The geocorona photo (Figure 134) shows that the Earth's atmosphere extends over 10 Earth radii, similar to the solar corona on a smaller scale. Jupiter has a Lyman-alpha corona, detected by Pioneers 10 and 11, and it is likely that Saturn does also.

We turn now to distant stars, and the many observations made by NASA's Orbiting Astronomical Observatories and the European satellites, as well as from rockets and manned-spacecraft missions. Background information on the nature of starlight and the various types of stars in the sky is given in Volume 5, Starlight, and Volume 6, The Evolution of Stars. The sequence of star types, O, B, A, F, G, K, and M, runs from 50,000°K surface temperature in O types to about 2000°K in M types. At temperatures of 10,000°K and higher, most of a star's light is in UV wavelengths (shorter than 3000A), and the cool stars of types K and M, less than 3000°K, have most of their light in the infrared (wavelengths longer than 10000A or 1 micron) and almost none in the UV. This color effect can be confusing: a far-UV photo of stars, such as Figure 135, shows only O, B, and A stars, and looks very different from a normal white-light photo or eyeball view of the same star field.

Each lettered type in the sequence has been divided into ten numbered segments so that a Bo star is the "earliest" of B types, B1 a bit later, and B9 stars just before Ao in the temperature sequence. Other designations have been added to describe the star's instrinsic brightness or absolute magnitude—Ia for very bright "supergiant" stars, II, III, and IV for medium to low, and V for "dwarf" stars of lowest luminosity. The small letters e and p are added, e for emission lines and p for "peculiar." Thus, a star may be designated type B5IIIe, all determined from lines in its spectrum, as described in Volume 5.

Some of the early observations of stars extended the study of spectra down to about 2000A in the UV.—TLP

Stellar Ultraviolet Spectra from Gemini 10

KARL G. HENIZE AND LLOYD R. WACKERLING

(*Sky and Telescope*, October 1966)

On July 19, 1966, astronauts John Young and Michael Collins opened the hatch of the Gemini-10 spacecraft and spent the next 40 min. taking the first ultraviolet spectra photographed from a manned spacecraft. Twenty-two frames were exposed on the southern Milky Way as the spacecraft's orbital motion caused the camera to scan from Crux to Vela, recording spectra of the brighter stars in the wavelength region 2200-4000A.

Begun on Gemini 10, this experiment will be continued on Gemini flights 11 and 12. Later, an f/3 objective-prism spectrograph of 6-in.

aperture will be flown on Apollo spacecraft to obtain stellar spectra from 1300A to 3000A.

Since the Gemini windows do not transmit ultraviolet light, the observations must be made through the opened hatch. The camera is fastened on a bracket that points it 5° above the roll axis of the spacecraft. The original operating plan called for the command pilot to point the undocked Gemini at the desired star field, and to nullify the vehicle's angular motions by using the pulse control system. With the craft thus stabilized, the pilot was to make six exposures on the field, ranging in length from 10 sec to 60 sec.

However, in Gemini 10 this plan could not be carried out for a variety of reasons, and we later discovered that, during assembly of the equipment, the grating was turned 17°.

By good luck, one of the most interesting regions of the Milky Way was scanned, and spectra were obtained for many more stars than expected. On the other hand, the displacement of the grating prevented proper widening of the spectra, since it caused motion of the stars to be in the direction of dispersion. The resultant loss in wavelength resolution was so great that the spectra show no emission or absorption features other than the Balmer continuum, 3300-3650A (due to hydrogen ions and electrons).

Exposures were about 20 sec long. In preliminary examinations of the film, we have identified the spectra of 54 stars. About half are bright enough to permit quantitative analysis of the ultraviolet energy curves. The limiting ultraviolet magnitude at effective wavelength 2200A for unwidened spectra (no motion of stars) is about +3.5.

On future flights, grating spectra will be taken of the regions of Lambda Scorpii, Alpha Eridani, Gamma Velorum, Sirius, and Epsilon Orionis.

In addition to gathering basic astrophysical data about hot, young stars, our Gemini experiment aims at clarifying the problems and techniques of ultraviolet photography of stars from manned spacecraft.

Space astronomers also had some bad luck with the first Orbiting Astronomical Observatory (OAO), an unmanned ground-controlled set of telescopes and spectrographs, but it eventually obtained a remarkable number of far-UV observations of stars, nebulae, galaxies, and comets. OAO A was the first design, built around a standard spacecraft frame, or "bus," suitable for all later OAO's. It was renamed OAO 1 after launch in April 1966, but suffered a battery failure two days later. In an orbit 500 mi. above the Earth, it carried two experiments: the SAO "Celescope," a group of four Vidicon TV cameras with selectable filters, to measure *UV* brightnesses of stars in a 3° field, and the University of Wisconsin Experiment Package (WEP), consisting of a 16-in. f/2 telescope, four 8-in. f/4 telescopes, and two grating spectrometers (see Glossary). The large telescope was to observe individual nebulae and galaxies of low surface brightness; the 8-in. telescopes were to observe individual stars in four different wavelength bands. All five telescopes were provided with selectable filters in the range 2100-4200A, and the spectrometers scanned the ranges 1000-2000A and 2000-4000A.—TLP

An Observatory in Space

RAYMOND N. WATTS, JR.

(Sky and Telescope, January 1969)

After several delays, the long-awaited OAO 2 is now aloft, circling the Earth at a height of about 490 mi. The 4400-lb cylindrical package was launched on Dec. 7, 1968, over 2½ yrs after NASA's first attempt in April 1966. The present flight was at one time scheduled for October 1968, but troubles with the guidance system caused further postponement. One of the two experiments aboard is the Celescope, with four f/2 telescopes feeding starlight to television cameras sensitive to four regions of the ultraviolet spectrum between 3200A and 1050A. The normal exposure on a star field will be 1 min., but a maximum of 5 min. can be obtained when necessary. Transmitted to ground by telemetry, the television pictures can be analyzed to yield the brightnesses of individual stars in the four wavelength bands. It is hoped eventually to map some 25 percent of the sky, with special emphasis on the hot young stars of types *O* and *B* (see p. 335).

The other major experiment is WEP, a cluster of seven telescopes developed by the University of Wisconsin. After engineering and communications tests, OAO 2 began working on December 11 and measured successfully the ultraviolet light from Beta and Iota Carinae.

Some Early Results
from Celescope

RAYMOND N. WATTS, JR.

(*Sky and Telescope*, May 1969)

The purpose of Celescope on OAO 2 is to measure the ultraviolet brightnesses of a large number of stars, using an array of four televisionlike cameras operated by ground control. The optical parts of Celescope weight 440 lbs, its electronics 77 lbs.

Cameras 1 and 3 cover 1600-3000A, while 2 and 4 are for 1050-1800A. Each camera has two filters, so that the spectral sensitivity of the upper half of each picture is different from that of the lower half. The filters divide each camera's wavelength sensitivity ranges roughly in half. When properly processed and displayed on a television screen, the signals from Celescope produce a picture that can be photographed for immediate inspection. Simultaneously, the incoming data can be recorded on magnetic tape for subsequent analysis.

The cameras were all checked out on Dec. 15, 1968. Just before midnight, Cameras 1, 3, and 4 made 30-sec exposures that were relayed to Earth with very little "noise." They showed measurable A-type stars (see p. 335) down to 8th magnitude in Camera 3, and to 6th magnitude in Camera 4.

A picture obtained by Camera 4 at 6:47 a.m. on December 16 provided the first of a series of scientific surprises. The half of the picture exposed through the barium-fluoride filter showed a normal, dark sky. The other half, through lithium fluoride, showed a sky brightness that drowned the stars. This was Lyman-alpha radiation of 1216A from the Earth's hydrogen corona (see p. 332).

By April, some 500 star fields had been photographed by Celescope, covering areas in the constellations Puppis, Orion, Taurus, Lyra, and Draco. Many of the stars observed have ultraviolet brightnesses differing significantly from expectations. For example, four sets of pictures of Orion's sword show that the multiple star Theta Orionis is 10 times brighter than anticipated in the two longest-wavelength ultraviolet intervals, and more than 100 times brighter in the two shortest!

On the other hand, the same pictures show Iota Orionis to be five times fainter than anticipated in the 1500-2700A region, and 15 times fainter at 2300A. Iota Orionis is a young, hot star of spectral class O9.

From other photographs, it seems that giant stars in general are fainter in the ultraviolet than predicted.

A series of observations of the geocorona was made between December 17 and the end of January. Preliminary measurements show that the Lyman-alpha radiation is most intense when the sun is near the horizon as seen from the satellite. The radiation is at minimum when the sun is 90° below the horizon, but that minimum is as much as half the maximum intensity.

By mid-March 1969, the Celescope experiment was still operating, although several difficulties have arisen as the cameras aged. For example, Cameras 1 and 3 have lost sensitivity, and Camera 2 has ceased operation, but Camera 4 still functions well.

By the middle of January OAO 2 had collected 20 times more ultraviolet information about the stars than had been accumulated in 15 yrs of rocket launchings.

Ultraviolet Photometry from a Spacecraft

ARTHUR D. CODE, THEODORE E. HOUCK,
JOHN F. MCNALL, ROBERT C. BLESS,
AND CHARLES F. LILLIE

(*Sky and Telescope*, November 1969)

Observational astronomy has taken a large step forward in the past year with the successful operation of NASA's Orbiting Astronomical Observatory, OAO 2. Apart from its "wings" covered with solar cells, the satellite is a stubby octagonal column 10 ft long. At one end the seven telescopes of the University of Wisconsin Experiment Package look out. Their operation differs radically from that of the SAO Celescope, looking out of the column's other end.

Traveling high above virtually all of the atmosphere, OAO 2 can observe celestial objects in the astrophysically important wavelength range between 3000A (the atmospheric cutoff) and 912A, where interstellar space becomes opaque because of the continuous absorption by hydrogen atoms.

The most sophisticated of all NASA unmanned satellites to date, OAO 2 has the ability to point to any position on the sky with 1 arc-min. accuracy, and then to maintain that position to about 1 arc-sec. Primary orientation is afforded by the spacecraft's six gimbaled star trackers, three or more of which are locked onto chosen guide stars.

On command, the craft is slewed around the gimbal axes to point at a specified object. Should a star tracker move off its guide star, an error signal is generated that starts up flywheels inside the spacecraft, moving it back toward the star.

The largest of the seven telescopes in the Wisconsin package is a 16-in. f/2 reflector, located on the central line of the spacecraft. Following the 30-arc-min. entrance aperture is a six-position filter wheel, with four interference filters affording a choice of spectral regions between 4250A and 1800A. The other two positions on the wheel provide a standard radiation source and a dark slide. Light passing through a filter is imaged by a field lens upon the photocathode of an EMI-6256B photomultiplier tube.

Grouped around this "nebular telescope" are four 8-in. f/4 off-axis reflectors, each with its set of four filters and a photometer. These "stellar telescopes" have smaller fields, either 2 or 10 arc-min. on command. The filters' wavelength ranges are 4250-2900A, 2900-2000A, 2500-1500A, and 1600-1200A. Since each of the five telescopes shares a range of wavelengths with another, it can be calibrated with respect to all the others. Objects as faint as magnitude 12 or 13 have been detected.

The other two instruments are scanning spectrometers, which view sky areas 2°.5 long (in the direction of the dispersion) and 8 arc-min. wide. Each dispersing element is a 6-by-8-in. plane diffraction grating. The dispersed bundle is passed to a parabolic mirror and focused through a hole in the grating, where two detector slits are located, one 10 times as wide as the other. The spectral resolutions are 10A and 100A for the short-wavelength spectrometer (1050-1800A), 20 and 200 for the longwave one (1800-3800A). The spectrum is scanned by shifting the grating in 10-A steps. The faintest objects measured to date with these spectrometers are of magnitudes 6 to 7.

To make a photometric observation, a two-word command is transmitted to the spacecraft. The first word identifies the type of command, where the command is to be stored, and the time of execution. The second word specifies photometer aperture, exposure time and gain, and filter-wheel settings.

The spacecraft clock provides timing signals. The observational data can be transmitted to the ground immediately or stored for later transmission.

Planning an efficient observing program for the Wisconsin telescopes is a rather complex and time-demanding task, involving considerations unfamiliar to ground observers. As seen from OAO 2, nearly half the sky is blocked by the Earth, which appears to move relative to the stars at over 3½°/min., and the Earth's phase changes from new to full in less than an hour. Also, observing is forbidden within 45° of the sun, to prevent direct sunlight from entering the instruments, and within 30° of the antisolar point (180° from the sun), to prevent excess heating of the Celescope in the other end of the satellite. Of course, the sunshade there is closed while we are observing.

There are other operational restraints; the slew from one object to the next can be no more than 30°, the number of observing commands that can be issued between radio contacts with ground stations is limited, and we avoid observing when the satellite is in the South Atlantic Anomaly where the Van Allen radiation belt dips close to the Earth's surface and affects the photometers.

With these restrictions in mind, we choose objects in the area of sky available to us and determine in advance the sequence of commands necessary to observe them properly. When the observing sequence for the day has been completed, it must be translated into computer language and combined with those for the other spacecraft subsystems, then checked against engineering and orbital parameters to find if the operation is safe for the satellite. Approximately 400 commands are required each day to carry out our observing program and to operate the satellite properly. Clearly, high-speed computers are essential to the astronomer in planning this work. Even so, we have had to assign one astronomer full-time to the OAO operation room at GSFC.

SOME OBSERVATIONAL RESULTS

The ultraviolet photometric and spectrophotometric observations of stars and galaxies show promise of giving much-needed information on the structure and evolution of stars.

The temperature scale of the youngest and most massive stars requires measures in the ultraviolet part of the spectrum. The OAO data indicate that stars earlier (bluer) than B_1 (see p. 335) are hotter than was previously thought. This result helps to resolve a discrepancy between the observed luminosities and the predicted thermonuclear-energy generation for these stars.

Evidence of compositional differences between stars is provided. Thus, in the WEP filter photometry, if a star turns out to be abnormally faint at 1700A, this deficiency in ultraviolet flux can be understood in terms of the additional opacity produced by silicon and magnesium ions. If late-type (cool) supergiant stars are brighter than expected in the ultraviolet, it is presumably a result of strong emission from their chromospheres.

The data from the scanning spectrophotometers have turned out to be of high quality and extremely informative. In Figure 136, the ultraviolet spectrum of Epsilon Persei, a main-sequence $B_{0.5}$ star (see Vol. 5), the

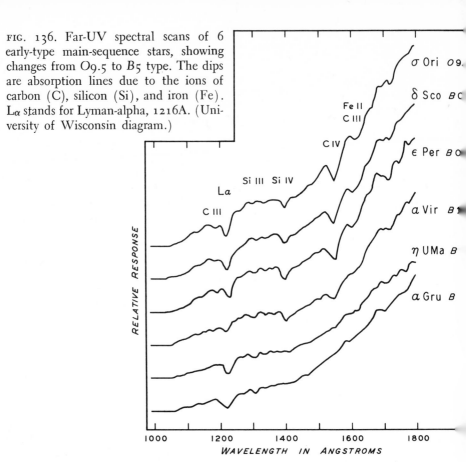

FIG. 136. Far-UV spectral scans of 6 early-type main-sequence stars, showing changes from O9.5 to B5 type. The dips are absorption lines due to the ions of carbon (C), silicon (Si), and iron (Fe). Lα stands for Lyman-alpha, 1216A. (University of Wisconsin diagram.)

three strongest features are the Lyman-alpha line at 1216A (due primarily to interstellar hydrogen), the resonance line of Si IV (triply ionized silicon) near 1400A, and the resonance line of C IV (triply ionized carbon) near 1550A.

In one Wisconsin OAO study, the Lyman-alpha line has been measured in 48 stars to study the distribution of neutral hydrogen gas in interstellar space. In various directions, there seems to be a good correlation between the amount of hydrogen so determined and the amount of neutral sodium, as inferred from the interstellar D lines at 5890A and 5896A.

In addition to stars, the OAO has observed planets, gaseous nebulae, the brightest x-ray source (Scorpius X-1), the brightest quasar (3C-273), and about 15 galaxies. The filter measurements of the great Andromeda galaxy, M31, show it to be abnormally bright at wavelengths shorter than 2700A. This feature is characteristic of most, but not all, galaxies. M33 in Triangulum shows it.

If this ultraviolet excess is general among galaxies, it implies a new distance effect resulting from the red shift (Doppler shift—see Glossary)

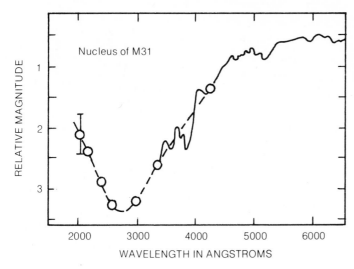

FIG. 137. OAO-2 filter measurements of M31's ultraviolet spectrum fitted to ground-based observations (solid line). M31, the "Andromeda Nebula," is the nearest spiral galaxy. (University of Wisconsin diagram.)

of very distant galaxies, and requires a revision of the distance scale. Some of the blue starlike objects found in searches for quasars may be just these red-shifted objects.

These are only some of the preliminary results from the rich harvest of data that the OAO 2 is continuing to collect.

Observing a Comet
from Space

(*Sky and Telescope*, March 1970)

In the first observations of a comet ever made from outside the Earth's atmosphere, a huge hydrogen cloud has been detected surrounding Comet Tago-Sato-Kosaka (1969g). This discovery was made with the WEP instruments aboard OAO 2.

Comet 1969g first came under observation on Jan. 14, 1970, by OAO-2's ultraviolet photometers and spectrometers. The head of the comet was bright in Lyman-alpha radiation of hydrogen out to $0°.5$ from the nucleus, and just measurable out to $1°.5$.

In addition, the comet radiated strongly at an ultraviolet wavelength of approximately 3070A. A. D. Code and his associates at the University of Wisconsin say this spectral feature is presumably due to hydroxyl (OH) molecules. A continuous spectrum (intensity at all wavelengths) was also present, which reached minimum intensity near 1900A, then increased again to as far as 1100A.

Until now, free hydrogen had not been detected in a comet, although molecules containing this element produce some of the more conspicuous features of cometary spectra. The hydrogen halo of Tago-Sato-Kosaka was huge, since on January 14 the comet was 0.429 AU from the Earth, so that 1°.5 corresponded to 1.7 million km. It was 1¼ times the size of the sun! On that same night, the visible head of the comet extended only about 8 arc-min., for observers with small telescopes.

This discovery of the hydrogen halo around a comet triggered far-UV observations of several later comets, including Comet Kohoutek, mentioned on p. 305. The Lyman-alpha emission grows in size and brightness as each comet approaches the sun, and can be used to measure the rate at which hydrogen boils off the comet's nucleus.

With these successes in mind, NASA went on to launch two more OAO's.—TLP

Another Orbiting Astronomical Observatory

RAYMOND N. WATTS, JR.

(*Sky and Telescope*, December 1970)

OAO 2 has amassed (to early September 1970) 5127 observations for the University of Wisconsin's Washburn Observatory and 3494 for the Smithsonian Astrophysical Observatory.

Now a further great advance is promised by OAO B, scheduled by NASA for launch late in November. This satellite bears a single telescope, a relatively fast 36-in. Cassegrain $f/5.2$ reflector, with a spectrometer operating in the spectral region from 1150A in the far-UV to about 3900A in the blue. With the exception of the quartz secondary mirror, all of the mirrors in the 1000-lb instrument are made of the light element beryllium, even the diffraction gratings.

The Ritchey-Chrétien optics (see Vol. 4) provide image quality better than 1 arc-sec over a 10-minute field of view, and the grating of 1219 lines/mm produces a dispersion of 0.18A/arc-sec on the sky.

Observations are being programmed by principal investigator Albert Boggess III, and associates at GSFC. OAO B has far better pointing accuracy than OAO 2. Yaw, pitch, and roll motions are controlled by gas jets and nine gyroscopes. The fine-guidance control, using about 10 percent of the starlight collected by the 36-in. primary mirror, will permit locking on to a particular light source to within 1-2 arc-sec. The precision gyros of the inertial reference system provide a drift rate of less than 10 arc-sec/hr.

With an instrument this large operating without the usual interruptions of daylight and bad weather, astronomers have to provide round-the-clock control. However, OAO B operates automatically for longer periods than OAO 2, since it has a command-storage memory of 1280 instead of only 256 two-word commands (see p. 340). The detectors' outputs are stored by a data accumulator for subsequent relay to the ground.

*Unfortunately, the well-designed OAO B failed during launch on March 30, 1970, because the protective shroud did not separate properly (p. 68), so it never became OAO 3. Astronomers had to wait another two years before getting a 32-in. telescope above the atmosphere for far-UV observations.—*TLP

An Astronomy Satellite Named Copernicus

RAYMOND N. WATTS, JR.

(*Sky and Telescope,* October 1972)

"Spacewarn 112. The U.S. National Academy of Sciences announces the launch of Copernicus (OAO C) from the Eastern Test Range on 21 August 1972 at approximately 10:28 UT. Period 99.7 min., apogee 751 km, perigee 739 km, inclination 35°.0. Tracking beacon transmits continuously on 136.440 MHz at 0.16 watts. Tracking beacon may be commanded on and off. Telemetry data are transmitted on command on 136.260 MHz at 2 watts and 400.550 MHz at 10 watts. Copernicus has been designated 1972-065A by World Warning Agency on behalf Cospar. Satwarn."

This teletype message from the World Data Center at GSFC, Greenbelt, Maryland, was transmitted at 17:47 UT on Aug. 23, 1972. A 32-in. reflecting telescope, the largest astronomical instrument ever sent into space by man, is now viewing the heavens as it revolves in a nearly circular orbit about 460 mi. up. Named Copernicus after the great Polish astronomer, who was born in 1473, OAO 3 is the heaviest unmanned satellite ever orbited by NASA (4900 lbs).

The primary experiment aboard OAO 3 is being carried out by Princeton University astronomers, with Lyman Spitzer, Jr., as principal investigator and John E. Rogerson, Jr., as the executive director. Their 32-in. ultraviolet telescope weighs about 1000 lbs, including the guidance optics, and is housed in the 10-ft central tube of the satellite. The $f/3$ fused-silica primary mirror—an "egg-crate" sandwiched between front and rear thin disks—weighs only 96 lbs. A conventional mirror of this diameter would weigh more than three times as much.

The aluminized mirror surfaces are protected with coatings of lithium fluoride, which is transparent to far-UV light. The fused-silica optics and special thermal design of the telescope allow it to operate at temperatures ranging from $-30°$ to $+15°C$ (about $-20°$ to $+60°F$).

The Cassegrain optics, with an effective focal length of 630 in. ($f/20$), feed a scanning spectrometer. The entrance slit is 3 mm long and may be adjusted to a width of either 24 or 96 microns. The spectrometer has a fixed concave diffraction grating (see Glossary) and two carriages, each moving a pair of exit slits and ultraviolet photometers along the Rowland circle, where the spectra are focused.

The first carriage has exit slits 24 microns wide and scans the first-order spectrum from 1623A to 3185A, as well as the second-order spectrum from 711A to 1492A. These scans are done simultaneously, with one phototube observing the first order at 0.05-angstrom intervals and the second photometer working at 0.025A intervals.

In a similar manner, the second carriage uses 96-micron exit slits for the first order, 1550-3300A in 0.4A steps, and the second order, 775-1650A in 0.2A steps. Two calibration sources are mounted behind their own slits on the Rowland circle.

The Princeton instrument has a fine-error sensor (FES) with a star-pointing accuracy of 0.1 arc-sec and the thermally-sensitive controls for the primary mirror maintain its alignment within 0.1 mm.

The FES is part of the over-all guidance system, controlling small errors in pitch and yaw, but only when the satellite is in its fine-pointing mode. The use of tilted slit jaws permits on-axis guiding without introducing lenses or prisms into the light path from the primary mirror to the spectrometer.

The Princeton observing program, allotted 90 percent of the viewing time, will obtain stellar spectra with a resolution hitherto impossible in the far-UV.

The spectrometer should be highly effective in measuring the very narrow absorption lines of interstellar molecules and atoms to help determine the abundances of elements in the interstellar medium. Such data are much needed in studying the origin of the stars and galaxies.

A second experiment aboard OAO 3, to be operated 10 percent of the time, studies celestial x-ray sources and x-ray absorption in interstellar space.

Although physically similar to the OAO 2, which is still operating after nearly four years in space (launched Dec. 7, 1968), OAO 3 has improved sun baffles (shields to keep sunlight off the telescope) and a newly developed gyro inertial-reference unit as the primary "attitude sensor," augmented by four star trackers to keep track of pointing directions. A unique on-board computer can handle 16,000 18-bit words and store up to 1024 ground commands. This will allow automatic operation of OAO 3 during intervals between its daily contacts with the primary ground station at Rosman, North Carolina. The design lifetime of Copernicus is one year, but the longevity of OAO 2 gives hope that the actual operating life will be many times longer.

A great deal more should be said about OAO 3, Copernicus. Many research papers have been published on the far-UV data collected by some 80 groups of astronomers in the U.S. and 10 other countries, including the U.S.S.R. Most of these studies are more technical than appropriate for Sky and Telescope or this book, but it should be stressed that they have vastly extended astrophysical research.—TLP

The Corona of Pollux

(*Sky and Telescope*, March 1975)

Until now, the sun has been the only normal star known to have a corona (p. ooo). The bright spectral lines characteristic of a possible corona are drowned out by the light of the star in the visible part of the spectrum. But in a cool star's far-ultraviolet spectrum, accessible only outside the Earth's atmosphere, the star's continuum may be faint enough to allow the coronal lines to be detected.

The K0 giant star Beta Geminorum (Pollux) was observed by the scanning ultraviolet spectrometer on board the Copernicus OAO-3 satellite. Near the strong Lyman-alpha line of hydrogen another, fainter emission line at 1218.4A is identified as due to oxygen four times ionized. This appears to be a coronal emission, since the ions that radiate it require a very high temperature.

The most plausible interpretation is that Pollux is surrounded by a corona which has temperatures near 260,000°K.

Results from the Utrecht Orbiting Spectrophotometer

THEO M. KAMPERMAN, KAREL A. VAN DER HUCHT,
HENNY J. LAMERS, AND ROEL HOEKSTRA

(*Sky and Telescope*, February 1973)

In March 1972, the Utrecht orbiting stellar spectrophotometer experiment S59 became operational on the ESRO TD-1A satellite. So far it has recorded with 1.8A resolution the ultraviolet spectra of more than 200 stars.

TD 1A travels a retrograde near-polar orbit, and the orbital plane precesses (turns) 1°/day around the Earth, while the satellite is locked onto the sun and the Earth. Thus, any instrument pointing away from the Earth scans the celestial sphere along meridians of the ecliptic coordinate system (perpendicular to the plane of the solar system), and covers the entire sky in half a year. Our spectrophotometer makes use of this scanning motion of the satellite.

Experiment S59 is programmed to observe 225 stars, chiefly along the Milky Way, and ranging in spectral type from early G (like the sun) through F and A to the hot stars of types B and O (p. 335). Our observing list includes all luminosity classes from subdwarf to supergiant, together with many abnormal objects, such as magnetic stars, peculiar stars, emission-line stars, contact binaries (Fig. 146), and one Wolf-Rayet (very hot) star.

THE INSTRUMENTATION

Experiment S59 was proposed in 1964 by C. de Jager, director of the Astronomical Institute.

The gimballed telescope-spectrometer has its own tracking system with a pointing accuracy of 1 arc-sec. The 26-cm (10-in.) primary mirror has an ellipsoidal figure. With a spherical secondary, the combination gives an effective focal length of about 100 cm (40 in.).

The spectrometer has a concave grating (1200 lines/mm) and the spectrum is scanned mechanically with three 0.030-mm-wide slits. The monochromatic light from each slit is measured by a photomultiplier.

When an object with an ultraviolet flux exceeding some threshold value (set by ground command) enters the field of view, the telescope

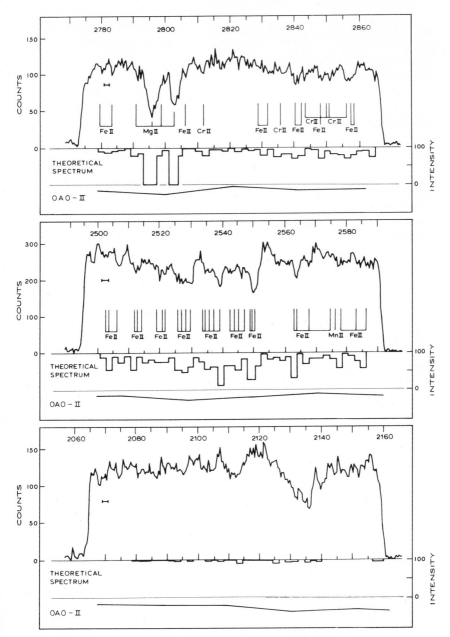

FIG. 138. Utrecht S59 spectra of Beta Aurigae, a main-sequence A2 star, in three sections between 2060A and 2860A. The dips are absorption lines due to ionized iron (Fe), magnesium (Mg), chromium (Cr), and manganese (Mn). (Astronomical Institute of Utrecht diagram.)

becomes locked on that object for 4 min. while three spectral regions are scanned simultaneously. They are each 100A wide and centered near 2100A, 2500A, and 2800A.

The field of view is rectangular, 50 by 10 arc-min., with its long dimension set at right angles to the scan direction. With the given precession of 1°/day and the satellite's orbital period of 95½ min., a star on the ecliptic may appear within the rectangle and be observed on 12 consecutive orbits. (This number increases with ecliptic latitude; Zeta Draconis, near the ecliptic pole, is observable 145 times.)

We must restrict our observations to Fo stars brighter than magnitude 2.5, Ao to 3.5, and Bo to 4.2. In rich parts of the sky, even higher threshold values can be adopted, so that important bright objects are not missed because too much time has been spent on fainter ones.

The first star we observed was Beta Aurigae, of visual magnitude 1.9 and spectral type A2V. By averaging the results of four orbits, we obtained the spectra shown in Figure 138. Thereafter the S59 equipment observed some 10 stars per week, obtaining at least 10 to 12 spectra for each one.

The strongest absorption features in the S59 spectra are found at 2795.5A and 2802.7A. By studying this Mg II (ionized magnesium) doublet, we obtain information on physical conditions, such as temperature, density, and motions in the top layers of stellar atmospheres.

A first analysis of two dozen stars has already shown that in spectral class B the lines are about three times stronger than predicted by model atmospheres. The most plausible explanation is that the XUV flux from the surfaces of these stars is considerably weaker than was expected. Hence, a large fraction of the magnesium atoms remain singly ionized, instead of becoming doubly ionized (losing two electrons) as they would if the XUV flux were as strong as expected. This is an example of how we can obtain indirect information on stellar radiation at very short wavelengths that cannot be observed directly because the interstellar hydrogen gas is opaque to wavelengths shorter than 912A.

NASA is supporting the research of foreign scientists in many of its space missions, including the Apollo series (p. 150), Skylab (p. 296), OAO 3 Copernicus (p. 347), and particularly in the International Ultraviolet Explorer Satellite (IUE), to be launched in 1977 as a joint undertaking with the British Science Council and ESA (formerly ESRO—p. 30). The small satellite will be put into geosynchronous orbit over the Atlantic Ocean, with telemetry to ground receivers in Europe and at GSFC. It will obtain spectra of stars, nebulae, and galaxies, with a 45-cm f/15 Ritchey-Chrétien telescope feeding one of two echelle spectrographs with

0.1A resolution, one from 1150A to 1950A, the other 1850-3200A. It will be possible to replace the echelle grating with a mirror to get low-resolution (6A) spectra from the cross-dispersion grating (used to separate orders in the echelle spectra) and measure stars of magnitude 15.

As this book goes to press, 150 investigators from 17 countries have asked for observing time. Meanwhile, the Soviet Union is also developing space astronomy. Their large number of low-resolution spectra matches Henize's results on Gemini flights (p. 335) and on Skylab (Table 13). —TLP

Ultraviolet Spectra of Faint Stars from Space

G. A. GURZADYAN

(*Sky and Telescope*, October 1974)

In December 1973, during the flight of the Soviet spaceship Soyuz 13, manned by Peter Klimuk and Valentin Lebedev, ultraviolet spectra of many stars were obtained with the space-observatory instrument "Orion 2."

It contained a 24-cm (9.4-in.) Cassegrain telescope of 100-cm (39.4 in.) focal length, which yielded a 5° field. In front of the telescope aperture was placed a 4° objective quartz prism. Our spectra could extend from 2000A to 5000A.

At the shortest wavelengths, the optical system transmitted 20 percent of the incident light, the dispersion of the spectra was 170A/mm, and the spectral resolution was 8A. At 3000A the corresponding numbers were 48 percent, 550A/mm, and 28A.

To obtain this good spectral resolution, the space observatory had to be held steady to ±5 arc-sec, by means of a six-wheeled gyro stabilization system guided on two stars, one of them in the center of the area being studied. Orion 2 and its stabilized platform were set up by the cosmonauts on the outside of Soyuz 13, but the guide star had to be acquired from inside by means of a special sighting mechanism.

The spectra were photographed on Kodak 103-O film, 10 cm wide, which was sensitized for ultraviolet light before use. This film was kindly supplied by NASA scientists. In Figure 139, where Capella's image is very much overexposed, even a cursory glance shows many hot stars of early spectral type with strong ultraviolet spectra in this region of the Milky Way.

FIG. 139. Objective-prism ultraviolet spectra of stars within 2°.5 of Capella (center), photographed with the Orion-2 telescope on Soyuz 13. Long "tails" (spectra) extending toward the upper left indicate hot blue stars, some extending to 2100A. Short "tadpoles" are cooler stars of types *F*, *G*, and *K*. The arrowhead at lower right points to a star of magnitude 12.6, showing that spectra of faint stars are recorded. South is up, east to the right.

Before the flight, the Orion-2 telescope had been calibrated for spectrophotometry, by means of synchrotron radiation in order to determine the energy distribution in the continuous spectra of several thousand stars. However, the quality of many spectrograms has turned out good enough for the strong absorption lines or bands to be distinguished. In the region of 2800-2150A, preliminary identifications have been made of a score of lines due to the neutral metals and their ions.

As noted in Chapter 7 (Table 13) low-resolution stellar spectra were also obtained from Skylab, primarily by Karl Henize's Northwestern

University So19 Experiment, which used a 6-in. f/3 camera with an objective prism in front of the camera, and recorded as many as 100 spectra in each 4° × 5° field. Note that OAO 2, OAO 3, TD 1A, and IUE obtain spectra one at a time, with slower but much higher resolution results.

It is to be expected that these many far-UV observations will change the astrophysicist's concepts of hot stars. Already, OAO 3 (Copernicus) has discovered huge clouds of highly ionized low-density, interstellar gas between us and hot stars of types O and B (see Astrophysical Journal letters 193, L121, 1974). Temperatures in these clouds may be 10 million degrees K or more. Individual stars are turning out to be hotter than we previously thought. Some O- and B-type stars observed by the Astronomical Netherlands Satellite, launched in August 1974, seem to have surface temperatures of 100,000°K.

The Apollo spacecraft on ASTP (see p. 32) carried a 50-lb extreme-ultraviolet (XUV) telescope designed by Berkeley astronomers led by Stuart Bowyer. Four concentric grazing-incidence mirrors coated with gold (Fig. 140) focus XUV light of 100-1000A wavelength in a 2-ft beam on a channeltron detector about 2 ft behind the mirrors. Photoelectrons from the channeltron cathode were multiplied and counted, the counts then telemetered to NASA Space Net receivers and recorded at JSC, Houston. The XUV telescope, mounted in the SM, was pointed by maneuvering the Apollo CSM during about 16 hrs of the 6 days following the docking with Soyuz. About 40 targets were observed—nearby

FIG. 140. Berkeley XUV telescope for the Apollo-Soyuz mission. The edges of four concentric rings (grazing-incidence mirrors) can be seen in the 2-ft aperture. (University of California photo.)

stars and nebulae which show low Lyman-alpha absorption in OAO spectra (p. 342), indicating less than the average amount of atomic hydrogen in the line of sight. (Atomic H absorbs XUV below 912A wavelength.) The XUV data are expected to detect 200,000°K coronas around a few of the stars (p. 347) and possibly hot flares (p. 317). Two planetary nebulae observed have very hot central stars (see Vol. 7), and four nearby pulsars observed may have very high temperatures. Jupiter was also observed, with the hope of detecting 584A emission from a very-low-density helium cloud around the giant planet. As this book goes to press, the XUV results show one strong source that coincides with a white-dwarf star called HZ43 in the constellation Coma Berenices, which may have an unusual hot corona like Pollux (p. 347). Only three or four other barely detectable XUV sources were found.

Infrared observations have been neglected in this chapter, partly for lack of space. Future developments of optical-observing techniques will be discussed in Chapter 10.—TLP

9

X-Ray and
Gamma-Ray
Astronomy

The electromagnetic spectrum extends from very long radio waves, kilometers in length, through high-frequency radio, very-high-frequency (VHF), millimeter waves, infrared, visible red to violet, ultraviolet to far-UV, and , theoretically, down to zero wavelength (infinite frequency). Because the Earth's atmosphere has only a few narrow "windows" where it is transparent (to visible light, near infrared, and high-frequency radio —shown in Fig. 1), astronomers have been hindered in their observations and were delighted with the space-age extension to the far UV and the extreme UV (300-1000A). The next shortest wavelengths, 1A to 300A, are called x-rays, associated with physicians' offices and atomic experiments early in the twentieth century. At this point in the spectrum, the index shifts, arbitrarily, from wavelength to energy because laboratory physicists used higher-energy beams to get shorter-wavelength x-rays. The voltage, V, needed in doctors' offices or physics labs to accelerate the electrons (striking a target) to produce x-rays of wavelength λ in angstroms is approximately $V = 12345/\lambda$, a formula easy to remember.

Actually, it is the kinetic energy of an electron, $(\frac{1}{2})mv^2 = eV$, which produces x-ray photons of energy $h\nu = hc/\lambda = eV$, where e is the charge on the electron, h is Planck's constant, c is the velocity of light, and ν is the frequency of the x-rays $= c/\lambda$. So the index of x-ray energy is electron-volts, proportional to 1/wavelength, and a 10A x-ray photon has energy 1234 eV or 1.234 keV (kilo-electron-volts). Farther down in wavelength—higher in energy—we come to gamma rays, well known to cosmic-ray physicists, with energies in the millions of electron-volts, or MeV—hence the name "High-Energy Astrophysics." The relative intensities of "soft" (low-energy) and "hard" x-rays from a source can be sorted out by barrier filters—strips of material such as beryllium and steel in front of the detector—which block low-energy x-rays longer than some specified wavelength.

As late as 1950, the average astronomer would have said: "No star can be hot enough to emit x-rays, and x-rays cannot get through the Earth's atmosphere, so leave them to the physics lab. Gamma rays are produced by primary cosmic rays striking the Earth's atmosphere. The primary cosmic rays come in from all directions and are not associated with astronomical bodies, so leave them to the cosmic-ray physicists." These views turned out to be incorrect; many x-ray and gamma-ray sources have by now been detected in the sky.

X-rays can be detected in several ways, one being the film used in doctors' offices. However, they cannot be focused by ordinary lenses or mirrors (x-rays are just absorbed), and special equipment had to be built to detect x-rays and gamma rays from one direction only. Early directional detectors had metal-tube collimators giving an angular resolution of about 20°, so it was difficult to identify the x-ray source with an optical object (star, nebula, or galaxy). Some of this background information is given in Volume 6, The Evolution of Stars, where it is noted that x-ray astronomy was started about 1962 by Riccardo Giacconi (AS&E, Cambridge, Massachusetts) largely by accident during rocket observations of the Moon, and by Herbert Friedman (NRL, Washington, D.C.). Giacconi discovered x-rays coming from the center of our Galaxy, and Friedman measured x-rays from the sun (in 1949), several stars, a nebula, and a galaxy (in 1963). The sources were named by the constellations where they are located, followed by X-1, X-2, etc. in order of discovery. Although the sun was found to be an x-ray source (p. 322), it would be a faint one at great distance.

Friedman scanned part of the sky with a 10°-resolution, 6keV x-ray detector on an Aerobee rocket and found Scorpius X-1 was close to no bright star or nebula, but Taurus X-1 was located in the Crab nebula, known to be a giant gas remnant of a supernova explosion in the year 1054. Theory predicts that such an explosion should leave a small, invisible neutron star, so Friedman assumed that both these x-ray sources are neutron stars. However, in 1964 he was able to watch Taurus X-1 from another rocket-borne x-ray detector while it was eclipsed by the Moon. Instead of "blinking out," as a stellar point source would, Taurus X-1 gradually declined, during 5 min., showing that the x-rays come from a region at least 2 arc-min. across.

On the other hand, Scorpius X-1 was near a faint star (magnitude 12.5), which flared in optical brightness at irregular intervals that match variations in the x-ray intensity. In 1967 and 1968, simultaneous measures of x-ray and visual brightness roughly checked the new theory that both were produced by "bremsstrahlung" from an electron cloud around a star about 1/100 the size and 1/10 the mass of our sun.

Bremsstrahlung (literally "brake-radiation" in German) simply means the radiation produced when high-speed electrons hit a target—in this case by falling under gravity into a small super-dense star (a white dwarf, neutron star, pulsar, or black hole—see p. 366, rather than being shot by electrical force as in a doctor's office. Two other mechanisms for the production of x-rays in space, other than in super-hot stars, are "synchrotron radiation" and "inverse Compton effect." The first comes from high-energy electrons passing through a magnetic field which makes them move in helical paths following the magnetic lines of force; these oscillations of the electrons produce polarized x-rays, the polarization depending on angle between the line of sight and the magnetic field direction. The inverse Compton effect, named after Arthur Compton, an American physicist, is an interaction between electrons, atoms, and radiation, whereby electron energy is added to radiation (photon) energy, giving it a shorter wavelength (higher photon energy).

These and other mechanisms can account for the high-energy gamma rays coming from some regions of space.—TLP

Polarized X-Rays from the Crab Nebula

(Sky and Telescope, July 1972)

The expanding Crab nebula is the debris from the explosion of a supernova in A.D. 1054, which also left the pulsar NP 0532.

About 20 years ago it was discovered that both visible light and radio emission from the Crab nebula are partially polarized. This indicates synchrotron radiation, arising from the spiraling of fast-moving electrons around magnetic field lines within the nebula.

Then, in 1963, the Crab nebula was discovered to be a copious source of x-rays (Tau X-1), possibly also of synchrotron origin.

Such an origin is to be expected if the energy radiated by the Crab nebula is provided by the pulsar embedded in it. Such a rotating neutron star slows down by imparting some of its energy to nearby electrons via its rotating magnetic field. As these high-speed electrons move outward into the general magnetic fields of the nebula, synchrotron radiation is produced, some of it in x-rays which should be plane-polarized.

This has now been confirmed by the detection of polarization in Tau X-1. A team of Columbia University physicists, headed by Robert Novick, flew two x-ray polarimeters aboard an Aerobee-350 rocket to an altitude of 100 mi. on Feb. 22, 1971. One unit measured x-rays with energies between 2.0 and 3.2 keV, the other between 7.0 and 17.0. Both were mounted to count x-rays as the payload, pointing toward the central part of the nebula, slowly rotated around the line of sight.

The results indicate that about 15 percent of the x-ray flux is polarized, the plane of polarization being in position angle 156°—very nearly the same as the 14 percent and 154° found for visible light. This observation has an important bearing on the origin of cosmic rays. If electrons can be accelerated to energies high enough to produce synchrotron radiation in the Crab nebula's magnetic fields, atomic nuclei should be accelerated also. These heavier particles would not interact strongly enough with magnetic fields to cause synchrotron radiation, but would retain most of their energy and escape from the nebula as cosmic rays. Thus, it is possible that cosmic radiation originates within supernova remnants like the Crab nebula.

Gamma-Ray Astronomy

CARL E. FICHTEL

(*Sky and Telescope*, February 1968)

Very-high-energy astronomy has great scientific interest because it can detect many of the major energy transfers occurring in the universe. Thus, it may help answer questions about stars, interstellar matter, galactic magnetic fields, cosmic rays, and their interactions.

Within our Galaxy and in intergalactic space, cosmic rays interact with interstellar matter to form unstable pi mesons (particles of weight between electron mass and proton mass). Approximately one-third of these are neutral pi mesons, which decay into high-energy gamma rays almost at once. The expected intensity of these photons can be calculated from the cosmic-ray flux near the Earth and the interstellar gas density.

Measuring the gamma-ray intensity in different parts of the sky will help to determine whether cosmic rays pervade intergalactic space, or are primarily restricted to our Galaxy.

In addition to gamma rays expected from the decay of neutral pi mesons in our Galaxy, they may result from the interaction of cosmic-ray electrons with galactic magnetic fields (synchrotron radiation) and with interstellar matter (bremsstrahlung). Pi-meson decay is expected to be dominant over the energy range 1-10,000 MeV.

Discrete sources of gamma rays may also exist, produced by the same mechanisms that produce x-rays: bremsstrahlung, the inverse Compton effect, and synchrotron radiation (p. 357). Gamma rays are also produced by nuclear interactions and in association with the shock wave of a supernova explosion.

Finally, there are gamma rays that result from the annihilation of a particle when it interacts with its antiparticle—100-MeV photons from proton-antiproton annihilation, and 0.5-MeV photons from electron-positron annihilation (p. 319).

Gamma rays are remarkably unaffected by the material along their paths; the Galaxy is essentially an open window up to energies of 10^5 MeV. At higher energies, a gamma ray has a high probability of interacting with a starlight photon to produce an electron-positron pair. This effect is important until almost 10^7 MeV, where again a window is open until 10^8 MeV. At this point, the 3.5°K blackbody photons (see Vol. 8) begin to be an important absorber.

The detecting systems used in high-energy astrophysics generally resemble particle detectors rather than optical devices. In the energy range below about 10 MeV, the detector system normally employs a scintillation counter, which absorbs the photons and permits a measure of their energy. Such a counter is surrounded by another detector, which serves as an active anticoincidence shield to discriminate against charged particles (cosmic rays). This shield can tell if a charged particle has passed through it, in which case the scintillator has not recorded a gamma-ray pulse, and the pulse is not counted.

The detector system shown in Figure 141 was flown on Explorer 11. The top sandwich of crystal scintillators acts both as a converter of gamma rays and as part of the telescope, defining the direction of the incident radiation to be measured. The other half of the telescope is a detector which views a piece of lucite in which Cerenkov light is formed either by the electron or positron (or both) of a particle pair

FIG. 141. Diagram of the gamma-ray detector on Explorer 11. The middle of the upper three photomultipliers counts flashes in the sandwich crystal if they are coincident with Cerenkov radiation detected by the photomultipliers below and not coincident with flashes produced by particles in the surrounding plastic.

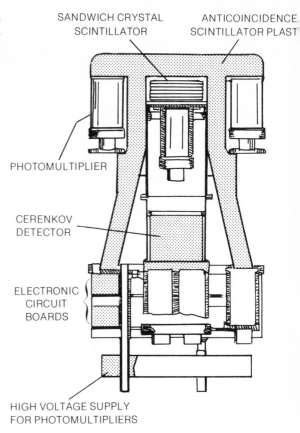

SANDWICH CRYSTAL SCINTILLATOR

ANTICOINCIDENCE SCINTILLATOR PLAST

PHOTOMULTIPLIER

CERENKOV DETECTOR

ELECTRONIC CIRCUIT BOARDS

HIGH VOLTAGE SUPPLY FOR PHOTOMULTIPLIERS

produced in the converter.[1] This detector is in coincidence with the scintillation sandwich, and is surrounded by a large, very efficient anti-coincidence dome to reject counts of charged particles.

Several investigators have turned to the *spark chamber* as a gamma-ray detector. This consists of a series of parallel plates in a gas. A high voltage difference may be applied to pairs of these plates. When a particle traverses the chamber, it leaves an ionized path in the gas, along which a spark may occur if a high-voltage pulse is applied quickly. The spark chamber has the advantage of providing a picture, allowing the investigator to distinguish between an electron-positron event and other events produced by neutral particles. Three distinct types of spark chambers are currently being considered. The *optical chamber* records on film two perpendicular, side views of the spark, providing a three-

[1] Cerenkov light is emitted by a charged particle traveling through a dispersive medium at a velocity greater than that of light in that medium. For detecting an electron or positron produced by a gamma ray, the medium may be lucite, lead glass, or a gas.

dimensional record. In a somewhat different technique, the *sonic chamber* uses microphones to determine the position of the spark with the aid of accurate timing signals. The *magnetic-core digitized spark chamber* has been developed at GSFC. Instead of two flat plates, it has paired gratings of parallel wires. In each pair the wire directions are perpendicular. A spark between the grids causes a current to flow from a wire in one to a wire in the other. With the aid of a magnetic core on the end of each wire, we can find which pair of wires is involved, hence the rectangular coordinates of the discharge. Thus, as the electron and positron traverse a stack of wire gratings, their three-dimensional paths are determined.

Progress in Gamma-Ray Astronomy

(*Sky and Telescope*, May 1968)

During the past decade many investigators have tried unsuccessfully to detect celestial radiation of wavelength shorter than x-rays. The dividing line between gamma rays and x-rays is taken as a photon energy of either 100 keV or 1 MeV, corresponding to a wavelength of 0.124A or 0.0124A. Because the strange nebula in Taurus is an intense x-ray source, it was a logical place to look for even shorter-wave radiations. Definite measurement of gamma rays from the Crab nebula has now been announced by R. C. Haymes and his co-workers at Rice University. Their scintillation-type detector was carried by balloon to 38 km above the Earth's surface on June 4, 1967, in an ascent from Palestine, Texas.

They obtained flux measurements in three bands: 0.36-0.124A, 0.124-0.049A, and 0.049-0.022A (0.035-0.1 MeV, 0.1-0.25 MeV, and 0.25-0.55 MeV). Counting existing radio, optical, and x-ray measures of the same source, its electromagnetic radiation has been recorded (with some gaps) over a range in frequencies of 10^{12} to 1! The Crab nebula has a power output of about 7×10^{37} ergs/sec.

In another balloon experiment with similar equipment on Aug. 29, 1967, Haymes' team succeeded in detecting gamma radiation from the constellation Cygnus, near right ascension 20^h 06^m, declination $+37°$. Since the uncertainty in each coordinate is about $\pm 3°$, the gamma-ray source may well be identical with the strong x-ray source Cygnus X-1, at 19^h 57^m, $+35°.1$.

Report from Rome:
X-Rays and Gamma Rays

GEORGE S. MUMFORD

(*Sky and Telescope*, August 1969)

Early in May 1969, some 170 astronomers, physicists, and other interested persons gathered in Rome, to discuss nonsolar x-ray and gamma-ray astronomy. Prof. L. Gratton of Rome University and the Dudley Observatory in New York was host to Symposium 37 of the International Astronomical Union. The participants came from many parts of the world, with large contingents representing Great Britain, India, Italy, Japan, and the United States.

Early searches for 100-MeV gamma rays were prompted by the prediction that they should originate from the interaction of cosmic rays with the interstellar medium.

On the other hand, early observations revealed the discrete x-ray sources and a diffuse x-ray background, both considerably stronger than had been expected. As a result, x-ray astronomy has evolved with remarkable speed. It is now well ahead of gamma-ray astronomy.

Almost all of the nearly 50 known discrete x-ray sources lie close to the equator of the Milky Way. The only accredited extragalactic x-ray source is the jet of Messier 87 in Virgo, which presumably is the result of an explosion in that galaxy. Other extragalactic objects that have been surveyed for x-rays by Friedman's group at NRL include the radio source Cygnus A, with negative results; the optical galaxy (NGC 5128) associated with Centaurus A (a radio source), with positive results; and the quasistellar object 3C-273, with no clearcut evidence.

However, at least three different groups reported additional sources at high galactic latitudes (see Glossary) in Cetus, Crater, and Virgo.

At least three x-ray sources are associated with supernova remnants (see Glossary)—the Crab nebula, Cassiopeia A, and Tycho's supernova of 1572.

A number of other remnants of old supernovae have been scanned for x-ray emission with negative results. It was suggested by R. Giacconi of AS&E that relatively youthful supernova remnants are x-ray emitters while older ones are not.

Two well-observed sources turn out to be entirely different. In the Crab nebula (Tau X-1), emission originates in an extended region.

The x-ray spectrum is "hard," that is, strong at short wavelengths, and the flux is constant with time. Scorpius X-1, however, appears pointlike, its spectrum is soft, and the x-rays undergo large intensity fluctuations.

Cygnus X-2 is fairly well identified with a bluish optical object, and Vela X-1 with a 17th-magnitude star 3 arc-sec away from one of 15th magnitude—possibly a double star.

In reviewing the optical properties of discrete x-ray sources, H. M. Johnson of Lockheed Research Laboratory said there is no evidence that any one of them is a very close binary star, but R. E. Wilson, of the University of South Florida, believes that only a binary system can explain the peculiar spectral features of Cygnus X-2.

Flare activity has been observed in many x-ray sources, the most striking example being the complete disappearance and subsequent reappearance of Centaurus X-2 shortly after its discovery. A flare of Scorpius X-1, described by L. E. Peterson, University of California at San Diego, was simultaneously observed by Johnson at optical wavelengths, suggesting that a rather uniform increase in energy occurred at all wavelengths.

It is clear from these reports that Western scientists got an early lead over the Soviet Union in high-energy astrophysics. We know that the Russians carried x-ray detectors in their orbiting laboratory, Salyut 4, in 1975—their press announced the confirmation of x-ray "flares" from Sco X-1—but no detailed data were released. This omission, and the lack of other announced Soviet missions devoted to x-ray and gamma-ray observations, probably indicates their backwardness in the necessary electronics technology described in this chapter.—TLP

X-Ray Results
from "Uhuru"

(*Sky and Telescope*, June 1971)

Significant observations have already been obtained by Explorer 42, the first artificial satellite entirely devoted to x-ray astronomy. It was launched on Dec. 12, 1970, by Italian scientists from the San Marco platform off the coast of Kenya.

Since that date was the seventh anniversary of Kenyan independence, this satellite was named *Uhuru*, "freedom" in Swahili. It is traveling in a nearly circular orbit about 335 mi. high, inclined 3° to the Earth's equator, with a period of 96 min.

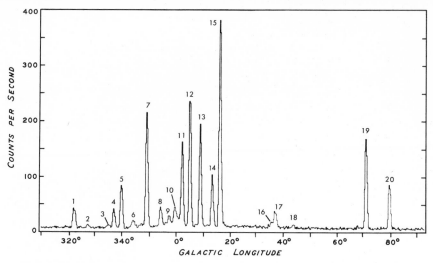

FIG. 142. Several scans along the galactic equator (center of the Milky Way) averaged to give x-ray "counts" per second (intensity) *vs* galactic longitude (angle eastward along the equator from the direction toward the center of the Galaxy). Each peak is a source; numbers 4, 10, 16, and 18 are newly discovered ones. (*Astrophysical Journal* diagram.)

Explorer 42 has two back-to-back x-ray telescopes developed at AS&E in Cambridge, Massachusetts, which is also responsible for analyzing the data telemetered from them. Riccardo Giacconi and seven collaborators presented preliminary findings in five papers at the American Astronomical Society meeting in Baton Rouge.

X-ray source survey. As the satellite spins on its axis every 12 min., its detectors sweep around the sky, recording photons with energies between 2.0 and 20.0 keV. By reorienting the spin axis, a new strip of sky can be observed, permitting an all-sky survey which was expected to be finished in May.

In 3 hrs of observing time, 29 discrete sources were detected, seven of them previously unreported. These results indicate that many more galactic x-ray sources remain to be discovered by Uhuru.

Positions of x-ray sources. Another purpose of Explorer 42 is to obtain accurate positions of x-ray sources to permit identifications with known optical or radio sources. Reliable identifications of this kind in our Milky Way are still very few.

For each occasion that a detector beam sweeps across an x-ray source, the maximum counting rate places the source somewhere within a narrow strip of sky that is perpendicular to the sweep direction. On a later orbit, when the same source is observed with a different spin-axis orientation, a second narrow strip is determined. The source position must lie within the "box" where the two bands intersect, usually about 25 sq. arc-min.

Extragalactic x-ray sources. The typical Seyfert galaxy is distinguished by a small, very bright nucleus and an optical spectrum containing strong, broad emission lines. Such galaxies are very bright at infrared wavelengths, and some of them are strong nonthermal radio sources (including NGC 1275, which is also known as Perseus A). The nonthermal emission led astronomers to anticipate that Seyfert galaxies might also be x-ray sources.

X-rays from NGC 1275 were first observed by NRL scientists with instruments on an Aerobee rocket launched on Feb. 28, 1970, from White Sands Missile Range.

Meanwhile, the Uhuru satellite measured the x-ray flux from NGC 1275 and from another Seyfert galaxy, NGC 4151.

The observed strength of NGC 1275 is surprising, exceeding many previously known sources. The intrinsic x-ray luminosity amounts to 2.4×10^{44} ergs/sec, which is comparable to that galaxy's total optical emission. For NGC 4151, the x-ray luminosity is about 200 times less. Thus, it is quite possible that in these two galaxies the x-rays are produced by different processes. No significant x-radiation was detected from NGC 1068 or from 16 other Seyfert galaxies.

Cygnus X-1. Perhaps the most spectacular discovery from Uhuru data is the very rapid variation of the x-ray emission from Cygnus X-1, a source in our Milky Way which has been known since 1966.

It takes a source about 20 sec. to pass through the wider Uhuru detector beam, the counts rising steadily to a peak and then falling. Unlike other strong sources, Cygnus X-1 shows a jagged rise and fall. Within a few seconds, changes amount to 25 percent of the average source intensity. No previously known x-ray source had ever displayed such large and rapid intensity oscillations.

On March 6, observations were made with a shorter counting interval, 0.096 sec, and again rapid pulses were observed, as well as slower changes in average intensity.

This led to the suspicion that the Cyg X-1 period might be shorter than the sampling interval of the detector, and a re-examination of the records indicated a period of 0.073 sec, with an amplitude of about 20 percent. On the assumption that the source's distance is 1000 parsecs, its average x-ray luminosity is about 3×10^{36} ergs/sec, summed over the energy range 0.5-100 keV.

While Cygnus X-1 can be regarded as the second known x-ray pulsar, it differs in several important respects from the Crab-nebula pulsar, NP 0532. For one thing, it is not a radio pulsar, nor is it detectable as a radio source. No supernova remnant is observable in its vicinity, and its very short period would imply an age of only 10^4 years—too little time for the debris of a supernova explosion to dissipate entirely.

The AS&E investigators note that the very short but persistent pulse period can be reasonably interpreted only by the rapid rotation of a collapsed star. This, they suggest, might be a "black hole" instead of a neutron star (see Glossary).

Black holes are objects of very high mass density, probably caused by the collapse of a star or large cloud of dust and gas pulled together by its own gravitational force. Einstein's General Relativity Theory explains the observed bending of a light ray as it passes close to the sun by a postulated "curvature of space-time" near any massive object, and a mathematical solution of the equations by Karl Schwarzchild in 1916 showed a "discontinuity" at distance R_s from a mass M, where $R_s = 2\,GM/c^2$; G is Newton's gravitational constant and c is the velocity of light.

The simplest explanation of this highly complex mathematical derivation is that when there is a very large mass within the Schwarzschild radius, R_s, a light photon cannot escape from M, as can be demonstrated with the much simpler laws of Newton (p. 17). The velocity of escape for a space probe (p. 12) is v_e, where $(\tfrac{1}{2})mv_e^2 = GmM/R$ and the space-probe mass, m, cancels out. Now, replace the space probe with a photon moving at velocity c; $(\tfrac{1}{2})c^2 = GM/R$, or $R = 2GM/c^2 = R_s$, or if M is made larger, the photon cannot escape, and from outside we would see no light from the black hole.

Of course, this derivation is over-simplified—the photon is not decelerated like a rocket—but distance and directions in Einstein's curved space-time are such that light cannot get past R_s.

Most recently, in 1975, D. H. Menzel at Harvard College Observatory has found other mathematical solutions than Schwarzschild's which eliminate the discontinuity at R_s; he predicts that photons can always escape from large M no matter how small R is. But the strong gravitational field has another effect—the so-called "gravitational red shift" which increases the wavelength or decreases the energy (p. 355) of the escaping photon. Hence, a photon of visible light from a collapsed star ends up as a far-infrared photon when it reaches us. A gamma-ray photon may come out as an x-ray if M (the "black-hole" mass) is large enough. Menzel's formula is

$$V_e = V \exp\left[-GM/Rc^2\right]$$

$$\text{or,}\quad \lambda_e = \lambda \exp\left[GM/Rc^2\right]$$

where V_e is the escaping photon energy (p. 000), λ_e is the escaping photon wavelength, V and λ are the original values at R distance from M, and "exp" stands for e raised to the power given by the ratio in brackets. Thus, at the Schwarzschild radius, $GM/R_sc^2 = 1$, and $V_e = V_e^{-1} = V/2.718$, $\lambda_e = 2.718\lambda$.

It was clear that higher resolution (for accurate positions) and greater detector sensitivity (for faint sources) are needed for x-ray astronomy. The detectors on Uhuru were proportional counters—cells filled with argon gas that conducts an electric current momentarily between charged electrodes when the argon absorbs an x-ray. Each such electric pulse is a "count." The larger the cell, the more counts are recorded from a beam of x-rays. Collimators are metal tubes fitted together in a "honeycomb" in front of the detector so that the only x-rays that can reach the cell come through the tubes. One of the Uhuru telescopes had a $5° \times 5°$ collimator, the other had a $\frac{1}{2}° \times 5°$ collimator, giving higher resolution but fewer counts.

What is needed is a true x-ray telescope that concentrates x-rays from a source into a small image.—TLP

Giant X-Ray Telescope

(*Sky and Telescope*, May 1969)

Sometime next year a giant balloon will rise to a height of 26½ mi. over the Southwest and float there for about half a day while a large, highly sensitive x-ray telescope of unique design makes astronomical observations.

Developed at Boeing Scientific Research Laboratories by R. Graham Bingham and Farrel W. Lytle, the instrument will measure x-rays of very high energy: 18-100 keV, corresponding to wavelengths of only 0.69A to 0.124A.

The "lens" of the instrument is pictured in Figure 143. The nested set of 19 concentric rings is 10 in. deep, with the outermost ring 50 in. in diameter, the smallest 10. But the operating principle is *not* that of the grazing-incidence total reflection of the x-rays.

Instead, the inside surface of each curved ring is covered with thousands of small thin crystals of lithium fluoride. These are machined to concentrate (by means of symmetric Bragg diffraction—the deflection of x-rays by crystals) a parallel beam of x-rays onto a 1-in. square of thallium-doped sodium iodide or lithium-doped germanium at the focus. This detector will be shielded by an anticoincidence counter (see p. 359) and so collimated that only photons diffracted from the lens will fall on its sensitive surface.

Each 10-in. ring is an annulus of a paraboloid. The curves were computed so that first-order diffraction will concentrate on the detector

FIG. 143. Boeing's 50-in. x-ray "lens," a set of 19 concentric rings, the inner side of each covered with accurately ground crystals of lithium fluoride which diffract incoming x-rays to a focus behind the "lens." (Boeing Laboratories photo.)

x-rays with energies between 18 keV and 87 keV. Second-order diffraction extends the useful range to 100 keV.

The system's diffraction efficiency is high, experimentally 30-50 percent over the entire energy range. The field of view is estimated at 0°.5.

About 110 lbs of LiF scrap were used to produce 150,000 individual crystals. They were affixed to the rings with epoxy cement and then machined. Each is 1/40-in. thick and half the size of a postage stamp.

In the 20-40 keV range, the system will detect x-ray fluxes as little as 1/500 that of the Crab nebula (Tau X-1)—much greater sensitivity than previously available.

Unfortunately, this giant instrument was never flown. The U.S. Air Force funds were cut off as a result of the "Mansfield Amendment," which prohibits Federal funds from being spent on "non-mission-oriented" projects. Since the Air Force and the Boeing Company are primarily concerned with other matters, this x-ray investigation was dropped.—TLP

Letter to the Editor

THOMAS R. LINDQUIST AND WILLIAM R. WEBBER

(*Sky and Telescope*, July 1969)

Readers of "Giant X-Ray Telescope" may be interested in the fact that a telescope of similar type has already been successfully flown by our group at the University of Minnesota.

Our lens consists of approximately 4000 rock-salt crystals mounted on a paraboloidal frame 3 ft in diameter. We chose Laue geometry (in which x-rays are reflected on passing all the way through a crystal) over Bragg geometry (first-surface reflection), as but one paraboloidal surface is needed. X-rays with energies of between 20 keV and 140 keV are focused on a sodium-iodide scintillator 1¼ in. in diameter. The telescope's angular resolution is approximately 1°.

This instrument performed well when flown from Palestine, Texas, in April 1968, and we are now working on data analysis.

X-Rays from Another Galaxy

GEORGE S. MUMFORD

(*Sky and Telescope*, May 1969)

Five scientists at the Lawrence Radiation Laboratory, University of California, report that they have detected x-rays from the Large Magellanic Cloud. Their observation was made with counters aboard a rocket flown to 208 mi. altitude above Johnson Atoll in the central Pacific Ocean, on Oct. 29, 1968.

This rocket was oriented so that one set of detectors scanned the far southern region that included the Magellanic Clouds while another set, mounted back to back with the first, scanned the northern Milky Way from Cygnus to Cassiopeia. This operation continued for 4 min. as the rocket made a slow roll. Then the scan plane was tipped 45° for another 1½ min. of observing. This procedure eliminated any ambiguity in the recorded positions of x-ray sources.

In the northern scans, three previously known sources were found: Cyg X-1, Cyg X-2, and perhaps Cassiopeia A. In the south, the Large

Magellanic Cloud was detected as a source about 12° in extent. The measured flux indicates that in the wavelength range of 1.2A to 8.3A (10-1.5 keV) the Large Cloud emits energy at the rate of 4×10^{38} ergs/sec. This corresponds to 25 or 30 x-ray objects having the same source intensity as the Crab nebula.

In the same wavelength range, the x-ray luminosity of our Galaxy is about 7×10^{39} ergs per second, according to H. Friedman and his associates at NRL—about 17 times that for the Large Magellanic Cloud. Since the Milky Way system has about 10 times the mass of the Large Cloud, these two galaxies have similar proportions of x-ray sources.

X-Rays from Rich Clusters of Galaxies

(Sky and Telescope, August 1972)

Clusters of galaxies can have more than a thousand members, and it now appears probable that all of the richest ones also contain extensive x-ray emission regions. Since last year, the Virgo, Coma Berenices, and Perseus clusters have been known to emit x-rays from an area about 0°.5 across in each case. At the adopted distances to these aggregations, these angular diameters correspond to linear dimensions between 800,000 and 3,000,000 light years.

From an examination of a catalogue of 125 x-ray sources detected by the Uhuru satellite, a team from AS&E and ESRO has identified more clusters with intense, widespread x-ray emission. These are numbers 401, 1367, and 2256 in the Abell Catalogue, a dense cluster in Centaurus, and galaxies associated with the radio sources 3C-129 and Cygnus A. The possibility of chance coincidence is very small.

In the Virgo and Perseus clusters, the centroids of x-radiation are close to the galaxies M87 and NGC 1275, which are powerful sources of nonthermal radio radiation. Furthermore, many single galaxies in Coma and Perseus have tails of radio emission pointing outward from the cluster centers, while the clusters themselves have associated halo-shaped radio-emission regions.

The average x-ray luminosity of the five richest clusters is 4×10^{44} ergs/sec, about 100,000 times that of the Milky Way Galaxy. If this intensity is typical of the very richest clusters, then more distant ones are probably below the detection threshold of Uhuru. No x-radiation is detected in sparse clusters, indicating that the emission is not proportional to the number of galaxies in a cluster.

New X-Ray Source

(*Sky and Telescope*, December 1969)

Two University of Wisconsin scientists, A. N. Bunner and T. M. Palmieri, have announced a previously unknown source of x-rays in the southern constellation Ara. This is one of the results from the flight of an instrumented Aerobee rocket on Sept. 21, 1968, from White Sands Missile Range in New Mexico.

In all, three x-ray sources were detected, two of them the previously known Scorpius X-2 and Ara X-1. The third and strongest, at 17^h 16^m, $-48°$ $42'$, is new. To account for it having been missed previously, the Wisconsin investigators suggest that its intensity may change with time.

The x-ray spectrum of the new source is described as "unusually soft."

Several more cases of variable x-ray sources have been discovered. These, and the very powerful sources in rich clusters of galaxies, disparate though they are, have recently been linked by the cosmological theorists' "white-hole" explanation (p. 385).

"Soft" x-rays, of wavelengths longer than 15A or energies less than 800 eV, are more difficult to detect. The proportional-counter window must be thin, and gas mixtures such as neon-methane are used.

By 1973 it was clear that there are many x-ray sources in our Milky Way Galaxy. It is shown in Volume 7, Stars and Clouds of the Milky Way, that stars, gas, and dust are spread along spiral arms in our Galaxy, which is very like a spiral of type Sb seen from far outside.—TLP

The Distribution of X-Ray Sources in Our Galaxy

FREDERICK D. SEWARD

(*Sky and Telescope*, April 1973)

The spiral structure of the Milky Way system is not obvious. Since we are located almost exactly in the galactic plane, close to or in a spiral arm, we must look through a large amount of material to see distant objects. The interstellar material—small particles of dust—have long made things difficult for optical astronomers. The average attenuation of visible light in the plane of the Milky Way is about one magnitude for each kiloparsec of distance (App. V). Only nearby objects can be seen clearly, and it is very difficult to determine the structure of our Galaxy using optical techniques.

The best insight into the structure of the Milky Way came from measurements of the 21-cm radiation of neutral hydrogen. The distribution of this gas has been mapped throughout the Galaxy, revealing the over-all spiral pattern. (See Vol. 7, *Stars and Clouds of the Milky Way*.)

With the discovery of x-ray sources in 1962, a second window was opened through which objects on the other side of the Galaxy might be observed. The interstellar dust is transparent to x-rays, although the gas in the hydrogen spiral arms absorbs low-energy or soft x-rays. Assuming about one hydrogen atom/cm³ in the centers of the spiral arms, calculations show that the Galaxy is transparent to photons with energies above about 3 keV (less than wavelength 4A).

The entire thickness of the interstellar gas, from galactic center to rim, is about equivalent to a sheet of paper in x-ray stopping power. At photon energies below 3 keV the gas starts to absorb appreciably, and below 0.5 keV we can see only out to about 1 kiloparsec distance in the galactic plane. (By way of comparison, a dental x-ray machine produces photons of about 100 keV—much more penetrating than the ones to be discussed here.)

An all-sky survey with Uhuru (p. 363) is 90 percent complete and has located 125 x-ray sources (Fig. 144). About two-thirds are probably located within the Milky Way system, and some of these may even be on the other side of the galactic center.

Although the Uhuru instruments can accurately determine the positions of these sources on the sky, no information is obtained about their distances. Here is where soft x-ray observations, extending from 3 keV down to 0.2 keV, can be most useful. At these low energies, x-ray absorption by the interstellar gas can be used to measure the amount of gas in the line of sight, on the basis of which distances of some objects can be calculated.

X-ray data were obtained by two sounding-rockets from the rocket range at Kauai, Hawaii. One rocket was launched in May 1970 and the other in May 1971.

The plane of the Galaxy was surveyed from Vela in the south, through the galactic center in Sagittarius, and on to Cassiopeia. X-rays were detected with a large-area proportional counter system recording energies from 0.2 to 2 keV with a field of view 1° wide and 30° high.

The rocket was oriented so that the long dimension was centered on and perpendicular to the galactic equator and scanned at the rate of

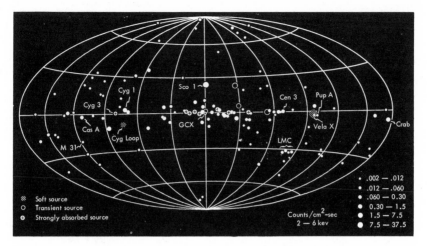

FIG. 144. Uhuru map of x-ray sources plotted on galactic coordinates with a 30° grid (north galactic pole at top, 0° longitude at center). Size of each symbol corresponds to source intensity in 2-6 keV x-rays. Well-known sources are labeled: "GCX" = galactic-center x-ray source, "LMC" = Large Magellanic Cloud, "Sco 1" = Sco X-1, etc. "Soft" sources emit mostly low-energy x-rays; "transient" = variable; an "absorbed source" has almost no soft x-ray component.

1°/sec along the length of the Milky Way. When a source passed through the field of view, the detector-counting rate would rise and fall and was recorded as a time plot.

A camera on board the rocket was recovered after the flight, and its star-field pictures were used to determine where the detector had looked at different times. Thus, the positions of the observed sources could be fairly accurately determined; they agreed quite well with Uhuru's.

Data from the two flights provide two continuous records, one for photons between 0.5 and 1.0 keV, which shows strong sources out to several kiloparsecs, and one for 0.3-0.5 keV, showing only nearby objects, less than about 1 kiloparsec (3260 light years) away.

The Cygnus Loop nebula and Vela X dominate the sky at 0.3-0.5 keV. They are big sources—the Cygnus Loop x-ray source is about 2° in diameter, Vela X about 5°.

Both of these sources are old supernova remnants (see Glossary), about 0.8 and 0.4 kiloparsecs distant, and since they were not detected by Uhuru at high energies, they are appropriately classified as soft x-ray sources. Another supernova remnant, Cassiopeia A, is just visible in the 0.5-1 keV energy range. It is not seen at lower energies because it is about 3.5 kiloparsecs from the Earth, and the weaker photons are absorbed.

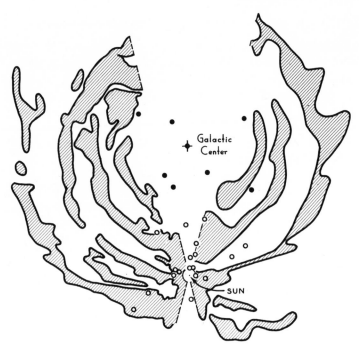

FIG. 145. X-ray sources plotted on a radio map of hydrogen spiral arms (shaded) in the Milky Way Galaxy. Our position ("sun") is indicated; dots indicate strong sources, circles weaker ones (in total x-ray energy radiated).

In this way, we have been able to assign distances to the two dozen brightest sources. They all lie close to the plane of the Milky Way, and their distribution in that plane is shown on Figure 145, where two types of x-ray sources are indicated.

First, there is the group of strong emitters clustered loosely around the galactic center. These are very luminous, radiating energy at a rate of about 10^{38} ergs/sec (25,000 times the sun). So strong are these sources that all of them in our Galaxy have been detected.

(Four other strong x-ray sources have been found by Uhuru in the Large Magellanic Cloud. Since the distance to the Large Cloud is known, the x-ray luminosities of these sources can be accurately calculated. They have about the same strength as one of the strong sources in our Galaxy; probably they are the same type of object.)

The second class consists of weaker sources, with luminosities between 10^{35} and 10^{37} ergs/sec. We detected these only to a distance of 4-5 kiloparsecs, and there are probably more, extending all the way around the rim of the galactic disk.

Peculiar behavior of Cyg X-1 (plotted on Fig. 144—see also p. 365) was observed by Harvard and Utrecht astronomers using the Astronomical Netherlands Satellite (ANS) to record soft x-rays of 1.5-7 keV. They found that the 1 to 2 keV x-rays increased suddenly by a factor of 10. At the same time, the National Radio Astronomy Observatory noted that radio emission from Cyg X-1 in the 3.7 to 11-cm wavelength band increased by a factor of 3. No explanation has yet been given. They also found that the bright star Sirius is a soft x-ray source, possibly evidence of a million-degree corona around that A1V star.

Another class of irregularly variable x-ray sources was recognized in 1975 by George Clark and co-workers at MIT using x-ray detectors on OSO 7 (p. 319). They identified five of the Uhuru sources with globular star clusters NGC 1851, 6440, 6441, 6624, and 7078, whose x-ray luminosities change by factors of 2 to 5 in short intervals (hours to weeks). These cluster sources are all nearby (within 50,000 light years).

Herbert Friedman and co-workers at NRL had a soft x-ray experiment on Apollo-Soyuz (p. 35) counting 0.2-keV (61A) photons. Unfortunately, the proportional counters (detectors) began to arc in a continuous discharge (no counts) two minutes after being turned on. However, the scientists got the astronauts to turn the counters on and off every two minutes, and thus obtained soft x-ray measurements of many of the Uhuru x-ray sources, which will add to Seward's distance measurements. They also discovered that a source in the Small Magellanic Cloud, SMC X-1, is a strong pulsar with period 0.715 sec.—TLP

AAS 75th Anniversary Meeting

THORNTON PAGE

(*Sky and Telescope*, November 1974)

In a paper that aroused much interest, C. T. Bolton, University of Toronto, summarized x-ray and optical observations of five binary systems where the unseen secondary is possibly a super-condensed black hole (p. 366). The problem is to interpret x-ray eclipses and optical Doppler shifts to estimate the secondary's mass and size. The x-ray source SMC X-1 in the Small Magellanic Cloud has a mass 2.5 times that of the sun, he says, and may be a black hole. Cygnus X-1 is a likely case, and 2U 0352 + 30, if it is really a comparison of X Persei, would have a mass of 70 suns and very probably be a black hole.

Edwin Kellogg of Harvard, who received the Newton L. Pierce prize, spoke on "X-ray Astronomy in the Uhuru Epoch and Beyond," discussing theoretical studies of Cygnus X-1 that show it can be neither a triple star nor an ordinary binary with a hot spot on the primary or secondary. The only model that fits consists of a distorted primary from which matter flows into a black-hole secondary, having a mass of 10 to 15 suns. Hydrogen and ionized-helium emission lines in the optical spectrum come from the mass-flow bridge, and x-rays from the gases disappearing into the black hole.

He started by surveying the remarkable success of the Uhuru satellite in locating 161 x-ray sources, of which 29 are identified with stars, 5 with globular clusters of stars (see Vol. 7), and 5 with supernova remnants in the Milky Way. Others are associated with galaxies and clusters of galaxies.

For instance, the Perseus cluster around NGC 1275 is an x-ray source $1°$ in diameter. This may be interpreted as a gas sphere at $84,000,000°K$ throughout, of mass 4×10^{14} suns (four times the combined mass of all the galaxies in the cluster), which would take 10 billion yrs to cool off. This model fits radio observations of NGC 1275, if that galaxy is moving through a tenuous intergalactic medium containing about 200 hydrogen atoms/m^3. It is also a strong radio source—Per A—and its x-ray emission is estimated to be 10^{44} erg/sec, more than five times the whole Virgo cluster of galaxies.

Friedman's estimate, ten years earlier, that our Galaxy is a strong x-ray source, has been partly confirmed. Unfortunately, NASA was not given enough money to finance the High Energy Astrophysical Observatory (HEAO) Program originally planned for more accurate and more sensitive detectors to be flown in 1975. HEAO was postponed to 1978, when its 1.2-m. x-ray telescope will greatly improve x-ray resolution and help identify the 122 unidentified x-ray sources found by Uhuru and earlier missions.

The various connections between x-ray astronomy and optical astronomy were reviewed in some detail by Harvard astronomers.—TLP

X-Ray Sources and Their Optical Counterparts

CHRISTINE JONES, WILLIAM FORMAN, AND WILLIAM LILLER

(*Sky and Telescope*, November and December 1974, January 1975)

X-ray astronomy came into existence in the early 1960's. By 1967, approximately 30 discrete sources had been catalogued, mostly by groups working at AS&E, MIT, NRL, Lockheed Research Laboratories, and Lawrence Radiation Laboratory. The main problem in trying to identify these sources was that the positions were not accurate enough. Only for the most intense sources, such as Scorpius X-1, did the rocket experiments achieve high positional accuracy.

In December 1970, NASA's Explorer 42, Uhuru, was launched off the coast of Kenya. This small satellite had on board two x-ray detectors which for the next three years scanned nearly the entire sky. The most recent Uhuru Catalogue[1] lists 161 sources (see Fig. 144). The best positions are accurate to 1-2 arc-min. and consequently several dozen new optical identifications were made.

There is a concentration of strong sources toward the galactic plane and particularly in the direction of the galactic center. It seems likely that essentially all the bright x-ray emitters within our Galaxy have now been detected and that most of the x-ray sources not near the plane of the Milky Way are extragalactic.

One tremendous advantage of the Uhuru satellite was that individual objects could be observed over extended intervals of time. Not only were very weak sources detected by repeated scans, but periodic fluctuations in brightness were found in some of the sources, with characteristics of eclipsing-binary stars. As of this writing, five sources are known to be eclipsing binaries (see Glossary), a sixth is a spectroscopic binary, and two other sources are possibly eclipsing systems. Table 15 lists some of the known characteristics of these eight binaries.

Because of the unique signature provided by a precisely repeating eclipse, it is possible to identify *with absolute certainty* the optical source responsible for the x-rays. Furthermore, we can derive the star masses, sizes, and densities, and begin to say something about their past and future histories. Of these eight systems, the first five are quite similar;

---

[1] R. Giacconi and others, *Astrophysical Journal Supplement* No. 237, 1974.

Hercules X-1 is remarkably different from the rest, and is one of the most puzzling star systems in the entire sky.

X-ray production in binary systems is generally believed to be caused by the accretion of fast-flowing material from a more or less normal star onto a compact x-ray companion.

TABLE 15. BINARY X-RAY SOURCES

Source	Period	Optical	Mag.	Spectral type
3U 0900-40, Vel X-1	8.97	HD 77581	6.9	B0.5 I
3U 1700-37	3.41	HD 153919	6.6	O6.5f
3U 0115-73 SMC X-1	3.89	Sanduleak 160	13.3	B0 I
Cyg X-1	5.60	HDE 226868	8.9	O9.5 I
3U 1118-60 Cen X-3	2.09	Krzeminski's star	13.4	B0 I
Her X-1	1.70	HZ Her	14	B8-F3
Cyg X-3	0.18	?	—	—
Cir X-1	12.3?	?	—	—

Period is the orbital period in days. Mag. is the average **V** (yellow) magnitude.

There are two general ways to transfer material from the optical star to the x-ray object. The first relies on the natural evolutionary process by which a star increases in size on its way to becoming a normal red giant or supergiant (very large, cool star). It grows until it fills its so-called Roche lobe, a gravitational-potential surface surrounding the star (Fig. 146). This and a similar lobe around the x-ray star are connected at one point, which provides a "nozzle" through which the escaping material from the primary star can flow toward the companion. The transferred gases probably form a disk around the compact star and gradually spiral in toward its surface. X-rays may then be produced as bremsstrahlung above the surface, where the accreting material's density increases very rapidly as it converges on the degenerate star.

A second process relies on stellar wind from the primary star, a fraction of which is captured by the gravitational field of the collapsed star, producing x-rays.

The amount of energy produced by material accreting onto four different kinds of stars—the sun, a white dwarf, a neutron star, and a black hole—is GMm/R, the final kinetic energy acquired by material of mass m falling on the surface of a star of mass M and radius R; G is the universal constant of gravitation. The fourth column of Table 16 gives the energy of a single hydrogen atom, and the last column shows the amount of material required to produce the observed energy (10^{36} ergs/sec) in a typical x-ray source. Note the great efficiency of a neutron star or black hole in producing x-rays through the accretion process.

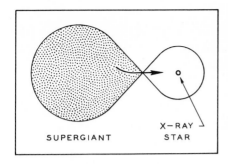

SUPERGIANT — X–RAY STAR

FIG. 146. Roche lobe in a binary-star system. Material inside each of the two egg-shaped areas is bound by gravitational force to the star in that area (actually a volume seen here in cross section). If gas pressure pushes material (gas or dust) past the X between the stars, as shown by the arrow, it will fall into the compact x-ray star. The size of the lobe depends on the mass of the star inside.

TABLE 16. HYDROGEN INFALL TO COMPACT STARS

Object	Mass[1]	Radius[2]	K.E.[3]	Infall[4]
Sun	1	700,000	0.002	10^{-5}
White Dwarf	1	10,000	0.1	10^{-8}
Neutron Star	1	10	100	10^{-11}
Black Hole[5]	10	4	1000	10^{-12}

[1] *Mass* is in solar units.

[2] *Radius* is in km.

[3] K.E. is the kinetic energy (in MeV) acquired by a hydrogen atom falling to the object's surface.

[4] *Infall* is the fraction of a solar mass/yr required to produce a typical x-ray source's output.

[5] See p. 366.

Now let us consider 3U 0900-40 as an example of one of these binary x-ray sources. As its name indicates, it is listed in the *Third Uhuru Catalogue* with right ascension 9h 00m and declination —40° (1950 coordinates). It is also known as Vela X-1. Its x-ray variations showed a periodicity of 8.95 ± 0.02 days. The optical counterpart of 3U 0900-40 seemed to be a 7th-magnitude star of spectral type B0.5 Ib, known since 1956 to have radial-velocity variations. Early in 1973, W. A. Hiltner, Jones, and Liller determined the light curve of this star, HD 77581, with photoelectric photometers, finding a range of about 0.1 magnitude.

The star goes through *two* maxima and *two* minima every 8.95 days, implying that one of the stars must be very faint yet moderately massive. Hence, we concluded that this star is an ellipsoidal variable, a binary where the primary is tidally distorted by an unseen secondary, something like Figure 146. As the x-ray source revolves around the *B* supergiant, the latter varies in brightness as we see it presented first broadside, then end-on—twice in each orbital revolution.

At the University of Michigan, L. Petro and Hiltner find that the period is 8.97 days, and that the velocity amplitude of the *B*-star primary is ±24 km/sec. The mass of the *B* star is inferred to be 12 times that of the secondary. The orbit of the x-ray source has a radius of 0.24 AU, and the radius of the supergiant is 62 percent that large.

The most interesting conclusion is that if the Bo supergiant has a mass of more than about 36 solar masses, the degenerate (highly compressed) companion is likely to be a black hole and not a neutron star. The best estimates for a Bo.5 star run about 15 or 20 solar masses, but if the luminosity class is Ia, the mass may be as high as 45 suns. Until more accurate data become available, we can only say that the x-ray object is at least one solar mass, and may be a black hole.

The results found for the other O- and B-supergiant systems in Table 15 are remarkably similar to those for HD 77581. The source SMC X-1 in the Small Magellanic Cloud (a nearby galaxy), also known as 3U 0115-73, is at a large distance from us; we believe that it has the highest intrinsic x-ray luminosity of the group. However, the star associated with it is faint and very difficult to study. The optical counterpart of the pulsating x-ray source Centaurus X-3 (3U 1118-60) is also a distant supergiant. Because of the regular pulsations of the x-ray source, its orbital radial velocity can be determined: (When the source is approaching, the pulsation period appears shorter than when it is receding, an instance of the Doppler effect—see p. ooo.) From this information, the x-ray source's mass is estimated to be not more than one sun.

The O-type star in 3U 1700-37 is probably the most luminous and most massive primary in this group, possibly 35 solar masses, implying 2.4 for its x-ray companion—a neutron star. None of the collapsed stars in the systems discussed so far has been shown to exceed the three-suns lower limit of mass for black holes.

The last of the x-ray sources identified with early-type supergiant binaries is Cygnus X-1. Its importance stems from the large mass of the compact x-ray object, implying that it is a black hole. However, the Cygnus X-1 source does not have periodic x-ray occultations, so its identification as the companion of the optical star HDE 226868 was at first not certain.

The first link in the identification resulted from a simultaneous x-ray and radio brightening in April 1971. The radio position could be measured with a precision of a few arc-seconds and agreed with that of the 9th-magnitude Bo I star HDE 226868, which was found to be a 5.6-day spectroscopic binary. (More accurate magnitude and spectral type are given in Table 15.)

The mass of the x-ray component could then be determined from optical observations of the spectral lines in the supergiant, giving its orbital motion and hence the ratio of its mass to that of the x-ray star. According to C. T. Bolton at the University of Toronto, the mass of

Cygnus X-1 is 14 suns, well over the line dividing neutron stars from black holes. Our picture of this odd pair is of a huge blue star swinging around a small, brilliant disk and spewing gas into it. Inside the x-ray-emitting disk is an invisible mass only 4-5 km in diameter!

The two remaining sources, Cygnus X-3 and Circinus X-1, remain puzzles for the most part. The Cygnus source has by far the shortest orbital period in Table 15. Circinus X-1 is highly variable at x-ray frequencies and tends to repeat its cycle every 12.3 days. As yet, no optical identification has been made.

The strangest of all the eight systems is Hercules X-1, now identified with the 14th-magnitude variable star HZ Herculis, located at right ascension 16^h 56^m 03^s, declination $+35°$ $25'.0$ (1950 coordinates), which is $38°$ from the plane of the Milky Way.

At x-ray frequencies, Hercules X-1 is a moderately intense source having pulsations with a 1.24-sec period, eclipses every 1.70 days, and a peculiar 35-day cycle. During this cycle the source is "on" and emitting x-rays for about 12 days, but "off" and not radiating detectable x-rays for the remaining 23 days. The 1.24-sec pulsations from Hercules X-1 are extremely regular; from the Doppler shift in these pulsations we find 3.954 million km for the orbital radius of the x-ray source.

To find the orbital velocity, we note that the circumference of its orbit (assumed circular) is 2π times the radius and it covers these 24.7 million km in the period 1.70 days (Table 15) giving an orbital velocity of 169.2 km/sec. From this velocity and the estimated mass of the primary star, it follows that the x-ray source is about as massive as the sun.

The size of the occulting primary star can be found from the duration of the x-ray occultation—about 0.265 day. In a circular orbit of radius 3.954 million km the radius of the optical star is 1.86 million km, or $2\frac{1}{2}$ times the radius of the sun.

The peculiar changes in Hercules X-1 on longer time scales are unique. Its x-ray emission can be detected only about one-third of the time. As Figure 147 shows, this source "turned on" abruptly Jan. 9, 1972, increased to maximum intensity in about 1 day, then gradually declined to nondetectability. After 23 days at this low level, the source reappeared. A similar pattern is repeated every 35 days.

The turning on is observed at only two phases in the orbit, and there are curious intensity dips just before the eclipses. One explanation for the dips and the abrupt turn-ons is a hypothetical small cloud of matter behind which the x-ray source passes just before it is occulted by the primary star. However, satisfactory explanation of the 35-day behavior must await further study.

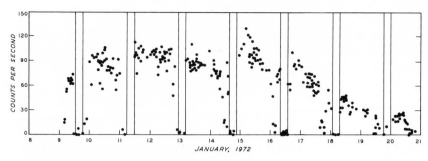

FIG. 147. X-ray intensity of Hercules X-1 measured by Uhuru during January 1972. The vertical strips mark the eclipses every 1.7 days; the interval from January 9 to 21 is the "on period" with sharp dips before each eclipse. (*Astrophysical Journal* diagram.)

The variable star HZ Herculis is not a typical well-behaved eclipsing binary (see Glossary). There is no secondary (shallow) minimum, and the primary (deep) minimum is very broad yet sharply pointed.

Apparently these peculiarities result from heating of the primary star by the x-ray companion. When the latter is between us and the primary, we see the heated side and the light curve shows a maximum. As the system rotates, we see less and less of the primary's bright side and the broad minimum commences.

Here is a new probe—x-rays from an external source—with which to study the structure and physical processes in the atmosphere of a star.

At minimum light, when we see the large star's hemisphere that is not irradiated by x-rays, the spectrum of HZ Herculis resembles a nearly normal Fo star. At maximum light, the spectrum is similar to a B8 star; only a few weak emission lines suggest the existence of free-flowing gas in the binary system.

The calculated radius is 2½ solar radii (p. 381), whereas that of a normal Fo star is only about 1½ times the sun. Hence, it appears that the occulting star is slightly bloated—starting to become a red giant.

This hypothesis is particularly attractive because the x-rays are believed to be produced by accretion of mass on the x-ray object coming from the primary. As shown in Figure 146, material from the more massive, larger star can pass through the nozzle connecting the two Roche lobes and fall onto the small x-ray star, producing x-rays. This accretion can be driven by the evolutionary expansion of the primary star, filling up its Roche lobe, and spewing out enough accreting material to produce an observable x-ray source.

How do the 1.7-day optical variations continue throughout the 35-day cycle, even when the x-ray source is turned off? Apparently some energy source continues to irradiate the primary star, causing the brightness changes. There seems to be some way by which x-rays can be beamed at a companion star continuously but not always at the Earth. Several schemes have been proposed. Most assume a very intense magnetic field (10^{12} times as strong as the Earth's field), which makes the accretion of material asymmetrical, and causes the outflow of x-ray photons from the compact neutron star to be beamed from its magnetic poles. Thus, the 1.24-sec x-ray pulsing we observe from Hercules X-1 would be caused by the rapid rotation of the neutron star, the beam of x-rays passing periodically through our line of sight.

A small change in the neutron star's rotational axis might divert the x-ray beam from the Earth's direction. Precession of the axis could move this beam back and forth slightly, thereby producing the observed 35-day effect in the x-ray emission.

A thorough search of the Harvard plates dating back to 1895 established that in some years the 1.7-day optical variation ceases and the star is "off," while in other years it is "on." Evidently, there are times when the F star is not irradiated. Possibly the F star contracts, cutting off the supply of material to the other star and temporarily ending the accretion.

At the present time, the 1.7-day cycle of HZ Herculis has its full optical amplitude of 1.5 magnitudes, as the F star shows us first its brighter, irradiated side and then its nonirradiated side. When will it next turn off at optical wavelengths? There is some tendency prior to 1956 for turnoffs to occur every 10 or 12 years.

With its pulsations, flickerings, its 1.7-day and 35-day variabilities, and its mysterious x-ray turning on and off, HZ Herculis is perhaps the most intriguing stellar system in the sky.

Of the 161 x-ray sources listed in the *Third Uhuru Catalogue*, approximately 40 percent are extragalactic. Their variety is remarkable. They include at least one quasar, several clusters of galaxies, radio galaxies, a normal spiral galaxy, the Magellanic Clouds, and a large number of still unidentified sources.

The unidentified sources are potentially the most interesting, because they may be extremely distant and therefore intrinsically very luminous. Since these sources do not have detectable optical counterparts, we can estimate that their x-ray emission is at least 100 times greater than their optical emission, a characteristic of considerable astrophysical significance.

Turning to the identified extragalactic x-ray sources, the Large and Small Magellanic Clouds are near enough to allow the resolution of individual x-ray sources. The Small Cloud is dominated by SMC X-1, which produces energy at about 10^{38} ergs/sec, a compact object revolving around a Bo supergiant star of apparent magnitude 13.3 (see Table 15).

The observed x-ray emission from the Large Cloud comes from four discrete sources. While none has yet been identified with an optical object, both LMC X-1 and LMC X-2 lie near type-B supergiants which appear to fluctuate slightly in brightness.

More distant than the Magellanic Clouds is Messier 31, a twin of the Milky Way system, which has detectable x-ray emission at a "normal" rate equal to contributions from all the discrete x-ray sources in our own Galaxy.

Approximately 10 times farther away is the giant elliptical radio galaxy Centaurus A (NGC 5128) with intrinsic x-ray luminosity roughly 10 times greater than a normal galaxy. An unusually large gas and dust lane extends across the central region of NGC 5128, and the spectrum of its x-ray emission reveals considerable absorption caused by material between us and the source. Presumably the x-rays originate in the nuclear region of the galaxy, and are partially absorbed as they travel through the lane of dust and gas.

Centaurus A shows two pairs of radio lobes (emission regions), which suggests that at times in the past violent explosions have occured in the nucleus, blasting relativistic (very high-speed) electrons out of the galaxy and producing the radio lobes. Presumably these explosions, nuclear radio emissions, and the x-ray source are all associated.

Still more remote is NGC 4151, one of the best examples of a Seyfert galaxy (pp. 7, 365), with a starlike nucleus whose spectrum shows very broad high-excitation emission lines—evidence that it is losing mass rapidly. It produces about 20 times more x-radiation than Centaurus A does, and it shows evidence of absorption, suggesting that the x-ray source does indeed lie in the compact nucleus behind interstellar material. Although NGC 4151 produces intense x-rays, another rather similar Seyfert galaxy, NGC 1068, has no x-ray brightness, although it is at roughly the same distance.

Another class of x-ray sources which extend over enormous regions of space—the giant clusters of galaxies—are the largest well-observed structures in the universe and some of them contain thousands of individual galaxies (p. 370). The total observed x-ray emission from a cluster is 10 to 100 times greater than would be expected from the combined contributions of the normal galaxies in it. The full extent of the region from which x-rays are received seems comparable to the size of the entire cluster—about a megaparsec (3 million light years) across. See page 376.

A still more distant and luminous x-ray source is the quasi-stellar object (quasar—see Glossary) 3C-273 of apparent magnitude 12. Its optical spectrum resembles that of the Seyfert galaxy NGC 4151, and it has an optical jet—a thin finger of light 20 arc-sec long—which extends from the bright starlike image. Other, fainter and more distant quasars may some day turn out to be x-ray sources also.

Since this article was written, a new x-ray source suddenly appeared and should be added to Table 15: Monoceros X-1, A 0620-00, discovered by the British Ariel-5 satellite in August 1975, and identified with a peculiar 12th-magnitude star. At Kitt Peak, Ted Gull obtained optical spectra of this star, and was able to determine its distance (from the interstellar sodium absorption line) as 2-3 kiloparsecs. Interpretation of the sudden increase in x-ray luminosity (by a factor of 500 between August 3 and August 11), and its very high level (5 times Sco X-1) is even more complex for Mon X-1 than for Her X-1 (p. 381), but it seems to be agreed that the compact companion star has a mass of 4-9 solar masses— therefore a black hole.

Another x-ray flare was detected on April 5, 1975, in the direction of our Galaxy's center, almost as bright as Sco X-1 for a brief period, and possibly associated with a black hole at the galactic center. Another nova was detected in Taurus, 4° from the Crab nebula, on April 21. By April 30 its x-ray intensity was twice that of the Crab. Apparently, the new high-energy satellites are finding that most of the x-ray sources are variable—some of them periodic, but most of them irregular or novalike.

In May 1975, Explorer 54 (also called "SAS 3," and similar to Uhuru) was launched from the San Marco Range off Kenya. Of its four x-ray telescopes, one is mapping the x-ray background, a second is measuring variations in Sco X-1, the third is surveying low galactic latitudes for nearby (strong) sources, and the fourth is surveying high galactic latitudes for x-ray galaxies. Sco X-1 has recently been shown to be a collapsed star in a close binary like the first five sources in Table 15.

It is clear from this chapter that high-energy astrophysics is developing very rapidly, that many more discoveries and interpretations will be made in the next few years. The existence of black holes—or something close to "black"—seems to be fairly well established.

Another theoretical speculation, "white holes," has recently been proposed by J. V. Narlikar, cosmologist at the Tata Institute in Bombay, India, and collaborator with Fred Hoyle on several cosmological theories. Narlikar's white holes are the opposite of black holes in that they are a source of energy and matter rather than a sink—a region in extreme conditions of density and temperature where matter "comes into our universe." The theory is supposedly based on field equations like Einstein's General Relativity, and undoubtedly involves speculative assumptions. Its proof will depend on fitting a wide variety of observations, including x-ray novae like Mon X-1.

As shown in Volume 8, Beyond the Milky Way, Seyfert galaxies and radio quasars are sources of energy so intense that theoretical explanations are very difficult. These radio sources, optical sources, the soft- and hard-x-ray sources, and gamma-ray sources imply a spectrum which Narlikar seeks to explain by "white-hole explosions" similar to the Big-Bang theory of Einstein and de Sitter for the origin of the universe. As the white hole first explodes into our space-time framework, its light would be extremely blue shifted, giving a hard x-ray or gamma-ray source (a "power-law" spectrum proportional to λ^{-3}). As time goes on, this source should "soften" to λ^{-2} or λ^{-1}, according to his calculations, very like observed changes in nearby x-ray sources and gamma-ray sources. Many distant white-hole explosions may account for the x-ray background extending over the whole sky, and also for high-energy cosmic rays, although various effects in intergalactic and interstellar space have yet to be taken into account.

It is not surprising that other cosmologists have criticized Narlikar's conclusions. Some of them argue that a white-hole explosion would immediately degenerate into a black hole because of the high density of matter. But then, if Menzel is right (p. 366), black holes cannot exist either! All this insures high interest in high-energy astrophysics for many years to come.—TLP

The Frontier
in Space

The next decade holds promise of many advances in space science and astronomy. The Space Shuttle with its Spacelab and Large Space Telescope, missions to a comet, Mars, and Uranus, and the High-Energy Astronomy Observatories—all mentioned in previous chapters. In addition, there will probably be another impact of the space age on human society. We are both old enough (62) to have conservative views on science fiction but also old enough to remember the beginning of the airplane age, when Lindbergh flew the Spirit of St. Louis across the Atlantic in 1927. It seems to us that the Apollo-11 flight to the Moon in 1969 confirmed the space age as Lindbergh's flight to Paris confirmed the airplane age. Ten or twenty years later airplanes were circumnavigating the Earth with passengers—in the 1980's spacecraft should be carrying passengers (for profit) around the Earth, and possibly around the Moon. Permanent bases on the Moon and permanent colonies on space stations are possible if we want them, as this chapter will show.

Of course, there have been a few "crazy ideas" that roused scientists' opposition.—TLP

"But Who Needs the Sun at Night?"

(*Sky and Telescope*, October 1966)

Under this headline, the Hartford (Connecticut) *Courant* on July 21 carried an Associated Press dispatch from Seattle, Washington, concerning a new government-sponsored study known as Project Able.

Prompted by Edgar Everhart, physics professor at the University of Connecticut, we inquired of the MSFC public affairs office, which confirmed the newspaper accounts:

"We have awarded to Boeing and Westinghouse contracts to study the feasibility of . . . an Earth-orbital light-reflective system for military operations at night. The studies will determine if a large reflective surface with mirrorlike qualities is feasible in providing a light source over land masses at night. The project studies are undertaken for the Department of Defense and NASA. These are phase-A study contracts costing about $125,000 each."

A diameter of perhaps 2000 ft is suggested for a large reflecting "dish" that would unfold after being placed in orbit. Upon command from the ground, the reflector would maneuver to reflect the sun's rays to a desired area. Half the breadth of the state of Florida could be illuminated at any one time.

"The satellites could be used for exposing enemy positions, for search and rescue operations, and for security purposes. Their large reflectors also would make them useful as navigation beacons and for radio astronomy experiments."

It is pointed out that three such satellites, if put in 24-hr orbits some 23,000 mi. high, could command the entire Earth. Professor Everhart comments:

"Entirely apart from the question of whether space projects should be used for warfare purposes, this program poses a most serious threat to observational astronomy. With such a reflector illuminating a large area to several times the brightness of full moonlight, deep-sky studies would be impossible by observatories anywhere within 500-1000 mi. of the aiming point. Furthermore, such a large object, unless very carefully blackened on the sides and back, is likely to raise significantly the illumination and resultant skyglow over an entire hemisphere of the Earth.

"Internationally, what right has this country, for its own purposes, to put up an artificial moon that would hamper astronomers in other countries, such as Japan and Australia? Since satellites are almost impossible to service, this reflector (and its successors in turn) would sooner or later become uncontrollable and scatter its sunlight at random —and forever.

"There is no indication in the newspaper accounts that the long-range consequences to astronomy have been considered. The precious darkness of the night sky is in grave danger unless astronomers the world over can effectively fight these evils."

Letter to the Editor

KURT ROSENWALD

(*Sky and Telescope*, November 1966)

The editorial "But Who Needs the Sun at Night?" is written from an astronomer's point of view. But the biological damage from satellites that illuminate the Earth will be far more extensive. As one with a doctorate in zoology, I would like to call attention to some major effects.

In their daily and yearly rhythms, both animal and plant life are geared to the amount and duration of daylight. While waiting for days to lengthen, satellite-illuminated plants might bloom prematurely and be damaged by frost. The loss of fruit would seriously affect the animal population. Trees might hold their leaves too long in the fall or fail to form proper buds for the ensuing spring growth.

Migrating birds, waiting for shorter days, might not leave in time, or not at all. The false illumination would misguide birds that use stars for navigation. Bird losses might become extensive enough to permit a disastrous increase in the number of insects.

The very least to expect would be a serious disturbance of the natural balance, with unforeseeable consequences. Much more must be known before a sunlight reflector can be permitted to orbit—if ever.

Many smaller orbiters (communications satellites, weather satellites, navigation-mapping satellites, the many abandoned hulks and third-stage rocket motors from past missions) litter up our night sky. Although their light cannot cause the biological damage that concerned Rosenwald, it often adds unwanted images on astronomical photographs. A "Space-warn" list of all Earth orbiters is maintained, as noted on page 345, so that such chance appearances can be identified.

The objections to "sun at night" were preceded, in 1961, by radio-astronomers' objections to another military development, Project West Ford, which would have scattered small "needles" of metal foil in a band around the Earth to improve long-range radio communication by reflecting radio waves downward. The needles would also have interfered with radio telescopes, and the astronomers' objections led to cancelling Project West Ford.

Now, over a decade later, sunlight on orbiters achieves a new importance, and the microwave (radio) transfer of solar power from orbiters to Earth may be an answer to our energy shortage—although it may have other adverse effects.—TLP

An Orbiting Solar
Power Station

J. KELLY BEATTY

(*Sky and Telescope*, April 1975)

Potentially, the sun's radiation is a limitless source of power, far in excess of the world's needs. In principle, sunlight can be collected on a vast scale and directly converted to electricity.

Unfortunately, the enormous flood of solar radiation reaches the Earth in a very dilute form. This means that any attempts to harness solar energy on a large scale will require collection devices of very great area. Only a few geographical regions favored by copious sunlight would be suitable, and even for them energy storage would be needed to compensate for the day-night cycle and cloudy weather.

The solution to these difficulties may be to place both the collector and the convertor in orbit around the Earth. This imaginative concept has been explored by a team of scientists and engineers headed by Peter Glaser of Arthur D. Little, Inc., Cambridge, Massachusetts.

They call their proposal a Satellite Solar Power Station (SSPS). It would move in a circular orbit over the equator at a height of 22,274 mi., in synchronism with the Earth's rotation. Thus it would hover over the same spot, remaining in almost continuous sunlight. Due to its equatorial orbit, the spacecraft will seasonally pass through the Earth's shadow in March and September, but never for longer than 72 min. each day. Photovoltaic (solar—see Glossary) cells would convert the sunlight to electricity, ready to be transmitted as microwaves to a receiving antenna on Earth.

Glaser and his collaborators suggest that the satellite must be large enough to produce between 2000 and 20,000 megawatts at ground sta-

FIG. 148. The proposed Satellite Solar Power Station, a 17-sq-mi. (45 km²) array in geosynchronous orbit 22,274 mi. above the equator. (Grumman Corporation drawing by Spero Kavases.)

tions if it is to be cost-competitive with other sources of power. The particular design for the SSPS (Fig. 148) would generate about 8000 megawatts in orbit and yield about 5000 on the ground. For comparison, the largest hydroelectric generating plant in the world, at Krasnoyarsk, U.S.S.R., has an output of 6096 megawatts. Grand Coulee, in Washington, now generates 2161 megawatts and has an ultimate capacity of 9771.

The SSPS as now envisioned is truly enormous, its two collector arrays having a combined area of 45 km² (17.4 mi.²). Even keeping this spacecraft in its proper orbit will be a major task. For a satellite in synchronous orbit to remain constantly over the same geographical point, it must travel in a circular path with constant 23-hr, 56-min. period and exactly in the plane of the Earth's equator. However, several kinds of perturbations strive to change the motion, causing the sub-satellite point to drift.

For example, the effect of solar radiation pressure on this large, relatively light structure will tend to make the orbit slightly elliptical, and the Moon's attraction will change the orbital inclination. If SSPS's inclination differs from zero, the oblateness of the Earth itself will introduce other perturbations in the satellite's orbit. To compensate for these effects and keep on station, the SSPS will have to expend about 30,000 lbs of propellants per year. [Ion propulsion may reduce this estimate—see page 276.—TLP]

THE ENERGY COLLECTORS

The 45 km² of collecting area will consist of high-efficiency solar cells and concentrator mirrors (Fig. 149). These inclined mirrors can increase substantially the amount of sunlight falling onto the cells. Suitable mirror coatings will reflect only the useful wavelengths of light, thus reducing the heating of the cells and increasing their efficiency. The weight and cost of a mirror is considerably less than for an equal area of solar cells.

Silicon cells for photovoltaic conversion of sunlight to electricity were first demonstrated in the laboratory in 1953, and now they are used on nearly every spacecraft (see p. 270).

At present, single-crystal silicon cells have an efficiency of 11 percent, about half of their maximum theoretical efficiency. Many improvements can be expected. Whereas single-crystal silicon cells are ordinarily wafers 0.5-1 mm thick that are exposed broadside to light, better results can

FIG. 149. Diagram of SSPS mirrors and solar cells. The plan view (top) shows the off-set position of the 1-km circular transmitting antenna inclined 6½° to transmit power to a ground receiving antenna in the U. S. The end-view (left) shows the V-shaped rows of mirrors concentrating light on the rows of solar cells. (A. D. Little drawing.)

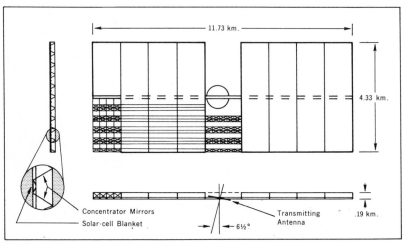

be obtained by stacking much thinner wafers so that light strikes them edgewise. . . . Silicon crystals eventually may cost only five times as much as plate glass. . . . Solar cells now have life expectancies of about 10 yrs . . . improvements in radiation-resistant cells are expected to increase the useful life to 30 yrs.[1]

Micrometeorite impacts should affect the satellite's performance only slightly, since they would destroy only about one percent of the solar cells in the course of 30 yrs.

Handling the Power

On the SSPS, the electricity generated by the collector arrays must be carried through a complex distribution network capable of handling 200,000 amperes of 40,000-volt direct current. Such high currents could result in internal stresses in the SSPS structure by the interaction of self-induced local magnetic fields, but this problem can be minimized by suitable circuit designs. Special switches and fuses or circuit-breakers will be needed to control sections of the solar-cell array for maintenance, and also to protect the system when entering or leaving the Earth's shadow.

The power produced and collected by the SSPS will be converted to microwaves and transmitted to Earth at a frequency tentatively selected as 3.3 gigahertz, corresponding to a wavelength of 9.1 cm. The actual frequency adopted will probably be set by international agreement, to minimize contaminating bands used by radio astronomers for their observations, and by defense radar.

Power transmission by microwaves was demonstrated as early as 1963, and its technology is well known. One kilometer (3280 ft) in diameter, the downward-transmitting antenna will consist of 800,000 Amplitron microwave generators, combined electrically in subarrays of about 25. Each subarray is to have an automatic phasing system so that the individual elements of the antenna will be in phase. These subarrays will radiate the microwave energy through a slotted waveguide, producing a beam of coherent radiation which is in phase like that of a laser, and has a beam spread of only 35 arc-sec.

One important engineering problem is to dissipate into space the large amounts of heat produced by the Amplitrons. The probable solution is to attach graphite fins to each unit for radiative cooling.

The 1-km dish will be able to transmit 86 percent of the power produced by the solar cells, since about 4 percent will be lost by the conductors aboard the SSPS and about 10 percent in the Amplitron conversion of direct current to microwaves.

[1] Late in 1975 NASA assigned $12 million through JPL for industrial development of cheap solar cells.—TLP

Additional losses suffered by the beam in traversing the Earth's atmosphere will be small. At a frequency of 3.3 GHz (see App. V), the ionosphere is nearly transparent. In the troposphere, the modest attenuation is largely due to water, being about 2 percent during light rain (0.1 in./hr) and 6 percent during moderate rainfall (1.3 in./hr).

RECEIVING THE POWER

The proposed receiving antenna on the Earth's surface is to be over 7 km (4.35 mi.) in diameter. This area is sufficient to intercept 90 percent of the energy in the transmitted beam. The ground antenna will use a phase-lock technique to keep the SSPS subarrays in phase and control the microwave beam from the SSPS, which will be virtually stationary in the sky. A pilot signal sent from the center of the ground antenna will control the phase of the microwaves transmitted from each of the subarrays. If the microwave beam were improperly pointed, the SSPS could not receive the pilot signal without which the coherence of the microwaves would be lost and the spread-out beam reaching the Earth's surface would be too dilute to pose a hazard. In effect, this assures that the beam could not be directed accidentally or deliberately toward any location other than the receiving antenna.

The 7-km antenna (Fig. 150) will consist of open metal mesh on which are spaced half-wave dipole antennas to capture the microwave energy and deliver it to solid-state diode rectifiers, which must convert high-frequency alternating current to direct current. Their output will then be fed into a high-voltage direct-current transmission network. At present, microwave rectifiers can attain 80 percent efficiency . . . about 68 percent of the output of the SSPS solar cells will finally enter the power grid.

Glaser's group looked carefully at the potential hazards of the microwave beam: at the edge of the 7-km antenna, a man would be exposed to as much radiation as if he were standing in front of a microwave kitchen oven with the door closed, meeting U.S. safety standards for microwave exposure (but not those of the Soviet Union). If a plane were to fly through the beam, the metal fuselage would probably shield the crew adequately.

The most obvious obstacle to the SSPS project is getting the gigantic solar panels, support structure, and transmission antenna into orbit. Glaser suggests first putting the component parts into a 5000-mile-high orbit for assembly, using a modified Space Shuttle. Obviously, many trips will be required to deliver SSPS's 25,000,000 pounds of parts. Once in that orbit, sub-units would be assembled into larger sections that

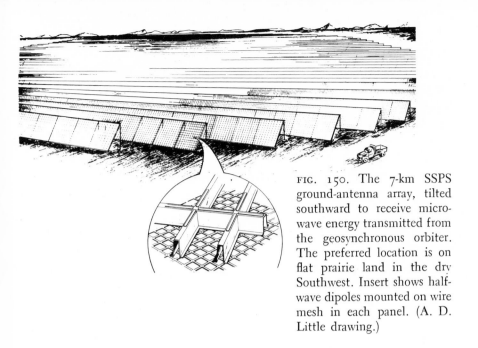

FIG. 150. The 7-km SSPS ground-antenna array, tilted southward to receive microwave energy transmitted from the geosynchronous orbiter. The preferred location is on flat prairie land in the dry Southwest. Insert shows half-wave dipoles mounted on wire mesh in each panel. (A. D. Little drawing.)

would be raised to the synchronous orbit. Advantage could be taken of solar-cell arrays to power ion-rocket engines for the last phase.

Fantastic? It is Glaser's considered opinion that a Satellite Solar Power Station may be operating in the 1990's, supplying electricity at a competitive cost.

The cost/effectiveness ratio for SSPS has not yet been worked out, but a competitive proposal by the Boeing Aerospace Company estimates that $4 billion per year for 15 years would build a manned solar-power station with 20 mi.² (51.5 km²) of mirrors to reflect sunlight on high-temperature "boilers" from which hot helium gas would drive 48 turbogenerators producing 14,400 megawatts of electricity, converted to microwaves and beamed to Earth. A crew of 75–100 men would live and work on this Power Satellite ("Powersat") in geosynchronous orbit 22,500 mi. from Earth. Boeing estimates that Powersat would pay off at $2 billion per year, starting as early as 1990. As this volume goes to press, it is too early to say which of these two solar-power systems is the better. Either one would light up the night sky.

Another cost-effectiveness problem is posed by the NASA-proposed Large Space Telescope. Its very high angular resolution, if achieved, would undoubtedly pick up faint astronomical objects that can never be detected from ground-based telescopes like the Palomar 200-in. (7.88-m.) reflector. The practical problem is whether so large a mirror can be accurately figured for o-g in terrestrial laboratories at 1-g, and then maintained in proper figure to achieve "diffraction-limited resolution" (the 0.04-arc-sec resolution theoretically possible with a 120-in., or 3.05-m. mirror). Smaller telescopes might first be tried out in Spacelab.—TLP

The Large Space
Telescope Program

C. R. O'DELL

(*Sky and Telescope*, December 1972)

The 1980's should see the establishment of the first major observatory in space, as a result of NASA's Large Space Telescope (LST) program. This observatory will contain a long-lifetime reflecting telescope of about 120 in. clear aperture, and will be repairable and refurbishable while in Earth orbit through visits by the Space Shuttle (Fig. 151). We expect that its high angular resolution, large auxiliary-equipment payload, and efficiency of operation will make this the most effective telescope ever built.

Observational astronomers are painfully aware that the irregularities in the Earth's atmosphere degrade star images to much below the performance potential of good instruments (see p. 3). A 120-in aperture is capable of resolving 0.04 arc-sec, but ground-based seeing problems cut this to 0.5-1.0 arc-sec.

The other great advantage of an orbiting telescope is that it is outside the atmospheric components that absorb the stars' ultraviolet radiations at wavelengths shorter than about 0.3 micron (3000A). This advantage is already being exploited by small telescopes aboard the OAO (see p. 337), but no instrument on these satellites has an aperture of even 1 m. or closely approaches diffraction-limited performance.

The advantages of finer images are hard to evaluate quantitatively. But we note that they will permit better studies of close binary-star systems, will provide badly needed trigonometric parallaxes (see Glossary) for stars at great distances, and will give much clearer views of already familiar objects.

We can only speculate about what will be seen when the LST is trained on presently known objects, but we know the improvement achieved in going from a resolution of 10 arc-sec to the presently attainable 1 arc-sec. Much fainter objects can be detected. By concentrating starlight into smaller images, contrast with the sky background will be improved, and the background is darker in orbit due to the absence of

FIG. 151. Cut-away drawing of the Large Space Telescope, with Shuttle in the background. The telescope tube has an extended sunshield at the front end (right), solar panels for power at the back end. Current plans include a vacuum-tight work space behind the primary mirror (left), not shown here. (NASA drawing.)

scattered light and airglow. For faint-object photography, the LST should go at least five magnitudes (100 times) fainter than can the same detection system on the ground.

Since the LST will be remotely controlled, it will use a television-type recording system of much higher sensitivity than photographic film. An additional benefit of this recorder is freedom from the limited storage range of an emulsion, thus permitting very long observing periods and even fainter magnitude limits.

The constancy of the telescope environment and of automation techniques will also allow long integration times, like those now employed to good advantage only by radio astronomers. Exposures of 10 hrs may not be uncommon with the LST and will be clearly worthwhile if they establish the scale of the universe through accurate photometry of variable stars in distant galaxies, or determine the true stellar composition of globular clusters (groups of about 100,000 stars).

Similar advantages also exist for astronomers interested in stellar spectra. The small images and darker sky will permit low-dispersion spectrographs to avoid more of the contaminating background and to go five magnitudes fainter. The crispness of the images also has potential for very efficient high-dispersion spectrographs. With only modest grating sizes, the observer will be able to use all of the light from the stellar image, instead of losing much of it at the entrance slit of the spectrograph, as is commonly the case on the ground.

A further advantage lies in the accessibility of all the sky and nearly around-the-clock observing, utilizing about 6000 hrs/yr. The best ground-based observatories obtain only about 2000 hrs/yr.

Design of the LST is dictated by many interacting elements. . . . At present, we are planning an $f/2.2$ primary and an over-all $f/12$ Ritchey-Chrétien system (see Vol. 4).

The scientific instrument package is located at the principal Cassegrain focus, with tertiary mirrors feeding several other instruments. Since stray-light suppression is extremely important in reaching faint light levels, the tube is closed (also serving as a meteoroid shield) and

FIG. 152. Diagram of LST design. The Support System Module (right) will permit astronaut or scientist to work on instruments in "shirtsleeve environment." The over-all length is about 75 ft, weight about 10 tons. (Perkin-Elmer drawing.)

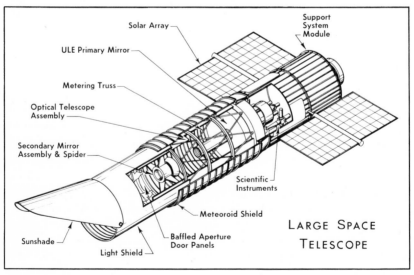

well baffled. There is also a sun shade (Fig. 152). The system is powered by two solar panels, and the supporting equipment and electronics are located in the instrumentation room.

The primary mirror will be of low-thermal-expansion material whose weight will be reduced by coring out most of the interior. The mirror will be heated during operation to about optical-shop temperatures to minimize variations from its original 1/64-wave accuracy. Pointing will be done by precision gyros, and fine guiding by means of bright field stars and an articulating secondary mirror.

In order to achieve a minimum LST lifetime of 10 yrs, we must plan on being able to visit it for repairs and for replacing equipment. This will be done with the Space Shuttle; the prime idea at present calls for docking the Shuttle on the aft end of LST, pressurization of the instrument room, followed by astronaut entry. Free of the need for a protective space suit, the astronaut can easily carry out adjustments and replacements. This repair capability will help keep down the total cost, since it is very expensive to achieve the high reliability necessary for a large one-chance satellite. Shuttle will also be able to retrieve the LST, return it to Earth, and relaunch it later—so the LST should approach the immortality of the major ground-based telescopes.

The instrumentation will be determined by the needs of the astronomy users, tempered by space and weight constraints of the spacecraft. The instruments will include a television camera capable of resolving the diffraction-limited (see Glossary) images, and low- and high-dispersion spectrographs, all to be modified or replaced from time to time.

The diffraction-limited camera must work at very long focal ratios to allow for detector resolution. Present detectors would demand focal ratios of about $f/96$, but it is hoped that improved resolution will reduce this ratio. Fortunately, much of the slowness associated with large focal ratios will be offset by the high sensitivity of the detectors. We expect that competition for new instruments will be opened every few years to guarantee that LST represents the best possible system— some instruments having long lifetimes, and others of a more specialized nature being changed during each revisit. Although the first launch date, in 1980, seems remote, our long lead time indicates the many developments that must be made in building a cost-effective new system.

The actual control of the observations with the LST will be from the ground, through the facilities at GSFC in Greenbelt, Maryland. MSFC is responsible for the over-all system, while Goddard is in charge of scientific instrumentation. At present, the major scientific effort is by the 12-man LST Science Steering Group headed by the writer.

In 1976, *LST* received a setback due to the reduced *NASA* budget; in effect, the development was delayed one year, much to the disappointment of many astronomers. However, $100 million was added to the solar-physics budget for the Solar Maximum Mission (SMM) in 1979-80, when solar activity is at its 11-yr periodic maximum.—TLP

Astronomy with the
Space Shuttle

GEORGE R. CARRUTHERS

(*Sky and Telescope*, September 1974)

We are now reaching the end of the first great era of space astronomy which began soon after World War II, with a variety of sounding-rocket experiments. It reached significant peaks with the successful OAO 2, OAO 3, and Skylab. Additional contributions were made by smaller unmanned satellites, and by nonsolar astronomy experiments carried out during Gemini and Apollo missions.

The main hope for future space astronomy rests with the new Space Transportation System (Space Shuttle), expected to become operational about 1980 (see p. 28). The Shuttle will be capable of placing a payload of up to 65,000 lbs in low Earth orbit, if it is launched from Florida, or 40,000 lbs in polar orbit if the launch is made from California. This capability is greater than that of any of the current expendable boosters except the giant Saturn-5 rocket, which can orbit more than 200,000 lbs. However, the cost of a Shuttle flight will be much less (about $10,000,000).

Even more important is the fact that the Shuttle payload may be *returned* from orbit, for refurbishment and reuse. It will also be possible for the crew to operate, monitor, maintain, and repair the instruments while the Shuttle is in orbit. Restrictions on size and weight will be greatly relieved, as well as the requirement for extreme reliability. It is these factors that are largely responsible for the high costs of present space experiments.

The Shuttle is intended to have three basic modes of operation:

1. To establish and maintain automated observatories in low Earth orbit.

2. To be the first stage of a combined payload and upper stage, for establishing an observatory in high Earth orbit, or in a translunar or escape trajectory.

3. To provide a platform for manned experimentation for up to 30 days in orbit—the Spacelab or "Sortie" mode.

Low Earth Orbit

Operations planned for the first category include LST. . . . Other low-orbit missions include the support of advanced HEAO (High-Energy Astrophysical Observatory—see p. 376) spacecraft, primarily for gamma-ray and cosmic-ray investigations which require large and heavy instrumentation. It will also include an x-ray telescope with a 1.2-m. aperture for detailed study of x-ray sources in the sky (see Chapter 9).

Shuttle Plus Second Stage

The second mode of Shuttle operation would be used for most of the solar-system probe missions after the early 1980's. It would also be used to establish astronomical observatories (and other satellites) in orbits higher than the 1100-km limits of the Shuttle itself. Present upper-stage rockets, such as Centaur and Agena, could be used for early missions, but the full capabilities of the Space Shuttle will be realized only with the upper stage specifically designed to match it. This upper stage, the Space Tug, could be reusable for near-Earth missions. As presently envisioned, the Tug could deliver 18,000 lbs payload to synchronous orbit (35,400 km above the equator) if the Tug is expended, or 8000 lbs if it is returned to the Shuttle for further use. The Shuttle-Tug combination may be the basis for a new program of lunar exploration.

The Space Shuttle could also serve as a launching platform for interplanetary spacecraft that use solar-electric or nuclear-electrc propulsion systems. Such systems are far more efficient than chemical engines, but they have very low thrust-to-weight ratios. Hence, they can only be utilized after the spacecraft has been carried up to Earth orbit by conventional rockets (see p. 276).

For astronomical observations, high orbits have important advantages over the lower ones used by present OAO spacecraft and directly accessible to the Shuttle. Particularly for nonstellar objects, such as nebulae, planets, and comets, the Earth's outermost atmosphere—the hydrogen geocorona (Fig. 134)—interferes with ultraviolet observations at or below the Lyman-alpha spectral line at 1216A (see p. 333). Extending out to more than 15 Earth radii, the geocorona scatters Lyman-alpha radiation from the sun and produces an intense "foreground" glow. It can also interfere with the observations of other planets, comets, and the like, by *absorbing* Lyman-alpha radiation. Therefore, effective studies of extraterrestrial Lyman-alpha sources, and spectrographic surveys in the 900-1230A range, require the observatory to be in a very high orbit—at least 10 Earth radii out. Such orbits are also nearly free of the Van Allen belts of particle radiation, which can deteriorate window materials transparent to far-UV light, create background noise in detectors, and fog film (p. 44).

An additional advantage of very high orbits, particularly synchronous ones, is that data can be taken in real time over long periods, since such satellites view almost half the Earth at a time. Low-orbiting satellites, on the other hand, can transmit data only when they are in range of a ground station.

Spacelab or Sortie Mode

Experiments in the third category will be similar to those carried out by astronauts in the Gemini, Apollo, and Skylab manned missions, in that the equipment will be operated on the spot, as in ground-based astronomical observations. Direct recording on film can be used, rather than television or radio-telemetry transmission of the data, since the film will be easily retrievable for return to Earth laboratories. For the images or spectra of celestial bodies, film-recording instruments are much simpler and have better resolution than similar ones using television readout.

FIG. 153. Space Shuttle in sortie mode with astronomical Spacelab (Sortie Lab) aboard. One or two telescopes will be operated by remote control on open "pallets," the astronomers in a pressurized module (Spacelab cabin). After 2-4 weeks in orbit, the Shuttle returns to Earth. (NASA drawing.)

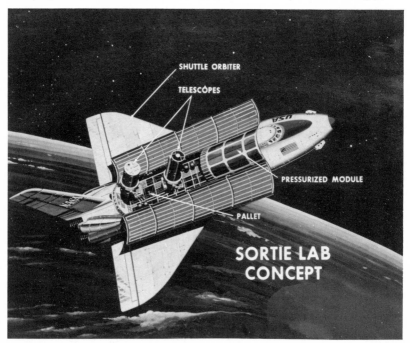

SHUTTLE ORBITER

TELESCOPES

PRESSURIZED MODULE

PALLET

SORTIE LAB CONCEPT

A particular advantage is that the Shuttle's large payload will allow scientists who are not trained as astronauts to go along to operate their experiments. This will be an extension of the present NASA airborne-astronomy program, in which infrared observations are made from high-flying jet aircraft. The nominal Shuttle crew consists of a commander, pilot, mission specialist, and payload specialist. Only the first three would have to be trained astronauts, the mission specialist being a scientist-astronaut. For Spacelab missions as many as four payload specialists might go along, making a total crew of seven.

In the Spacelab mode (a contribution of ESA—see p. 30), the Shuttle will be operated with the instruments (mounted in the payload bay) controlled by the crew in the orbiter's cabin and a second Spacelab cabin. Carried in the payload bay along with the instrument-mounting pallets outside, the second cabin would permit scientists to control the instruments in a shirtsleeve environment. All the crew members would ride to and from orbit in the orbiter cabin, but the payload specialists would work mainly in the pressurized Spacelab cabin (see Fig. 153).

Sky surveys at wavelengths inaccessible from the ground are particularly well suited to the Spacelab mode of Shuttle operation. Except for the Celescope experiment on OAO 2 (see p. 338), all experiments concentrate on photometry or spectrometry of individual stars or positions in the sky. Because of small fields of view, they cannot survey

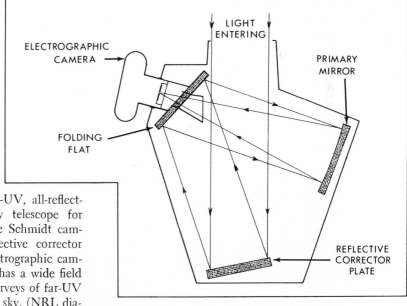

FIG. 154. Far-UV, all-reflecting sky-survey telescope for Spacelab. The Schmidt camera with reflective corrector plate and electrographic camera (p. 331) has a wide field suitable for surveys of far-UV sources in the sky. (NRL diagram by G. R. Carruthers.)

large parts of the sky to find out "what is there," in the manner that a wide-field camera or objective spectrograph can.

An all-reflecting Schmidt telescope has been proposed by a group at the University of Texas for a Spacelab mission (Fig. 154). The proposed $f/4$ instrument would be of 75-cm aperture (about 30 in.) and cover a field 5° square with a resolution better than 1 arc-sec. Reaching stars down to 20th magnitude in 15-min. exposures, it could extend the Palomar Sky Survey (48-in. Schmidt) to shorter wavelengths with equal or better resolution. The sky would be surveyed in three wavelength ranges (tentatively 1050-1600A, 1250-2000A, and 1800-2800A, using electrographic cameras (p. 331) both for photos and moderate-resolution spectra (by means of an objective grating in front of the telescope).

A *medium-size reflector* has been proposed by a group at JSC. This Astronomical Observatory for Shuttle (AOS) would have a 100-cm (about 40-in.) $f/15$ Cassegrain telescope mounted on the pallet, with controls inside the Spacelab cabin. EVA would be required to service the auxiliary instrumentation or to retrieve and replace film.

The telescope could provide a significant improvement over ground-based resolution, and would observe in the ground-inaccessible ultraviolet. Used with a variety of image detectors, spectrographs, and photometers, it would supplement LST by observing objects not requiring the full LST capability. It would have a larger field of view than the LST, and use photographic film.

Infrared astronomy will also benefit greatly from Spacelab operations. Good IR observations require that not only the infrared detectors, but also the entire telescope and optics, be cooled to the temperature of liquid hydrogen (about 20°K) or below. The short lifetimes of the necessary low-temperature material would limit the duration of the mission to a few weeks in any case. Hence, the Sortie Mode (a few weeks in orbit) is well suited to infrared astronomy.

The instrument sketched in Figure 155 would work best for observations in the 10-50-micron range of wavelengths, but could be used for broad-band photometry of extended sources over a wider range, roughly 5-200 microns.

Other Studies. Spacelab will also benefit solar astronomy, extending the results from the Skylab ATM (see p. 302). The backup ATM, which presently exists and would require minimal time and money to prepare for flight, has been recently proposed for an early Space Shuttle mission.

In Spacelab, the Atmospheric, Magnetospheric, and Plasmas-in-Space (AMPS) program could follow up work done with the Orbiting Geophysical Observatories, as well as the Apollo-16 far-ultraviolet camera and spectrograph (p. 331).

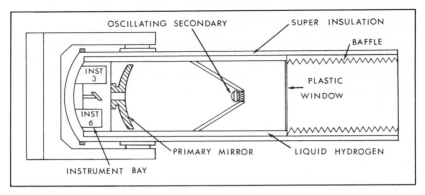

FIG. 155. Proposed 1.5-m. infrared telescope for Spacelab. The entire instrument is to be cooled with liquid hydrogen (near 20°K) to minimize thermal IR from the component parts. (NRL diagram by G. R. Carruthers.)

Since the Space Shuttle can be launched into polar orbit from Vandenberg Air Force Base, California, or into orbits of lower inclination from KSC, it can also undertake geophysical studies, including meteorology and Earth resources.

In summary, operational use of the Space Shuttle during the coming decades should make possible great advances in astronomy. Its development well deserves the required time and money.

Space Shuttle is being designed to allow great flexibility in operation. The astronomy described by Carruthers involves four different Spacelabs: Solar, Infrared, UV-Optical (the 1-m. telescope is now called SUOT for Shuttle Ultraviolet Optical Telescope), and HEAO (for High-Energy Astronomical Observatory). To complete the record, we list in Table 17 all seven types planned in August 1975, including Life Sciences, Advanced Technology Lab (ATL), and AMPS, together with the size of cabin ("module" on Fig. 153) for payload specialists, the number of pallets, instruments, study goals, and probable number of flights per year.

The "building blocks" of Spacelab include a long cabin, a short cabin, and outside pallets for mounting telescopes and other instruments. The cabins will be pressurized with normal air at 1 atmosphere, and each equipped with two or more computers, tape recorders, display screens, and telemetry to ground. There is room on Shuttle for

> *Long Cabin and 2 pallets, or*
> *Long Cabin, Short Cabin, and 1 pallet, or*
> *Short Cabin and 3 pallets, or*
> *(No Cabin) 5 pallets.*

By 1981 there will be four small Instrument Pointing Systems (IPS) which will stabilize a telescope on a pallet to point accurately (better than 1 arc-sec). The normal Spacelab flight will last 7 days in orbit, although 28-30 days will be possible. Each Spacelab will have Common Operations Research Equipment (CORE), the most elaborate being some 21 instruments, containers, and tools for Life Sciences. The 3 or 4 payload specialists, who operate this equipment and make observations round the clock, must spend 6 months in training before launch.

. In January 1976, JSC started 7-day tests of men and equipment in full-scale mock-up of Spacelab and the Shuttle crew cabin in which a scientist-astronaut and two qualified payload specialists are confined. During each simulated mission the men followed a tentative schedule of experiments, checking whether equipment and space are adequate.

Most of the Spacelabs will be flown twice a year, starting in 1981. Where compatible, a short cabin or single pallet for one discipline will be flown "piggyback" with another, such as Life Sciences with AMPS. The ATL is for trying out new instruments and designs—an engineering facility—and will probably go on Spacelab 1 in 1980. Spacelab 2 will probably carry HEAO with one payload specialist in the astronaut-crew space.

These many possibilities for astronomy and other research experiments aboard the Space Shuttle will allow scientists greater flexibility than in the past. It will be possible for an astronomer with a bright new idea to get observations from Spacelab, or even to accompany his own special equipment in orbit within a year following his proposal. There are major advantages in having the scientists aboard to recognize and correct malfunctions in the instruments—even more, to recognize an unexpected discovery and capitalize on it. Of course, this means that the scientist must have access to the data collected—being able to see what is being photographed, or even to develop photographic film on board.

Although Shuttle will receive the largest support, and be launched as many as 50 times per year, NASA hopes to maintain several other programs through the next decade, including those listed in Table 18. These cover a wide variety of goals; some described in previous chapters, some involving other government agencies, and some other countries. Many involve services or applications rather than astronomical research, and all of them utilize advanced space technology. Table 18 does not include international cooperative flights such as the planned (1977-78) Electrodynamic Explorers (U.S.-U.K.), three of which will explore the Earth's magnetosphere simultaneously in order to separate time changes from space changes that have been confused in the past, Lunar Orbiter (U.S.-U.K.) in polar orbit around the Moon, Helios (U.S.-W.Germany) in 1976, SIRIO (U.S.-Italy) in 1976, one flight per year with ESRO, and others for the armed forces.

TABLE 17. SPACELABS FOR SHUTTLE SORTIE MODE

Name	Cabin	Pallets	Major instruments	Study goals	Flights/yr
Life Sci	Long	none	Centrifuge Mass spectrometer	Biology Physiology Medicine Biomedical	1-2
Life Sci Piggyback	Short	none	(as needed)	(some of above)	1
ATL	Long	2	(as needed)	Space technology	2
AMPS	Short	3	Laser radar Radio generator Mass spectrometer	Earth's upper atmosphere Solar wind	1-2
Solar	Short or none	3-5	1-m. $f/35$ telescope EUV, XUV, and x-ray detectors X-ray spectrometer X-ray polarimeter Gamma-ray detector Neutron telescope Coronagraph Skylab ATM	Sun (6-7 experiments per flight)	2
Infrared	Long	2	Cryogenic 20°K telescope 2.6-m. telescope	Planets Stars Galaxies	1-2
UV-Optical	Short	2	1-m. $f/15$ telescope, or 75-mm $f/4$ Schmidt	Far UV observations Survey of stars, planets, galaxies	2
HEAO	none	5	X-ray concentrator Bragg spectrometer Gamma-ray detectors Cosmic-ray spectrometer	High-energy sources	1-2

TABLE 18. TENTATIVE NASA SPACE PROGRAMS FOR THE NEXT DECADE

Program	Pre-1976 Launches	(Text page ref.)	Planned Launches	Instruments	Major Program Goals
High-altitude balloons	(many)	(7)	5 to 10 per year	Far-IR, x-ray, gamma-ray detectors; high-resolution cameras	Heavy instruments above most of the atmosphere
Sounding rockets	Many Aerobees	(312)	20 to 40 per year	Far-UV, x-ray, gamma-ray detectors	Short-exposure detection and study of peculiar sources
Meteorological satellites	ATS 1 to 5 Nimbus 3, 5 SMS 1, 2	(24) (328)	1 SMS per year	Pointable cameras	Storm warnings, long-range weather forecasting, studies of the atmosphere
Earth-resources satellites	ERTS 1 LandSat 1, 2	(330)	1 LandSat per 2 yrs	Cameras and filters	Studies of crops, water conditions, geology
International Ultraviolet Explorer	Explorers 1-41	(350)	IUE in 1977	45-cm f/15 telescope and echelle spectrometer	Far-UV spectra of stars, nebulae, and galaxies
High-energy observatories	Explorers 42, 54	(363) (376)	HEAO 1, 2 1978-80 (3 more on Shuttle)	Directional x-ray and gamma-ray detectors, filters	High-energy spectra of x-ray and gamma-ray sources

Interplanetary space-probes	Pioneers 1-11	(246)	2 Pioneers in 1978	Cameras, spectrometers, and Lander	Land on Venus, explore atmosphere on day- and night-side
	Mariners 4-10	(168)	Mariner in 1980	Cameras, spectrometers, magnetometer, sampler?	Fly by Comet Encke for close-up view of nucleus
	Vikings 1-2	(233)	Mariner in 1980	Cameras, spectrometers, magnetometer, S-band	Fly by Jupiter and Uranus for close-up views and measures
Space Shuttle (manned)	None	(402)	Several per year 1980-85	(1) ESA Spacelab with survey telescope, 1-m. reflector, 1.5-m. IR telescope, solar telescopes (SMM), and/or high-energy telescopes	Survey of peculiar objects; immediate follow-up with detailed observations in a flexible program
		(396)		(2) Automated, separable, 2.4-m. diffraction-limited telescope (LST), with TV cameras and spectrometers	Detailed studies of very distant stars, nebulae, and galaxies between manned visits and returns to Earth
		(401)		(3) Various instruments on separable space-probes	Launch of deep-space probes to Moon and planets
Satellite Solar Power Station	None	(390)	?	(solar cells, microwave generators, etc.)	5000 megawatts of clean power (industrial)
Lunar base	None	--	?	Telescopes, detectors, spectrometers, and other equipment	Lunar observatory and manufacturing plants
Space colony	None	(418)	?	(sunlight mirrors, solar power station, "solar gardens," etc.)	10,000 people in 2 rotating cylindrical space stations

In June 1975, NASA appointed a 20-scientist panel to suggest broad future projects for the next 20 years of the space program. Their report, "Outlook for Space," was delivered to NASA Administrator James Fletcher, who was interviewed by J. Kelly Beatty for Sky and Telescope (see November 1975 issue). The report was organized around 11 "themes" with many subsidiary "objectives," scientific or engineering problems, and the instruments needed to solve them. NASA provided estimated costs in 1980 dollars; for example, a surface sample from Mars or Mercury will cost $1 billion, a Lunar Base $20 billion. (Fletcher considers it unlikely that the U.S. Congress will provide more than the current $3.5 billion/yr during the next few years, therefore precluding such major projects.)

Some of the more interesting ideas from the panel's discussion:

• LST aperture (p. 396) should be reduced from 3 m. to 2 m. (78 in.) to allow cooling by liquid hydrogen for infrared work (p. 404).

• The Space Tug for Shuttle (p. 401) should be powered with solid propellant.

• Solar-electric propulsion (p. 276) should be used for Venus and Mercury space probes.

• Major physiological problem for men spending months in 0-g is calcium loss, which was 1 percent/month on Skylab.

• A "Manned Orbital System" should be assembled in Earth orbit from two or more separately launched modules, where crews of three or more men can work continuously on engineering developments.

• High priority should be given to a Mercury orbiter for experiments on General Relativity and gravity.

• Planetary exploration should continue with "penetrometers" (costing about $300,000 each) to plunge into a planet's surface and telemeter data back to Earth.

• Planetary engineering proposals should be considered, such as altering Venus' atmosphere by dumping algae there (proposed by Carl Sagan).

• Manned exploration of planets will be limited by psychological problems on long flights.

Fletcher, who probably represents high-level NASA policies, said that NASA will emphasize Earth-applications satellites (such as described in Chapters 2, 7, and 8) for practical data on crops, mineral resources, pollution, and weather, although the U.S. Department of the Interior will take on LandSats and much of the work. He wants to continue planetary exploration and studies of the sun—hopes to have samples returned from Mars by 1981, and later Soviet cooperation on Shuttle. He said that over half of the NASA launches in 1975 were for foreign governments or other (non-NASA) U.S. agencies.

*To close this volume, we turn to two of the most speculative aspects of space science: interstellar travel and space colonization. In both, there are grounds for accurate scientific and technological studies.—*TLP

Visual Aspects of
Trans-Stellar
Space Flight

SAUL MOSKOWITZ

(*Sky and Telescope*, May 1967)

The idea of men traveling for years through the depths of space to another star holds a special fascination, even though we know that today it is still beyond the realm of feasibility. Trans-stellar space navigation is now being studied, and we are developing theoretical and practical concepts that may be of value in more immediate kinds of celestial navigation. Of particular interest among these pioneer studies of trans-stellar flight are the visual appearance of the celestial background and its changes during the journey to another star.

For a spacecraft in orbit around the Earth, the navigator can determine his distance from Earth by triangulation from objects of known size, or by measuring the angular heights of stars above the horizon with a special sextant. In the case of a flight to the Moon, the navigator will rely mainly on angular measurements between terrestrial or lunar surface features and the reference background of "fixed" stars.

A new situation is encountered in navigating a trans-stellar space vehicle. The reference background of stars can no longer be regarded as infinitely distant. Rather it must be employed as it is, a three-dimensional distribution of suns at various distances, each with its own motion. Also, it is necessary to consider spacecraft velocities that are a large fraction of the speed of light, c, so great that the aberration of light will distort the apparent shape of the universe. As the spacecraft moves onward, the stars ahead of it will appear bluer, those behind it redder, because of the large Doppler shifts. And because of the wavelength shifts in the peaks of their emission curves, some stars will appear to the human eye brightened, others dimmed. An observer aboard such a vehicle will see a spectacularly changing universe.

As viewed from within our solar system, the stars form stable groupings. From their positions in our sky, and apparent brightnesses, it is possible to "name" individual stars. However, those making up a constellation are at very unequal distances from us. Therefore, a voyage of only a few tens of light years into space will make some noticeable changes in the arrangement of the stars.

This effect has been investigated quantitatively at Kollsman Instrument Corporation where a computer program was prepared to generate star maps for any observation point in our Galaxy. The computer inputs included the right ascension, declination, parallax, and absolute magnitude (see Glossary) of all well-known stars in the region of interest. By inserting the coordinates of a particular displacement from our solar system (on an interstellar trip), a new right ascension, declination, and apparent magnitude could be obtained for each star as seen from the spaceship. From these outputs new star maps could be plotted.

For purposes of illustration, the star 45 Eridani was chosen as the destination of a trans-stellar spaceflight. The 1950 right ascension and declination of this K4-type red giant are 4^h $29^m.3$, $-0°$ $09'$, and its parallax of 0.007 arc-sec corresponds to a distance of about 143 parsecs, or 466 light years. As seen from Earth, its apparent visual magnitude is 4.97.

After the spaceship has traveled two-fifths of the way, or 186 light years, toward 45 Eridani, the constellation Orion is little changed, but the 1st-magnitude star Aldebaran (only about 65 light years from the sun) is no longer seen out the front window, having drifted astern.

After the spaceship has traveled 186 light years farther, the target star has brightened to magnitude $+1.5$, and when the spacecraft has attained the immediate vicinity of 45 Eridani, that star (now a "sun") dominates the scene.

Looking through the rear window during this trip, we would see the sun receding from sight, fading to below 5th magnitude at a little more than 30 light years. (The sun is of absolute magnitude $+4.8$, whereas 45 Eridani is of absolute magnitude -0.8—see Glossary.) Figure 156 shows the rear-window view, $45° \times 45°$, at 93 light years out, where the sun is only magnitude 7.

On this imaginary flight our spaceship has been moving away from the central plane of the Milky Way, for the galactic latitude of 45 Eridani is $-38°$. Thus the stars are actually thinning out, as seen from both the front and rear windows (in computer views not reproduced here). However, our maps are also becoming increasingly incomplete because they only include stars for which parallaxes are known and omit stars fainter than 5th magnitude.

DOPPLER SHIFTS

For small relative velocities (up to a few hundred kilometers per second), the change in wavelength can be regarded as proportional to the velocity (p. 98). However, an interstellar spacecraft will presum-

FIG. 156. View of our sun (cross) and nearby stars from an interstellar space-craft 93 light years on the way toward 45 Eridani, without the effects of spacecraft motion. (Kollsman Corporation computer map.)

ably travel at very much higher velocities, and hence a formula from relativity theory must be used to calculate the Doppler shifts of stars seen from the moving ship.

Figure 157 shows the distribution of energy at different wavelengths in the continuous spectrum of the sun, as viewed by an observer at rest with respect to the sun. The peak for him lies in the visual region of the spectrum. But for another observer who approaches the sun at 90 percent the speed of light ($v/c = 0.9$), the bodily shift of the spectral curve toward shorter wavelengths displaces the peak into the far ultra-violet. As the diagram implies, to this observer the sun would appear (in visible light, between the two dashed lines) about 40 times fainter than it does to an observer at rest.

On the other hand, consider a very cool star whose rest spectrum (no Doppler shift) has a peak in the infrared. As seen from a spacecraft approaching at extremely high speed, such a star would appear not only bluer but also brighter.

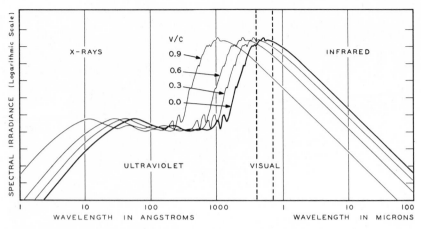

FIG. 157. The effect of large Doppler shifts on the spectrum of the sun. The curve labeled $v/c = 0$ is the spectrum we measure from above the Earth's atmosphere. Approaching the sun at $v = 0.9c = 167,000$ mi/sec, one would see this curve shifted toward shorter wavelengths (leftward). The wavelength scale changes from angstroms (A) to microns at 10,000A = 1 micron. (Kollsman Corporation diagram.)

Evidently, individual stars can deviate greatly from the general rule that stars seen through the forward window of the spaceship will look bluer and those through the rear window redder. In both fields, some stars will appear abnormally bright or faint, depending on the shapes of their spectral emission curves. At $v/c = 0.3$ there is not much effect, but at 0.6 the high approach velocity ($v = 180,000$ km/sec) has Doppler-shifted the light of many blue stars into the ultraviolet, thus diminishing the brightness of Rigel and Orion's belt. But red Betelgeuse and Aldebaran are much enhanced. In general, our constellations are no longer recognizable.

RELATIVISTIC ABERRATION

If a voyager is moving obliquely to the line of sight to a star, then its observed direction is shifted toward the direction of motion.

As the observer's velocity increases, this angular shift becomes larger and larger. Figure 158 shows the normal ($v = 0$) view toward 45 Eridani —a projection of the entire hemisphere centered on the target star.

At 60 percent the speed of light, many more stars have entered the visible hemisphere. Both Big and Little Dippers show, and many southern Milky Way stars. Orion has moved halfway from its normal position toward the center. (Doppler-shift effects are not included—so that the constellation can be recognized.) At $v/c = 0.9$, 96 percent of the stars in the back hemisphere have "moved" around to the forward side of the spacecraft, and no constellations are recognizable.

These results represent only one aspect of our current investigations of the problem of trans-stellar navigation. It shows that the greatest problem may well lie in the location and identification of specific navigation stars from a very fast spacecraft.

This shows some of the complexities of interstellar flight. Although NASA very sensibly avoids involvement with UFO's (Unidentified Flying Objects), it must be noted that a substantial fraction of Americans believe that extraterrestrial visitors aboard interstellar spacecraft are the cause of some unexplained UFO sightings—see UFO's, A Scientific Debate by Sagan and Page, Cornell University Press, 1972. Patterns of stars seen in a UFO that allegedly landed with a homonoid crew in New England have been compared to views like Figure 156 in support of this idea. The comparison yields no valid identifications.

It is true that terrestrial mankind can fire (and has fired—see p. 249) space probes out of the solar system. Moreover, nuclear-powered probes could now be made that would accelerate up to 0.52 c—96,800 mi/sec or 156,000 km/sec. But the believers overlook a major difficulty in carrying a crew or any living organisms on such an interstellar mission. Astrophysicists have ample evidence of hydrogen, dust, and other gases in interstellar space (p. 353), the hydrogen at a density of about 1 atom/cm^3. Thus, the fast-moving interstellar spacecraft (as pointed out by Freeman Dyson, at Princeton University) will be bombarded with hundreds of billions of H atoms every second, impacting at half the speed of light or more. Such impacts have been studied by physicists, using proton beams from high-energy accelerators, and are well known to produce high-energy gamma rays that can penetrate any feasible shielding and would kill any living organism in a matter of minutes. Another problem is where to go; there is no evidence of planetary systems where extraterrestrial life might exist within 50 light years, despite several astronomical attempts to find one, and several studies of alleged radio messages. Until these problems are solved, it is unlikely that terrestrial men will go on interstellar spaceflight.

Writers of science fiction ignore such practical limitations—and may, like Jules Verne and H. G. Wells, force space-age developments 50 years later. An excellent example is Arthur Clarke's Rendezvous with Rama (Harcourt Brace, N.Y., 1973) in which a huge, interstellar space ship is discovered in A.D. 2130 approaching the solar system. When visited by a large team of astronauts, this intruder is found to be a cylindrical structure over 50 km long and 16 km in diameter, with 1-km-thick walls, spinning about its long axis with a 4-min. period. With clever

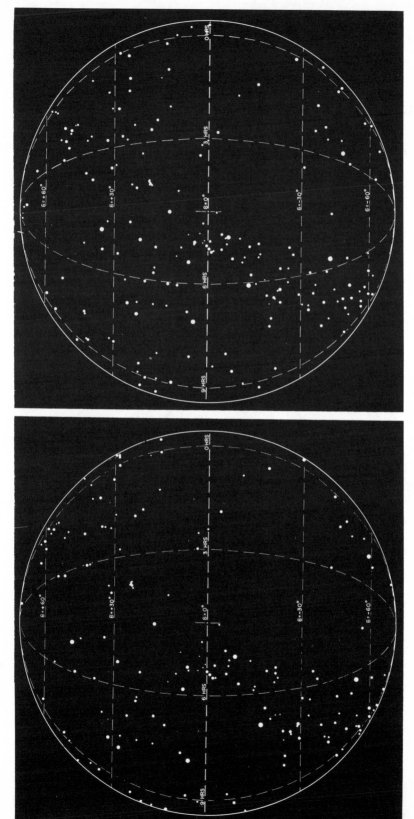

FIG. 158. On the left is a normal view of bright stars in half of the sky centered on 45 Eridani. The constellation Orion is about ½ in. (15°) left of center. On the right is the distorted view for an observer moving at $v = 0.3c = 55,750$ mi/sec toward 45 Eridani. Aberration shifts all the stars toward center, Orion by about 7°. (Kollsman Corporation computer map.)

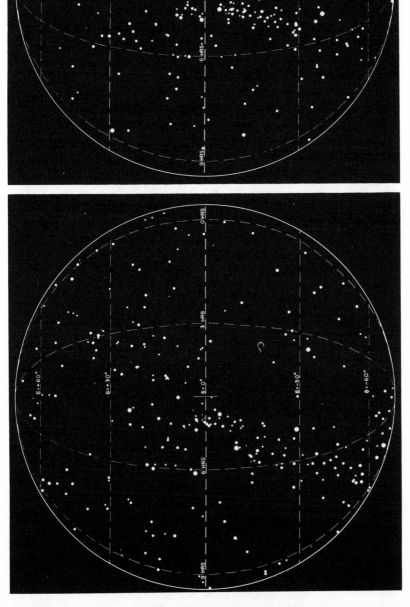

FIG. 159. Distortion of the view for observers moving at $v = 0.6c$ (left) and $v = 0.9c$ (right) toward 45 Eridani. Orion and all the stars in Figure 158 have been shifted farther toward the center, and other stars have moved in from beyond the edges of Figure 158. (Kollsman Corporation computer map.)

parodies of everything from Mission Control to the United Nations, Clarke gets his astronauts inside Rama where they find an oxygen atmosphere, a cylindrical sea, and 0.5-g gravity on the inside cylindrical walls. Many of his sci-fi descriptions are accurately worked out, including the "deep freeze" of interstellar travel. Rama's thick nose cone might even provide shielding against gamma rays from interstellar-gas impacts. But Clarke leaves unspecified the construction, launch, and control of such a large spacecraft.

In 1974, this general design was given a practical twist by Gerard K. O'Neill, a professor of physics at Princeton University. He first published an article, "Space Colonization," in the September 1974 issue of Physics Today, explaining his initial doubts but showing that no one had yet found an insurmountable difficulty in the construction of a large space station, whose planning and design he had assigned his students as a classroom project.

The success of Skylab (Chapter 6) had spurred suggestions like those of NASA's Christopher Kraft and William Schneider in February 1974 to have a permanent astronaut crew in an Earth-orbiting space station. O'Neill's went a good deal farther—to build a very large space station at a stable (Lagrange) point equidistant from Earth and Moon suitable for thousands of people to live normal lives in an unpolluted atmosphere, and served by space technology (low-cost solar energy, fast-maturing crops, artificial gravity, artificial day-night, and artificial seasons). Since the subject of space colonization has never been covered by Sky and Telescope, (as not yet being within its scope), we reproduce an article from Time Magazine, recounting the latest developments.—TLP

Colonizing Space[1]

(*Time*, May 26, 1975)

When Princeton Physicist Gerard K. O'Neill made the proposal that space colonies be established to relieve the Earth's overcrowding, increasing pollution and energy shortages, many of his more skeptical colleagues dismissed the scheme as one more exercise in scientific fantasy. But, unlike many other far-out proposals, the idea has not faded into oblivion.

For almost a year, O'Neill has continued to expand his vision with imaginative new details. He has become so convincing an advocate

[1] Reprinted by permission from *Time*, The Weekly Newsmagazine; Copyright *Time*, Inc.

that this month, at a three-day conference in Princeton, 100 scientists, engineers, international lawyers and social scientists agreed that space colonization is not only possible but eminently feasible. They even discussed such basic questions as what kind of meat the colonists will eat (the conferees were told that rabbits, chickens and pigs would be easier to raise in space than cattle) and what types of legal and social structures might be set up in their extraterrestrial world.

Huge Cylinders. This summer two dozen specialists, including O'Neill, will convene for ten weeks at NASA's Ames Research Center in Mountain View, Calif., to study the practical problems of getting the enormous project into orbit. Meanwhile, O'Neill, 48, is energetically continuing to press the idea in scientific articles, television and radio talks, and in campus lectures at the rate of at least two a week. Says he: "The whole thing is exploding so fast that I am beginning to worry about how to make time for my work in physics."

FIG. 160. A colony in space, proposed by Gerard O'Neill and Princeton students, a cylinder (one of two) 1 km long and 200 m. in diameter, rotating to produce artificial gravity on the inside walls. The three long mirrors provide sunlight through three segments of transparent wall ("sky" for the inhabitants) and the power-station mirror at the far end concentrates sunlight on solar cells. The "sister station" (not shown) rotates in the opposite direction. (*Time* magazine drawing reproduced by permission; copyright *Time*, Inc.)

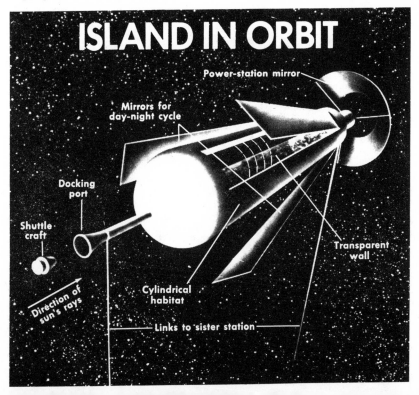

As it was first spelled out, O'Neill's scheme called for assembling in space large aluminum cylinders that would house self-contained communities. The cylinders would be built at the constantly moving "libration points," where the gravity of the Earth and of the Moon cancel each other out. Permanently in orbit at those positions, each pair of huge cylinders (1,100 yds long and 220 yds in diameter) would support 10,000 people; they would contain an atmosphere like Earth's, water, farm land and a variety of flora and fauna. The cylinders would rotate slowly, thus simulating gravity and holding people, buildings and soil "down" on the inner surfaces. For power, the space colonizers would rely on ever-present sunlight, captured by large external mirrors that could be controlled to create the effect of night and day and even of seasonal change.

At the Princeton gathering, O'Neill and others discussed the establishment of the first colony at a libration point called L5, which lies in the Moon's orbit at a spot equidistant from Earth and Moon. Initially, the space colonizers would set up a small mining base on the Moon. Its purpose: to provide most of the building blocks for the colonies. Rich in aluminum, titanium, iron, and other essential materials—including oxygen—lunar rocks could be fired off by a continuously catapulting device. Slowing as they climb out of the Moon's gravity, these building blocks would eventually arrive at the construction site in free space. That would be much cheaper than carrying the materials from Earth, where mineral-rich ores are already scarce and stronger gravity makes it necessary to use more powerful and costly rockets for launching.

Future Shock. As an added incentive, says O'Neill, the early colonies could be devoted to space manufacturing—for example, the construction of large turbogenerators driven by sunlight. Much easier to build in the gravity-free environment around the colonies, these giant machines could be towed back to the vicinity of the Earth, parked in fixed orbit and then used to relay the captured solar power down to Earth as a beam of microwaves (p. 390).

O'Neill concedes that such conceptions are "very rich in future shock" and larger than anything yet attempted in space or even planned on the drawing boards of space scientists. But he is firmly convinced that they could be achieved with technology that is either already available or almost perfected. In fact, says O'Neill, the first space habitat—he thinks the word colony connotes exploitation—could be functioning by the start of the next century. Its early inhabitants would probably be "hard-hat types," O'Neill says, but after the initial construction is finished almost anyone with a spirit of adventure could live at L5. The cost would be somewhat more than that of the $25 billion Project Apollo, which placed men on the Moon, but no more than a fifth of the estimated minimal $600 billion tab for Project Independence, the U.S. effort to free itself from dependence on foreign energy sources long before the year 2000.

The second, oppositely rotating cylinder is necessary to prevent precession of one cylinder's axis. Once the cylinders are "aimed" at the sun, the large mirrors reflect the sun as a stationary object in the colony's sky; it is made to "rise" and "set" by flapping the mirrors open (more in "summer" than in "winter") then shut. At the down-sun end, a rigid column between the two cylinders' axles keeps them apart; at the up-sun end (front, in Fig. 160) a cable is necessary to keep the axles together. O'Neill, his students, and conferees plan a small first pair of cylinders in 1988 A.D.—1 km long and 200 m. in diameter—for a few thousand construction workers who would build Colony 2, with cylinders 3.2 km long and 640 m. in diameter for 100,000 people or more in 1996 A.D. Now things are easier, and Colony 3, to be completed in 2002 A.D., would be 10 km long and 2 km in diameter, followed by Colony 4 in 2008 A.D., 32 km long and 6.4 km in diameter, the pair housing over a million people. By 2074, he says, 90 percent of the world population could be put in space colonies, and the Earth could return to its 18th-century beauty. The cost of a one-way trip to a colony would be about $3000.

A new cult of space colonists has formed the "L5 Society" (for the libration point, or L-5 location where the colonies would be built) with offices at 1620 N. Park Avenue, Tucson, Arizona 85719, from which further information can be obtained.

These bold plans triggered another NASA-supported conference and derision from conservative scientists, but no major holes have been found in the reasoning. After the conference (at the NASA Ames Research Center during the summer of 1975), O'Neill wrote about some of the conclusions (Science 190, 943, 5 Dec., 1975). In the long run, he says, orbiting solar power stations (p. 390) can be built more cheaply at L5 in what he calls "Space Manufacturing Facilities" (SMF) than on Earth. In fact, for an investment of about $100 billion, the first SMF could be built, from which several solar power stations could be constructed and moved to geosynchronous orbits to provide cheap electrical power on Earth. The profits would pay off the over-all investment in about 24 years. This would require a hard-working Lunar Colony, supplied with 10,000 to 20,000 tons of material from Earth and firing some 40,000 to 80,000 tons of lunar material to L5.

Dr. Tom Paine, Chief Administrator of NASA from 1968 to 1971, wrote for Newsweek magazine (Aug. 25, 1975) that he doubts some of O'Neill's cost estimates, but finds the proposals realistic, and more practical than colonizing other planets. He thinks that a Moon Colony may be possible, using Space Shuttle for support, in the 1990's. "Far out," he says, "is not far-fetched."—TLP

The Origin of
Sky and Telescope

In March 1931, publication of a small quarterly magazine, *The Telescope*, began at Perkins Observatory of Ohio Wesleyan University in Delaware, Ohio, with the director of the observatory, Harlan T. Stetson, as editor. By July 1933 the magazine had become a larger, bimonthly periodical. After Stetson moved to the Massachusetts Institute of Technology, the Bond Astronomical Club, a society of Cambridge amateur astronomers, and Harvard College Observatory assumed sponsorship of the magazine. Loring B. Andrews became editor, and in 1937 Donald H. Menzel succeeded him. *The Telescope* carried stories of important astronomical discoveries, reviews of current astronomical work, and articles on the history of the science.

In the meantime, the first issue of the small *Monthly Bulletin of the Hayden Planetarium* (New York City) appeared, in November 1935, edited by Hans Christian Adamson. In addition to a review of the current show at the planetarium, it contained other astronomical notes and articles. The interest and encouragement of its readers led, in November 1936, to the enlargement of its size and scope. Its name was changed to *The Sky*, and while retaining its planetarium ties, it became the official organ of the Amateur Astronomers' Association in New York City, replacing the magazine *Amateur Astronomer*, which had been published from April 1929 to the spring of 1936.

The Sky grew in reputation and circulation. In February 1938 Clyde Fisher, curator-in-chief of the planetarium, became editor. On November 1, 1939, the Sky Publishing Corporation was formed, owned by Charles A. Federer, Jr., who for four years had been a planetarium lecturer. He and his wife, Helen Spence Federer, edited and published *The Sky* through its fourth and fifth volumes.

Then, encouraged by Harlow Shapley, director of the Harvard College Observatory, Sky Publishing Corporation moved to Cambridge, Massachusetts, and combined *The Telescope* and *The Sky* into *Sky and Telescope*, born with the November 1941 issue. The ties with Harvard have been strong. Until the middle 1950's the magazine's offices were in the observatory—now they are located less than a mile away.

Mr. Federer was editor-in-chief until January 1975; he continues to participate in the magazine's preparation and to serve as president-treasurer of Sky Publishing Corporation. Joseph Ashbrook is now editor, with William E. Shawcross as managing editor, and Leif J. Robinson as associate editor. J. Kelly Beatty, Mollie Boring, Dennis di Cicco, Dennis Milon, and Roger W. Sinnott are assistant editors. Unsigned material in the magazine is prepared by this group.

During its thirty-five years, *Sky and Telescope* has been a distinguished and increasingly well-received publication, with two overlapping purposes. It has served as a forum where amateur astronomers can exchange views and experiences, and where they are furnished with observing data. It has brought to an ever-widening circle of scientists and educated laymen detailed and reliable information on new astronomical developments, and through its pages, has introduced them to the important figures of modern astronomy.

APPENDIX II

Chronology of the
Space Age

ca. 600:	Chinese development of military rockets.
1883–1935:	Tsiolkovsky, in Russia, wrote on the theory of rocket motors.
1914–45:	Goddard tested and flew rockets in Massachusetts, Maryland, and New Mexico.
1923–:	Oberth, in Germany, wrote books on space travel.
August 1942:	First V-2 rocket-bomb fired at London by German Nazis.
1946–:	Spectrographs carried above the atmosphere by NRL rockets.
October 1957:	First Earth orbiter, Soviet Sputnik I, 370 mi. high, 184 lbs.
February 1958:	USAF 31-lb Explorer I in orbit, discovered Van Allen belt.
	U.S. Congress Space Act established NASA.
September 1959:	First impact on Moon by 358-lb Soviet Luna II (see also Table 2).
October 1959:	Lunar far side photographed by 614-lb Soviet Luna III (p. 84).
August 1960:	NASA Echo I, first communications satellite (passive, 124 lbs).
February 1961:	First Venus fly-by by Soviet Venera I (see also p. 000).
April 1961:	First man in orbit: Cosmonaut Yuri Gagarin for 90 min. in Vostok I.
February 1962:	First NASA astronaut in orbit: John Glenn for 5 hrs in Mercury 6.
March 1962:	NASA OSO. 1 got far-UV observations of the sun from Earth orbit.
April 1962:	NASA 730-lb Ranger 4 got TV close-up pictures of Moon.
July 1962:	NASA Telstar I, first TV communications satellite (active, 171 lbs).
November 1962:	First Mars fly-by by Soviet Mars I.
July 1963:	NASA 80-lb Syncom II, first geosynchronous communications satellite (p. 48).

November 1963:   NASA 265-lb Imp A measured magnetosphere bow-shock front 60,000 mi. up-sun from Earth.

March 1964:   ESRO established in Europe with 10 member nations (p. 25).

July 1964:   NASA 806-lb Ranger 7 got TV close-ups of lunar surface before impact (p. 77).

August 1964:   NASA 830-lb Nimbus 1 in polar orbit got complete world cloud cover each day.

November 1964:   NASA 575-lb Mariner 4 got TV pictures of Mars on 6100-mi. fly-by (p. 188).

March 1965:   First spacewalk by Cosmonaut Leonov from Soviet Voskhod 2 with Belyayev.

April 1965:   NASA launched first commercial communications satellite, 87-lb Intelsat 1, in geosynchronous orbit over 27° W longitude.

July 1965:   First cosmic-ray measurements outside the atmosphere by Soviet 25,000-lb Proton 1.

November 1965:   First impact on Venus' surface by Soviet 2100-lb Venera 3.

January 1966:   First soft landing on · the Moon by 3520-lb Soviet Luna 9 (p. 87).

March 1966:   First docking in space by Astronauts Armstrong and Scott in NASA Gemini 8 with Agena rocket (p. 52).

March 1966:   First lunar orbiter, Soviet Luna 10 (3600 lbs).

August 1966:   First TV picture from lunar orbit by NASA Orbiter 1 (846 lbs).

April 1967:   Soviet Soyuz 1 crashed due to parachute failure, killing Cosmonaut Komarov (p. 64).

June 1967:   First soft landing on Venus by Soviet Venera 4 (2439 lbs).

September 1967:   First chemical analysis of lunar soil by NASA 619-lb Surveyor 5.

April 1968:   First orbiting astronomical observatory, Soviet Cosmos 215 (195 lbs).

September 1968:   First round-trip to Moon by Soviet Zond 5 with animals aboard (p. 54).

December 1968:   NASA OAO 2 orbiting astronomical observatory (4436 lbs—p. 337).

December 1968:   First manned round trip to Moon by NASA Apollo 8 with Astronauts Borman, Lovell, and Anders.

February–March 1969:   TV pictures of Mars by 838-lb NASA Mariners 6 and 7 in 2300-mi. fly-by (p. 193).

July 1969:   NASA Apollo 11 (109,565 lbs) led to the first manned landing on the Moon by Astronauts Armstrong and Aldrin (with Collins in CSM), returning 48 lbs of lunar samples from Mare Tranquillitatis (p. 99).

February 1970:	First Japanese Earth orbiter, 58-lb Lambda (see also Table 4).
April 1970:	First Chinese Earth orbiter (386 lbs—p. 26).
September 1970:	Soviet unmanned Luna 16 returned 3.5 oz. of lunar soil after soft landing in Mare Fecunditatis.
November 1970:	First unmanned lunar vehicle, Lunokhod 1, soft landed by Soviet Luna 17 for 11-months' operation in Mare Imbrium (p. 112).
December 1970:	X-ray survey of the sky started by NASA-Italian 320-lb Explorer 42 (Uhuru—p. 363).
December 1970:	France launched Earth orbiter.
May 1971:	Soviets launched Mars 2 and 3 for first impact on Mars by Mars 2 and first soft landing on Dec. 2, 1971, by Mars 3 (p. 208).
May 1971:	NASA launched 2271-lb Mariner 9, first orbiter of Mars, which telemetered high-resolution TV pictures of the surface (p. 211).
June 1971:	First space lab, Soviet 41,580-lb Salyut 1, visited for 22 days by Cosmonauts Dobrovolsky, Patsayev, and Volkov, who died in Soyuz 11 on return to Earth (p. 68).
July 1971:	First manned Lunar Rover used by Astronauts Scott and Irwin on Apollo-15 mission near Hadley Rille (p. 115); first subsatellite launched from CSM by Astronaut Worden.
March 1972:	NASA launched Pioneer 10 for first Jupiter fly-by in December 1973 (p. 246).
April 1972:	First astronomical telescope on the Moon with NASA Apollo 16 and Astronauts Young and Duke (p. 136).
July 1972:	First color-TV coverage of Earth from NASA 1800-lb ERTS 1.
August 1972:	32-in. telescope in Earth orbit on 4850-lb OAO 3 (Copernicus—p. 345).
December 1972:	Last manned landing on Moon by Astronauts Cernan and Schmitt in NASA Apollo 17; 250 lbs of lunar samples returned from Taurus-Littrow (p. 137).
1973–74:	NASA 190,000-lb Skylab in Earth orbit, launch damage repaired, many astronomical observations and engineering experiments performed by three crews in visits as long as 85 days (p. 296).
July 1973:	Soviet Union launched Mars 4, 5, and 6, of which Mars 5 soft landed in March 1974 (p. 231).
November 1973:	NASA 1108-lb Mariner 10 launched for Venus and Mercury fly-bys (p. 168).
December 1974:	German Helios 1 launched by NASA to orbit sun at 0.3 AU (p. 326).

June 1975:   Soviet Union launched Venera 9 and 10 to orbit Venus; Venera-9 lander reached the surface October 22 and transmitted the first pictures from another planet's surface (p. 167).

July 1975:   NASA Apollo docked with Soviet Soyuz in joint ASTP mission (p. 36).

August 1975:   NASA launched 7500-lb Vikings 1 and 2 to visit Mars in July 1976 with 2293-lb landers to make first analysis of martian soil (p. 233).

1980–:   NASA Space Shuttle to be used repeatedly (p. 400).

APPENDIX III

Notes on the Contributors

ASHBROOK, JOSEPH (1918–), the Editor of *Sky and Telescope* since January 1975, and on its editorial staff from 1953. His articles on people and events in the history of astronomy are a regular feature of the magazine. An astronomer specializing in variable stars, he was on the Yale faculty from 1946 to 1953. ("Results from Lunar Orbiter 2," "Apollo-15 Pictorial")

BEATTY, J. KELLY (1952–), geologist, assistant editor on the *Sky and Telescope* staff since 1974; from 1971 to 1974 he was a technical assistant at CalTech, analyzing Gemini and Apollo Earth-orbital photos and Earth-based photos of the terrestrial planets, and making mosaics of Mariners-9 and -10 photographs. ("Mariner-10's Second Look at Mercury," "Europe in Space," "An Orbiting Solar Power Station")

BLANCO, VICTOR M. (1918–), astronomer, born in Puerto Rico; Director of Cerro Tololo Inter-American Observatory, La Serena, Chile, since 1967; Director, Astronomy and Astrophysics Division, U.S. Naval Observatory, 1965–67; on the astronomy faculty at Case Institute of Technology, 1951–65. His chief interests in astronomy are galactic structure and stellar statistics. ("An Astronomer Looks at Space Travel")

BLESS, ROBERT C. (1927–), astronomer, since 1958 on the faculty of the University of Wisconsin and in the Space Astronomy Laboratory there; physicist at the Naval Research Laboratory, 1947–48. His chief fields of interest are photoelectric spectrophotometry and space astronomy. ("Ultraviolet Photometry from a Spacecraft")

CANNON, PHILIP JAN (1940–), planetary geomorphologist; Assistant Professor of Geology, University of Alaska; Research Scientist, Texas Bureau of Economic Geology, 1972–74; Geologist at the U.S. Geological Survey's Center of Astrogeology, 1967–72; his chief interests concern the origin and evaluation of landforms on Earth, Moon, and planets, the interpretation of side-looking radar imagery, and the applications of remote sensing to environmental geologic problems. ("Lunar Landslides")

CARRUTHERS, GEORGE R. 1939–), astrophysicist; Research Physicist at the Hulburt Center for Space Research at the Naval Research Laboratory, where he has been since 1964. He designed and built the far-UV electrographic camera, which was flown on several rockets and on Apollo 16 and Skylab. His interests are in far-UV spectroscopy, the Earth's upper atmosphere, and space-probe instrumentation. ("Astronomy with the Space Shuttle")

CODE, ARTHUR D. (1923–), since 1958 at the University of Wisconsin, where he is a member of the Space Astronomy Laboratory (OAO team) and Professor of Astronomy at Washburn Observatory; from 1956 to 1958 he was a staff member of Mount Wilson and Palomar Observatories and on the faculty of CalTech. His chief interests are photoelectric photometry of stars and nebulae, stellar spectroscopy, and satellite astronomy. ("Ultraviolet Photometry from a Spacecraft")

ERÖSS, BOTOND G. (1942–), Hungarian-born Systems Programmer, now at Systems Control, Inc., and from 1971 to 1975 at Stanford's Artificial Intelligence Laboratory; Technical Developments Officer at U.S. Air Force Solar Forecast Center, 1969–71; from 1967–1969 weather forecaster at Hamilton Air Force Base. ("Mariner-9 Picture Differencing at Stanford")

FICHTEL, CARL E. (1933–), physicist, Head of the Gamma-Ray and Nuclear-Emulsion Branch at NASA's Goddard Space Flight Center, where he has been since 1959; among his special interests are solar-particle composition, galactic and solar cosmic rays, and gamma-ray astronomy. ("Gamma-Ray Astronomy")

FORMAN, WILLIAM (1947–), astronomer at Smithsonian Astrophysical Observatory since 1973; interested chiefly in x-ray astronomy and optical observations of x-ray sources. ("X-Ray Sources and their Optical Counterparts")

GEHRELS, TOM (1925–), Netherlands-born American astronomer; Professor and member of the Lunar and Planetary Laboratory at the University of Arizona since 1961; Research Associate at the University of Chicago and Indiana University between 1956 and 1961. His chief interests are minor planets, photopolarimetry, and outer-planet missions. ("The Fly-By of Jupiter")

GIBSON, EDWARD G. (1936–), Scientist-Astronaut since 1965, the Science-Pilot of the third and final Skylab crew which spent 85 days in space in 1973–74; a Ph.D. from CalTech, where he studied solar physics; Senior Research Scientist at the Applied Physics Laboratory of Philco Corporation, 1964–65, studying lasers and the optical breakdown of gases; author of a textbook, *The Quiet Sun*. ("Comet Kohoutek Drawings from Skylab")

GURZADYAN, G. A., astronomer, at the Garny Space Astronomy Laboratory, Erevan, U.S.S.R. ("Ultraviolet Spectra of Faint Stars from Space")

HABER, FRITZ (1912–), German-born engineer, Vice-President of Marketing at the Avco Lycoming Division, where he has been since 1954; from 1949 to 1953 he was a research scientist for the U.S. Air Force, assigned to the Department of Space Medicine at the Air Force School of Aviation Medicine at Randolph AFB. ("G Forces and Weight in Space Travel")

HARTMANN, WILLIAM K. (1939–), Senior Scientist, Planetary Science Institute, Tucson, Arizona, studying the early history of the solar system; Co-investigator on Mariner-9 TV Experiment, 1971–73; his chief astronomical interests are the formation of planetary systems and the evolution of planetary surfaces; author of *Moons and Planets*, co-author of *The New Mars*. ("The New Mariner-9 Map of Mars")

HENIZE, KARL G. (1926–), astronomer; since 1966 a Scientist-Astronaut at NASA's Johnson Space Center and, since 1967, on the faculty of the University of Texas; previously he was at Lindheimer Research Center of Northwestern University, and at the Smithsonian Astrophysical Observatory. His chief astronomical interests are emission-line stars and planetary nebulae. ("Stellar Ultraviolet Spectra from Gemini 10")

HILLENBRAND, ROBERT (1944–), Director, Daytona Beach, Florida, Planetarium of the Museum of Arts and Sciences since 1971; his chief astronomical interests are planetary and comet observations. ("The First Men on the Moon")

HOEKSTRA, ROEL (1938–), in charge of the development of space-research instrumentation for ultraviolet stellar spectroscopy at the Space Research Laboratory at University of Utrecht, The Netherlands; from 1963 to 1970 he was at the Zeeman Laboratory in Amsterdam, studying the spectra of the rare-earth elements; his special interest in astronomy is high-resolution stellar spectroscopy. ("Results from the Utrecht Orbiting Spectrophotometer")

HOUCK, THEODORE E. (1926–1974) was Director of the Space Astronomy Laboratory at the University of Wisconsin; a pioneer in the adaptation of image-tube and television techniques to astronomical observing; he played an important role in instrumenting OAO 2 for ultraviolet photometry of stars, and in operating the Wisconsin Experiment for over two years. ("Ultraviolet Photometry from a Spacecraft")

HUCHT, KAREL A. VAN DER (1946–), Assistant Professor at the Space Research Laboratory at the University of Utrecht, The Netherlands; his chief interests in astronomy are the atmospheres of early-type stars and stellar evolution. ("Results from the Utrecht Orbiting Spectrophotometer")

JONES, CHRISTINE (1949–), astrophysicist; since mid-1975 a Harvard Junior Fellow associated with the Center for Astrophysics; a Post-Doctoral Fellow there from 1974–75; her chief astronomical interest is x-ray and optical observations of galactic x-ray sources. ("X-Ray Sources and their Optical Counterparts")

KAMPERMAN, THEO M. (1946–), Staff Member for Development and Research at the Space Research Laboratory of the University of Utrecht, The Netherlands, where he has been since 1970; his chief interests are the ultraviolet spectroscopy of stars, data processing, and electronic engineering. ("Results from the Utrecht Orbiting Spectrophotometer")

KOVAR, NATALIE S. (1938–1975), astrophysicist; Associate Professor at University of Houston and astronomer at the Manned Spacecraft Center, 1967–73; a devoted worker for the advancement of women in scientific and academic fields; her professional interests centered on comets, gas dynamics, and interstellar and interplanetary matter. ("Atmospheres Surrounding Manned Spacecraft")

KOVAR, ROBERT P. (1937–   ), astrophysicist; now Physicist with the U.S. Air Force; from 1967 to 1973 he was Staff Astronomer at the Manned Spacecraft Center and Assistant Professor at Rice University; his chief interests are infrared astronomy and comets. ("Atmospheres Surrounding Manned Spacecraft")

KUIPER, GERARD P. (1905–1973), Netherlands-born American astronomer, widely known for his studies of the planets and the Moon; he was Director of Yerkes and McDonald Observatories, and later head of the Lunar and Planetary Laboratory of the University of Arizona; editor of *The Atmospheres of the Earth and Planets*, *The Solar System*, and *Stars and Stellar Systems*, three University of Chicago series of astrophysical summary volumes. ("Lunar Results from Rangers 7 to 9")

LAMERS, HENNY J. (1941–   ), Research Staff Member of the Space Research Laboratory, University of Utrecht, The Netherlands; recently a Research Assistant at the Princeton University Observatory; his chief interests are the spectra of stars, the formation of coronae around hot stars, and the problems of mass loss. ("Results from the Utrecht Orbiting Spectrophotometer")

LILLER, WILLIAM (1927–   ), astronomer, at the Center for Astrophysics, Harvard and Smithsonian Astrophysical Observatories, a member of the Harvard astronomy faculty since 1960; from 1952 to 1960 at McMath-Hulburt Observatory, University of Michigan; his chief interests are photoelectric photometry of planetary nebulae and hot stars, the investigation of x-ray sources, and spectrophotometry. ("X-Ray Sources and their Optical Counterparts")

LILLIE, CHARLES F. (1936–   ), Associate Professor of Physics and Astrophysics at University of Colorado, where he has been since 1970; Research Associate at University of Wisconsin's Space Astronomy Laboratory, 1968–70; his chief research interest is far-UV spectroscopy. ("Ultraviolet Photometry from a Spacecraft")

LINDQUIST, THOMAS R. (1942–   ), mathematician; Applied Computer Scientist at Philips Medical Systems, Inc.; Systems Programmer at Honeywell, Inc., 1972–75, Berkeley Scientific Laboratories, 1971–72. ("Letter to the Editor")

LIPSKY, YURI N. (1909–   ), Head of the Department of Physics of the Moon and Planets, Sternberg State Astronomical Institute, Moscow University, U.S.S.R.; co-author of *Atlas of the Invisible Half of the Moon* (in Russian). ("Zond-3 Photographs the Moon's Far Side," "What Luna 9 Told Us About the Moon")

MARAN, STEPHEN P. (1938–   ), astrophysicist, now at the Laboratory for Solar Physics and Astrophysics at NASA's Goddard Space Flight Center, where he has been since 1969; from 1964 to 1969 he was astronomer in charge of the automated telescope project at Kitt Peak National Observatory; his special fields of interest are pulsars, gaseous nebulae, and infrared and solar-research satellites. ("What Two Sun-Observing Satellites Tell Us," "The OSO-7 Year of Discovery," "The Last OSO Satellite")

McNALL, JOHN F. (1930–), electrical engineer; Associate Director of the Space Astronomy Laboratory at University of Wisconsin, which supervised OAO-2 WEP; formerly Assistant Professor, Computer Science Department, University of Wisconsin; his chief work in astronomy has been instrumentation in ground and space astronomy and the use of computers in control and data collection. ("Ultraviolet Astronomy from a Spacecraft")

MELIN, MARSHALL (1917–), biochemist, amateur astronomer; an expert on visual observations of artificial satellites; he conducted the department "Observing the Satellites" in Sky and Telescope from 1958 to 1962. ("Space Age: Year One," "Man's Farthest Step into Space," "First Two Years of the Space Age," "Man in Space")

MOSKOWITZ, SAUL (1933–), since 1970 an instruments-engineering consultant and President of Historical Technology, Inc., of Marblehead, Massachusetts; at NASA's Electronics Research Center, 1967–70; from 1961 to 1967 he was Chief of the Navigations and Guidance Systems for the Space Division of Kollsman Instrument Corporation. ("Visual Aspects of Trans-Stellar Space Flight")

MUMFORD, GEORGE S. (1928–), astronomer; Dean of the College of Liberal Arts at Tufts University; from early 1965 to mid-1967 he conducted the "News Notes" department of Sky and Telescope; his current research is on cataclysmic variables and related objects. ("Surface Conditions on Venus," "Report from Rome: X-Rays and Gamma Rays," "X-Rays from Another Galaxy")

O'DELL, C. R. (1937–), astronomer; at NASA's Marshall Space Flight Center; formerly Director of Yerkes Observatory and Chairman of the Astronomy Department at University of Chicago. His chief astronomical interests are the evolution of planetary nebulae, diffuse nebulae, and comets. ("The Large Space Telescope Program")

ORDWAY, FREDERICK I., III (1927–), Consultant, National Science Foundation, and analyst, Research Analysis Corporation since 1974; Professor of Science and Technology Applications, University of Alabama, 1967–73; President, General Astronautics Research Corporation, 1965–67; Chief, Space Information Systems Branch, NASA's Marshall Space Flight Center, 1960–65; Saturn Systems Office, Army Ballistic Missile Agency, 1955–1960; co-author History of Rocketry and Space Travel; author, Pictorial Guide to Planet Earth. ("Rocket Propellants—The Key to Space Travel")

PAGE, THORNTON (1913–), Research Astrophysicist at Naval Research Laboratory, stationed at Johnson Space Center; Professor of Astronomy at Wesleyan University, 1958–69, at Yerkes Observatory and University of Chicago, 1938–40 and 1946–51, where his chief interest was in galaxies; editor of Stars and Galaxies, co-editor of UFOs: a Scientific

Debate and *The Sky and Telescope Library of Astronomy*. ("The Third Lunar Science Conference," "Notes on the Fourth Lunar Science Conference," "Notes on Lunar Research," "AAS 75th Anniversary Meeting")

QUAM, LYNN H. (1944–), Research Associate, Computer Science Department and Cardiology Division, Medical School, Stanford University; interested in image processing and robotics. ("Mariner-9 Picture Differencing at Stanford")

ROSENWALD, KURT (1901–1975), German-born Scientific Research Librarian at the Naval Research Laboratory. ("Letter to the Editor")

SAGAN, CARL (1934–), Director of the Laboratory for Planetary Studies at Cornell University since 1968; at the Harvard-Smithsonian Astrophysical Observatory, 1961–68; Investigator for the Mariner-2 Venus Probe, the Mariner-4 Mars Orbiter, and the Viking Mars Landers. His chief interests are planetary atmospheres and surface conditions, the production of organic compounds in astronomical environments, the origin of life, and extra-terrestrial biology. Author of *The Cosmic Connection, Communication with Extraterrestrial Intelligence*; co-author of *Intelligent Life in the Universe* and *Mars and the Mind of Man*. ("Mariner-9 Picture Differencing at Stanford," "Viking to Mars: The Mission Strategy")

SEWARD, FREDERICK D. (1931–), physicist; at the Lawrence Livermore Laboratory of the University of California since 1958; from 1967 he has been Leader of the High Altitude Physics Group there, which has made many observations of cosmic x-ray sources with rocket-borne detectors, discovered and identified the brightest sources in the soft-x-ray sky, studied the absorption of x-rays in the interstellar medium, the x-ray spectra of supernova remnants, and the structure of the x-ray source within the Crab nebula. ("The Distribution of X-Ray Sources in Our Galaxy")

SINNOTT, ROGER W. (1944–), an Assistant Editor of *Sky and Telescope* since 1971, co-conductor of its "Gleanings for ATM's" section; in the Navy Seabees, 1968–71; his chief astronomical interest is making and using telescopes as an amateur astronomer. ("Visit to Taurus-Littrow")

STROM, ROBERT G. (1933–), Associate Professor, Lunar and Planetary Laboratory and Department of Planetary Science, University of Arizona, where he has been since 1963; his chief interests in astronomy are lunar and planetary geology and geophysics. ("The Planet Mercury as Viewed by Mariner 10")

THOMAS, ROGER J. (1942–), Astrophysicist in Solar Studies and Assistant Project Scientist for the OSO Program at NASA's Goddard Space Flight Center, where he has been since 1970; between 1961 and 1970, a Research and Teaching Fellow at University of Michigan; his chief interests in astronomy lie in the field of x-ray and EUV studies of solar active regions and flares. ("What Two Sun-Observing Satellites Tell Us," "The OSO-7 Year of Discovery," "The Last OSO Satellite")

TUCKER, ROBERT B. (1940–   ), Programmer, Instrumentation Research Laboratory, Genetics Department, Stanford University; his chief astronomical interest is digital image processing. ("Mariner-9 Picture Differencing at Stanford")

VEVERKA, JOSEPH (1941–   ), Czech-born astronomer; since 1970 at the Laboratory for Planetary Sciences at Cornell University, where he is Assistant Professor of Astronomy. ("Mariner-9 Picture Differencing at Stanford")

WACKERLING, LLOYD R. (1937–   ), Staff Astronomer in the Department of Astronomy at Northwestern University, 1964–72; his chief interest in astronomy is stellar spectroscopy, especially of early-type stars with emission-line spectra. ("Stellar Ultraviolet Spectra from Gemini 10")

WATTS, RAYMOND N., JR. (1932–   ), Manager of the Geoastronomy Division, Center for Astrophysics, Harvard and Smithsonian Astrophysical Observatories; formerly Chief of Editorial and Publications Division there. From 1962 until early 1975, he reported regularly in *Sky and Telescope* on spacecraft and space research. He is the author of 54 articles covering most of the space technology in this book.

WEBBER, WILLIAM R. (1929–   ), astrophysicist, Professor of Physics and Director of the Space Science Center at University of New Hampshire; from 1961 to 1969 he was on the faculty of the University of Minnesota, and from 1963 to 1966 was a member of NASA's fields and particles subcommittee; his chief interests are cosmic rays and x-ray astronomy. ("Letter to the Editor")

APPENDIX IV

Glossary

A Angstrom(s); 1A = 10^{-8} cm. See App. V.

absorption lines, bands Gaps in the spectrum of planet, sun, or star due to atoms or molecules in its atmosphere. See **spectrum.**

Aerobee A class of U.S. sounding rockets. See Fig. 122.

airlock A vacuum-tight door in the cabin's pressure hull which can be opened when a canister is sealed to it on the inside so that equipment can be placed in the outside vacuum.

albedo Ratio of the amount of light reflected and scattered by a surface to the amount falling on that surface. The same ratio for a whole sphere (planet or moon) is called geometrical albedo.

alpha particles Helium atoms with two electrons removed (the helium nucleus), ejected by radioactive elements during their spontaneous disintegration.

ALSEP Apollo Lunar Surface Experiments Package, including a different set of experiments on each Apollo mission, deployed by astronauts with RTG power source and telemetry to JSC.

anorthosite A rock composed almost entirely of calcium-rich feldspar, which crystallized from magma.

aperture Diameter of lens or mirror in a camera or telescope.

apogee Point farthest from Earth in an elliptical orbit around the Earth. In order to enlarge the orbit, a spacecraft's apogee motor is turned on at apogee to give it an "apogee kick."

Apollo A NASA series of manned spacecraft and 12 missions, six of them landing on the Moon.

AS&E American Science and Engineering Company of Cambridge, Massachusetts, expert on x-ray detectors.

Atlas-Agena Rockets used to launch NASA space probes.

ATM Apollo Telescope Mount, an integrated set of instruments for observing the sun, and solar-cell panels for power, combined with Skylab (pp. 289, 302).

atmospheric drag Frictional retarding force on spacecraft or meteoroid moving through a gas.

ATS Applications Technology Satellites: a series of NASA Earth orbiters for weather, communications, and other applications, p. 328.

basalt A dark-colored, fine-grained rock, which crystallized from lava; of the same composition as gabbro.

436

beam splitter A flat plate of glass or other transparent material, placed at an angle in the beam of light, reflects a portion of the beam to one side, while the remainder goes on through.

binary star A double star—two stars in orbit around one another. When widely separated, they can be followed around orbits similar to, but larger than, planets' orbits. See **orbit**. Many binaries are so close together (and far away) that they appear as one star, even in large telescopes, but can be detected by the changing (orbital) Doppler shifts in their spectra, and are called "spectroscopic binaries." In a few cases, we see the orbit edge-on, and one star eclipses the other each half-revolution. These are called "eclipsing binaries." See **Doppler shift; spectrum;** Vol. 6.

bits Radio transmissions to and from spacecraft are in binary code bits, each bit being a 0 or a 1. These are spaced in "words" of nine bits each. Such a "word" has a value between 1 and 256 (2^9) each (just as, in the decimal system, nine digits have a value between 1 and 10^9). Thus, one word of nine bits can specify one of 256 values (such as the brightness of light in a small patch—"picture element" or "pixel"—of a TV picture). The TV picture is transmitted pixel by pixel, one word describing the brightness of each pixel.

black hole Einstein's General Relativity Theory explains gravitation by a curvature of space-time near a mass. If the concentration of mass in a very dense object exceeds a certain value, space is effectively curved back into itself, forming a separate universe, from which no light or material can pass into our universe, although the gravitational field is still with us. That is, the invisible "black hole" can be one component of a binary star, with a mass comparable to our sun's (and a diameter of 10 mi. or so), or it can be the nucleus of a galaxy with 10^{10} solar masses. The evidence for black holes is still uncertain and the theory was questioned in 1975 (pp. 366, 376, 385). See **neutron;** Vol. 8.

breccia A rock made up of angular, coarse fragments of different sorts of rock.

CalTech California Institute of Technology in Pasadena, manager of JPL.

Cassegrain telescope A reflector with concave primary mirror and convex secondary mirror producing an image (picture) at a focus through a hole in the primary. See Vol. 4.

celestial sphere The sphere of the sky, centered on the observer, with co-ordinates (right ascension and declination, like longitude and latitude on the sphere of the Earth) that define directions in space. See Vol. 1.

chromosphere Layer in the sun's atmosphere between photosphere and corona. See p. 000 and Vol. 3.

Command Module (CM) A component of the Apollo spacecraft attached to the Service Module (CSM) until re-entry into the Earth's atmosphere. See Figs. 43 and 111.

contour map A map with (contour) lines connecting points of equal value: elevation, brightness, etc. See Fig. 124.

corona Outermost layer of an atmosphere. The solar corona (p. 315) extends to 4 million mi. above the photosphere. See Vol. 3. The Earth has a geocorona (Fig. 134).

cosmic rays Charged particles (ions) moving at enormous speeds and striking the Earth from all sides. Some originate in the sun, and most are absorbed high in the Earth's atmosphere after some deflection in the Earth's magnetic field. Non-solar cosmic rays may originate throughout the Milky Way Galaxy (p. 356).

Cosmos A Soviet series of unmanned spacecraft used primarily for Earth observations. See Table 4.

COSPAR Committee on Space Research. An international committee advisory to space agencies in all countries (p. 55).

CSM Command and Service Modules, two parts of the Apollo spacecraft which remained linked together until just before splashdown (Fig. 29).

declination Angle north or south of the celestial equator. See celestial sphere.

Deep Space Net NASA's three 210-ft radio dishes in California, Australia, and Spain, used for receiving and broadcasting communications with distant spacecraft (p. 279).

diffraction Deflection of light waves passing through a narrow slit or reflected off a narrow strip of mirror. The interaction of waves from many parallel slits or strips forms a series of spectra from a parallel beam of light—the basis for most modern spectrographs. Light is also diffracted slightly at the edge of a telescope aperture so that even perfect optical parts must give slightly fuzzy—diffraction-limited—images. See Vol. 4.

dispersion The separation between different wavelengths (colors) in a spectrum, usually expressed in angstroms per millimeter (A/mm). See spectrum.

Doppler shift A change of frequency of radio, or sound, or light waves, or of any periodic emission, such as flashes. It is caused by approach or recession of the source relative to the observer (see p. 97).

down-link Radio or TV transmissions from a spacecraft to the control center on Earth.

dynamo action Motion of electrically conducting material in a fluid core of a rotating planet or moon; it produces a magnetic field.

echelle spectrograph Similar to a grating, the echelle gets high dispersion by using high-order spectra from fairly wide mirror strips. See dispersion, grating, spectrograph.

EREP Earth Resources Experiment Package carried on Skylab (p. 296).

ERTS Earth Resources Technology Satellite, a NASA series of Earth orbiters carrying downward-looking cameras (p. 330), later renamed LandSat.

ESA European Space Agency, a cooperative organization of 10 countries formed from two earlier organizations, ESRO and ELDO, in 1975 (p. 30).

EVA Extra-Vehicular Activity, time spent by one or more astronauts in space suits outside of their spacecraft; the longest EVA was on the lunar surface (pp. 74, 99, 137).

Explorer A series of NASA space probes, including Explorer 42, "Uhuru" (p. 363).

fault A fracture in bedrock along which there has been displacement of the sides relative to one another parallel to the fracture. In a "thrust fault," caused by compression, the rocks on one side of the fracture slip over those on the other, at a low angle.

feldspar One of the most important minerals on Earth: a group of silicates of aluminum with potassium or sodium and/or calcium.

flux Amount passing through a unit area each second. Light flux is the intensity in energy per square centimeter per second (ergs/cm^2 sec); particle flux (in the solar wind, for instance) is the number of protons/cm^2 sec.

focal length (f) The distance from lens or mirror to the image (picture) it forms in a camera or telescope. Larger focal length gives a larger image. Focal ratio is the ratio of focal length to aperture; an $f/4.5$ lens has focal length $4\frac{1}{2}$ times its aperture. An $f/2$ lens is "faster" than an $f/4.5$, requiring shorter exposure time. In a Cassegrain telescope, secondary mirrors increase the primary mirror's focal length. See Vol. 4.

FY Fiscal Year (for U.S. Government agencies), running from July 1 to June 30 (6 months earlier than the calendar year).

gabbro A dark rock which crystallized from magma, made up chiefly of calcium feldspar and certain dark minerals; its volcanic equivalent is basalt.

galactic latitude Directions from Earth to objects in the Milky Way Galaxy are measured in galactic latitude north or south of the Milky Way, and galactic longitude along the Milky Way, starting in the constellation Sagittarius. See Vol. 7.

galaxy Galaxies are groups of about 100 billion (10^{11}) stars in disk-shaped regions about 100,000 light years across. There are three types: ellipticals, spirals, and irregulars, the last two being young, and the ellipticals old. We are in the Milky Way Galaxy, a spiral, with two nearby irregulars, the Magellanic Clouds, about 300,000 light years away. Other galaxies are known by M-numbers and NGC-numbers, and are millions of light years away. Some are strong radio sources (radio galaxies) and some are undergoing vast explosions (Seyfert galaxies). See Vol. 8.

Geiger tube A radiation detector that counts cosmic rays, high-energy electrons and protons, and gamma rays passing through a low-density gas between charged electrodes. See p. 255.

Gemini (Twins) A series of NASA two-man spacecraft and missions. See Table 3 and p. 50.

granite Light-colored rock, consisting mainly of potassium and soda feldspar and quartz (SiO_2), which crystallized from magma. Extremely large bodies of granite were produced deep underground from other sorts of rock.

grating A mirror, plane or concave, with many fine, parallel scratches (rulings) that diffract incident light into several orders of spectra. The rulings can be shaped to throw most of the light into one order. Plane gratings require parallel light and a camera lens or mirror to focus the spectrum on film or detector. Concave gratings, properly placed, form spectra from divergent light (from a slit) on a film without a camera. Transmission gratings are similar but ruled on transparent plates. See Vol. 4.

gravity anomaly The force of gravity, or gravitational field, would be constant at the surface of a uniform spherical mass. The Earth, Moon, and planets approximate this ideal, but show deviations caused by mascons (p. 97) or low-density regions. These deviations are gravity anomalies.

GSFC Goddard Space Flight Center in Greenbelt, Maryland, near Washington, D. C.

gyro A spinning wheel, which resists change of the direction of its axis of rotation. Two or three gyros are used in spacecraft to steady its orientation during astronomical observations. By "twisting" the gyro axle, or changing its spin, the spacecraft can be rotated to a new orientation.

gyroscope A small rotating wheel maintains its axis fixed in space (except for slow drift due to unwanted forces). Two or three gyroscopes provide a fixed reference frame from which the spacecraft orientation is measured. Gyroscope drift is checked by observations on stars.

high-gain antenna A dish antenna on a spacecraft that amplifies the received radio transmission, if pointed at the distant transmitter. The 210-ft antennas on Earth (Fig. 108) are high-gain for down-link transmissions from spacecraft, and can be used as radio telescopes. See Vol. 4.

hydrostatic equilibrium Material in a star or planet or moon can be "in balance" between downward gravitational force and the outward pressure within. This hydrostatic equilibrium implies zero structural strength (no rock layers supporting heavy mountains) and no turbulent currents within that would cause bulges or depressions in the surface. Hydrostatic means "fluid at rest."

IAU International Astronomical Union, a cooperative organization of astronomers from 43 countries. Its various commissions collect data, adopt time standards, review research progress, and authorize official names (p. 184).

ilmenite An iron-titanium mineral ($FeTiO_3$) which is slightly magnetic.

IMP Interplanetary Monitoring Platform, a series of NASA space probes that collect data on Van Allen belts, magnetosphere, and near-Earth environment. See Table 4.

infrared (IR) Light of wavelength 0.7-100 microns. The near infrared (1-3 microns) is strongly absorbed by water vapor and carbon dioxide in the Earth's atmosphere (Fig. 1). The far-IR (10-100 microns) is difficult to observe telescopically because any warm body (including the telescope) is radiating strongly at these long wavelengths. Hence the telescope and accessories must be cooled, during measurements, with liquid air (90°K or −297°F) or liquid hydrogen (20°K or −423°F).

interstellar matter Both optical and radio telescopes show several gases (mostly hydrogen, with small amounts of H-C-O compounds, sodium, calcium, and neon) spread thinly (1 atom/cm³) between the stars in spiral galaxies. The gas is mixed with dust, commonly concentrated in clouds, and all of it moving in orbit around the galaxy's center. See Vol. 6.

ion An atom with one or more electrons removed. Atoms are ionized by heat, or light, or electrical forces, or bombardment of other particles. Ionization changes the spectrum of a gas. If hydrogen is 100 percent ionized (to protons), it has no spectrum lines or bands.

ionosphere A very high layer in the atmospheres of planets where the gases are mostly ionized.

isotope The atoms of many chemical elements have several different forms (isotopes) having the same characteristics except for different mass. Most of them occur naturally; some are products of radioactive disintegration of other atoms; and some isotopes are themselves radioactive ("radioisotopes"). Isotopes of one element can be separated by a mass spectrograph.

JPL Jet Propulsion Laboratory in Pasadena, California, operated by CalTech under NASA contract.

JSC Johnson Space Center in Houston, Texas; formerly MSC.

KPNO Kitt Peak National Observatory near Tucson, Arizona, financed by the U.S. National Science Foundation, with several large telescopes used by visiting astronomers.

KREEP A set of elements (K = potassium, REE = rare earth elements, and P = phosphorus) found in small but regular amounts in lunar rocks and soil. This set serves as a tracer or indicator of the amount of highland material mixed with the other material on the lunar surface.

KSC Kennedy Space Center at Cape Canaveral, Florida.

laser An intense coherent (all waves in phase) source of monochromatic light (one wavelength only), usually pulsed, and used (through telescopes) to measure accurate distances to spacecraft and Moon (p. 128).

LM Lunar Module, detachable part of the Apollo spacecraft that descended to land on the Moon (Fig. 33).

LRL Lunar Receiving Laboratory at MSC (now JSC), Houston, Texas, where lunar samples were unpacked and catalogued, and are now stored (p. 103).

LST Large Space Telescope planned by NASA (p. 396).

Luna A series of Soviet space probes to the Moon.

Lunar Orbiter A series of NASA space probes used to photograph the lunar surface (p. 91).

Lunokhod Soviet unmanned vehicle operated on the lunar surface (p. 112).

Lyman-alpha (Ly-α) The fundamental spectrum line of atomic hydrogen at wavelength 1216A; this line in emission or absorption is a sensitive detector of H atoms.

Magellanic Clouds Two irregular-type galaxies close to the Milky Way, visible only from the southern hemisphere of Earth's surface. See galaxy and Vol. 8.

magma A solution of chemical constituents which, when cooled sufficiently, crystallized to form the minerals that make up the various types of crystalline rocks. Magma which reaches the surface before hardening is called lava.

magnetic field Region in space where a small test magnet (compass needle) is forced to a particular direction. The force is a measure of the field strength measured in gauss (see Appendix V). This field extends around magnetized material in the direction from N pole to S pole, smaller at greater distance. The Earth's field is about 0.5 G at the surface. It is changed by electric currents and magnetic materials like iron, even when the iron is not magnetized. The source of the Earth's magnetic field is probably "dynamo action" in a molten nickel-iron core caused by electric currents in the spinning core. The field extends throughout the magnetosphere (Fig. 7).

magnetic star Many stars have strong magnetic fields, detected by the (Zeeman) effect on the star's spectrum. See Vol. 5.

magnetometer An instrument to measure the magnetic field at its location. Two sensitive types used by NASA are the fluxgate and helium-vector magnetometer, both sensitive to fields of 1 gamma or less (see Appendix V). Readings must be taken far from artificial magnetic materials in spacecraft or LM.

magnitude The brightness of a star in the sky is measured in magnitudes; the brightest stars are of magnitude 0, the faintest visible on a dark night about 5. These fifth-magnitude stars are 1/100th as bright as a 0-magnitude star. In a 2-in. telescope, 10th-magnitude stars can be seen—100 times fainter. Each increase of 1 magnitude refers to stars $2\frac{1}{2}$ times fainter (a "logarithmic scale" such that $2.512^5 = 100$). These are all "visual magnitudes" referring to visible (white) light. Photographs record blue light, and "photographic magnitudes" refer to blue light; infrared magnitudes refer to infrared light, and so on. The brightness (magnitude) of a star depends on both distance and luminosity, so a faint star may be a very luminous one at large distance. "Absolute magnitude" measures the luminosity; it is the magnitude (visual, photographic, IR, etc.) that the star would have if it were at a standard distance of 10 parsecs = 33 light years. The sun's absolute visual magnitude is $+5$, whereas its visual magnitude is -26 (because it is so close). Giant and supergiant stars have absolute magnitudes of -5 to -10 (see Vol. 5). The magnitudes of planets and comets depend on distance from both sun and Earth. The absolute magnitude of a comet is for both these distances equal to 1 AU. See Vols. 1, 2, and 5.

mare (pl. **maria**) A large, fairly flat plain on the Moon, originally called a "sea" because the surface looked waterlike in early telescopes. The maria probably resulted from giant impacts where lava flows filled in the impact craters (p. 78 and Fig. 16).

Mariner   A series of NASA spacecraft designed and built by JPL, used for planet fly-by missions (pp. 168 and 190).

mass spectrometer   An instrument for separating and measuring molecules of differing mass by deflecting a beam of the ionized molecules.

meteoroid   The small solid particle or chunk of iron, stone, or waxy material that flares to produce a bright meteor ("shooting star") when it hits the Earth's atmosphere. A large stone or iron may last long enough to reach ground as a meteorite. The numbers of meteoroids of different sizes in interplanetary space are larger for smaller and smaller sizes. See p. 44. Millions of micron-size micrometeoroids (or micrometeorites) strike Earth and Moon each second. See Vol. 2.

Milky Way stars   A band of stars, visible only on a clear, dark night, stretches right around the celestial sphere (sky). From the stars' distances, astronomers can plot out the Milky Way Galaxy, a disk-shaped group of over 100 billion stars, including the sun. See **galaxy** and Vol. 7.

Mission Control Center (MCC)   Where manned spacecraft of Mercury, Gemini, Apollo series, Skylab, and ASTP were controlled from MSC (later JSC) in Houston, Texas. See p. 304.

Mission Requirements Document (MRD)   A list of crew activities, experiments, and communications expected to be performed on a NASA manned space mission, with a schedule (p. 305).

MIT   Massachusetts Institute of Technology in Cambridge, Massachusetts.

MSC   Manned Spacecraft Center in Houston, Texas (changed to JSC in 1973).

MSFC   Marshall Space Flight Center in Huntsville, Alabama.

NASA   National Aeronautics and Space Administration, with headquarters in Washington, D. C., founded by the U.S. Congress in 1958. See p. 10.

neutron   An atomic particle with the mass of a proton (hydrogen ion) but no charge. In effect, the neutron is a tight combination of proton and electron, very much smaller than a hydrogen atom. "Neutron stars" are collapsed stars, consisting almost entirely of closely packed neutrons, and therefore having extremely high density (10^{17} gm/cm^3). When a neutron star cools it collapses further into a "black hole." See **black hole** and Vol. 8.

NORAD   North American Air Defense, a radar network covering U.S. and Canada.

NRL   Naval Research Laboratory in Washington, D. C.

OAO   Orbiting Astronomical Observatory (pp. 68, 337, and 345).

OGO   Orbiting Geophysical Observatory, a series of NASA space probes for studies of the Earth's atmosphere (pp. 57, 62, 404).

olivine   A green mineral which is an important constituent of dark crystalline rocks; a mixture in varying proportions of silicates of magnesium and iron.

orbit   Spacecraft, moons, planets, and comets all move in orbits around a massive primary—moons around planets, planets and comets around the sun. They follow Newton's laws of gravitation and motion (p. 17). Except for minor perturbations, these closed orbits are exact ellipses, each exactly in a plane, from perigee closest to Earth (or perihelion closest to sun) to apogee farthest from Earth (aphelion farthest from sun). Each orbit can be described by six numbers, called "orbital elements." The "semi-major axis," $a = \frac{1}{2}$ distance from perigee to apogee, the "eccentricity," $e = 0$ for a circle and nearly 1.0 for long, thin ellipses, the "inclination," $i =$ angle between orbit plane and equator of Earth, the "period," $P =$ time to go around once (related to a) and two angles that define the direction of the major axis. A spacecraft flying by a planet is in open orbit—an hyperbola instead of an ellipse—it has infinite a and no P. See Vol. 1.

OSO   Orbiting Solar Observatory, a series of NASA space probes for studies of the sun (p. 316).

parallax   Shift in a star's direction due to a 1-AU change in the Earth's position as it moves along its orbit; parallax is the basis of astronomical distances. A nearby star at 1 parsec (3.26 light years) has 1-arc-sec parallax; at 10 parsecs, the parallax is 0.1 arc-sec. See Vol. 5.

perigee   Point closest to Earth on an orbit around the Earth. Similar point on an orbit around the sun is called "perihelion," around the Moon "perilune," around Mars "perimartium," and around Jupiter "perijove," etc.

photometer   An instrument to measure the intensity (brightness) of light by an electrical voltage or pulse counter. There are several types: a "photomultiplier" amplifies the electron-current from a photoelectric surface, "photo tubes" and "channeltrons" work only in the ultraviolet, measuring ionization of a gas in the tube. A "photopolarimeter" measures both intensity and polarization. See Vol. 5.

photon   A quantum of light—the smallest separable amount of energy in a beam of light. Photon energy is inversely proportional to wavelength (see p. 355). High-energy photons of x-rays or gamma rays are counted by photometers; low-energy photons of visible light and infrared are measured continuously.

photosphere   The visible surface of the sun, at temperature 5700°K, where sunspots occur and flares originate (p. 313). See Vol. 3.

PI   Principal Investigator, the individual recognized by NASA as responsible for a space experiment.

plasma   Low-density, ionized gas (mostly protons and electrons) in motion, carrying a complex magnetic field, high in the solar atmosphere, in interplanetary space, in interstellar space, and in intergalactic space.

polarized   Ordinary light and radio waves are unpolarized; they oscillate in all directions around the line of sight. "Plane polarized" waves oscillate in one direction (plane) like a clothesline jiggled up and down only, or right-left only. (Circular polarization corresponds to "cranking" one end of the clothesline round and round.) Instruments to detect polarized light are called "polarimeters," using Polaroid or crystals to select one plane. See Vol. 5.

prominences Flame-like clouds of gas shot outward from the sun's chromosphere (Fig. 128). See Vol. 3.

proton A positively charged particle, the nucleus of the hydrogen atom, and component of all atomic nuclei. Ionized hydrogen is separated protons and electrons.

pulsar A rotating, condensed star, of a type first detected by regular (about one per second) pulses of radio waves. These, and the optical pulses, can be used to measure a pulsar's distance from us. See Vol. 7.

quasar A very distant, highly luminous galaxy, probably undergoing a violent explosion. See Vol. 8.

radial velocity Motion of a source toward or away from an observer in the line of sight. See **Doppler shift.**

radiation pressure The very small force exerted by light when it is absorbed or reflected from a surface. Sunlight, falling almost continuously on orbiters, slowly changes their orbits by radiation pressure.

radioactive decay Some chemical elements are radioactive. The "parent" nuclei of their atoms spontaneously eject alpha particles, protons or electrons, leaving a different "product" atom. Laboratory measurements have established the rate of such decay, so the relative amounts of parent and product in a rock give its "radioactive age." Some products are gases (helium from alpha particles, argon from radioactive potassium) and are released into the atmosphere or to space.

Ranger An early NASA series of space probes, pp. 47 and 77.

regolith The blanket of rubble on the Moon's surface created by lunar impacts.

resolution "Spatial" resolution of a photograph is the separation of the smallest features that can be distinguished (such as two close stars). For telescopes this "resolving power" is given in angle (arc-sec). See **diffraction;** Vol. 4. "Spectrographic resolving power" is the difference in wavelength between two spectral lines that can just be distinguished— given in angstroms (A).

retrorocket A rocket fired to oppose (reduce) spacecraft motion, causing it to descend in orbit, or to slow up in free fall.

right ascension Angle in the sky east of the Vernal Equinox, a point in the constellation Pisces. See **celestial sphere.**

Rover (LRV) Powered vehicle used on the lunar surface by astronauts on the Apollo-15, -16, and -17 missions (Fig. 45).

RTG Radioactive Thermoelectric Generator. A power source used on NASA space probes to Mars and Jupiter, and for ALSEP. See Fig. 106.

S 4B Third-stage rocket used on Apollo missions.

Salyut ("Salute") Class of large Soviet spacecraft-workshops (p. 293).

SAO Smithsonian Astrophysical Observatory in Cambridge, Massachusetts.

Saturn First-stage booster rocket used on Apollo missions.

Schmidt camera or telescope A concave spherical mirror with a smaller-aperture "corrector plate" and a curved focal surface, giving good focus over a wide field. The corrector plate, thicker in the middle and at the edges, eliminates spherical aberration and coma that would give foggy, deformed images with spherical mirror only. A Schmidt-Cassegrain has a secondary mirror to reflect light through a central hole in the primary to a focus behind. See Vol. 4.

sedimentary rocks   Rock fragments or particles deposited in layers. After burial they are cemented or compacted into rock layers.

seismometer   An instrument that measures earthquake (or moonquake) waves—"seismic waves" that move the ground up-down, north-south, and east-west, at the seismometer location. Seismic waves travel through solids at speeds depending on the material (p. 129). There are two types of wave: "P" (pressure, like sound waves) and "S" (shake or transverse waves like light waves). S-waves are slower and cannot pass through liquids.

Shuttle   NASA's reusable spacecraft (p. 400).

signal-to-noise ratio   All measurements are subject to error, and some are swamped in background that must be subtracted to get the desired data. Measures of the brightness of a faint star against strong background may have errors plus background light ("noise") equal to twice the star's actual brightness—signal-to-noise ratio (S/N) is then 0.5. Eliminating the background with filters would improve S/N.

Skylab   Large NASA spacecraft-workshop (p. 296).

solar cells   These convert sunlight to electric power by crystals of silicon with an efficiency of 10-20 percent. "Solar panels" of many such cells are used on almost all spacecraft as a source of electric power (pp. 270 and 391).

solar nebula   The large cloud of hot gas and dust from which the sun and planets formed.

solar wind   A stream of ionized gas, mostly protons and electrons, blown out of the sun at high speed (300 km/sec) on all sides. The wind spirals outward because of the sun's rotation (1 turn/month), so it is not parallel to sunlight. There are "gusts" and longer term variations, but the solar wind is fairly continuous, deforming the magnetosphere (Fig. 7), geocorona (Fig. 134), and outer atmospheres of the planets.

Soyuz   A class of Soviet manned spacecraft, pp. 33, 64, and 68.

Spacelab   ESA-designed part of Shuttle (p. 402).

Spacewarn   A list of spacecraft and jettisoned parts in orbit near the Earth, maintained at GSFC.

spectrum (plural **spectra**)   Light from a star or other source spread out according to wavelength from very short gamma rays through x-rays, ultraviolet, visible light, infrared, to long radio waves. Gaps are called "absorption lines" or "bands," due to "cool" atoms or molecules in the light beam. Atoms or molecules at high temperature (or "excited" by electron impacts, etc.) give "emission lines" or "bands" in the spectrum, each line or band characteristic of the atom or molecule causing it. Optical spectra are photographed in "spectrographs," in which the light to be analyzed passes through a prism or is reflected from a grating or echelle. See **diffraction, grating**; Vol. 4. A "spectrometer" uses a photometer to scan the optical spectrum. Radio spectra are measured by tuning the receiver in a radio telescope. For x-ray and gamma-ray spectra, see Chapter 9.

supernova Very massive stars (10 times the sun) "burn" their hydrogen fuel rapidly. Near the end of its life, such a star can suffer a tremendous explosion, increasing its luminosity by a billion times for several months. About 90 percent of the star is blown outward to form a "supernova remnant" like the Crab nebula (p. 357) and the central core remains as a collapsed neutron star or pulsar. At least four such supernovae have been seen from Earth in our region of the Milky Way Galaxy. It is estimated that there is one such supernova each 100 years in the average spiral galaxy. See Vol. 8.

Surveyor A class of NASA space probes (p. 47).

telemetry Most of the data obtained on spacecraft are sent by radio transmission to ground receivers. This "telemetry" involves coding the measurements in binary numbers (0's and 1's) which can be transmitted rapidly with few errors. Counts of cosmic rays and x-ray photons are easily coded in bits, but "smooth" (analog) data such as light intensity in a TV picture element (pixel) must be "digitized" in discrete levels for telemetry. See also **bit.**

Uhuru (Explorer 42) The first NASA high-energy-detector satellite (p. 363).

ultraviolet Light of wavelengths shorter than about 4000A, the limit of human vision. See **spectrum** and Fig. 1.

Universal time The standard time at Greenwich, near London, England, used by astronomers to avoid confusion in the time of an event. 12:00 UT is noon at Greenwich, 7 a.m. EST, 4 a.m. PST, and 9 p.m. in Tokyo, Japan.

UT Universal time.

Van Allen belts A doughnut-shaped region around the Earth, 200-20,000 mi. above the magnetic equator, where high-speed protons and electrons oscillate north-south in the Earth's magnetic field. See Fig. 7 and Vol. 2. Jupiter has even more intense radiation belts (Fig. 99).

Viking A NASA pair of spacecraft with landers to land on Mars (p. 233).

Voskhod ("Sunrise") An early Soviet class of manned spacecraft.

Zond A class of Soviet space probes (p. 54).

Units, Measures, and Powers of 10

1 metric ton = 1.10 tons = 2205 pounds
1 kilogram (kg) = 2.20 pounds (lbs)
1 gram (gm) = 0.035 ounces (oz)
1 microgram (μg) = 10^{-6} gm

1 parsec = 3.26 light years
1 light year = 9.46×10^{12} kilometers (km)
1 astronomical unit (AU) = 93,000,000 mi. = 150,000,000 km = 1.5×10^8 km
1 mile (mi.) = 5280 feet (ft) = 1.61 km
1 kilometer (km) = 1000 meters (m.) = 0.62 mi. = 3280 ft
1 meter (m.) = 3.29 ft = 39.4 inches (in.) = 100 centimeters (cm)
1 inch (in.) = 2.54 centimeters (cm)
1 centimeter (cm) = 10 millimeters (mm) = 0.39 inch (in.)
1 micron (μ) = 0.001 millimeter (mm) = 0.00004 inch (in.)
1 angstrom (A) = 10^{-8} cm = 10^{-4} micron (μ)

Area measured in square miles (mi.^{2}), square kilometers (km^2), etc.
Pressure measured in pounds per square foot (lbs/ft^2), kilograms per square centimeter (kg/cm^2)
Volume measured in cubic inches (in.^{3}), cubic centimeters (cm^3), etc.
Density measured in pounds per cubic foot (lbs/ft^3), grams per cubic centimeter (gm/cm^3)

10^9 (giga) = 1 billion = 1,000,000,000
10^6 (mega) = 1 million = 1,000,000
10^3 (kilo) = 1 thousand = 1000
$10^{1/2}$ = square root of 10 = 3.16
10^{-2} (centi) = 1 hundredth = 0.01
10^{-3} (milli) = 1 thousandth = 0.001
10^{-6} (micro) = 1 millionth = 0.000001

Angle around a full circle = 360 degrees (360°) = 2π radians
1 degree (°) = 60 arc-minutes (arc-min)
1 arc-min = 60 arc-seconds (arc-sec)

Temperature is measured in degrees Fahrenheit (°F), or degrees Celsius
(°C), or degrees Kelvin (°K)
The unit degree Fahrenheit = 5/9 of the units C or K
Freezing water = 32°F = 0°C = 273°K
Boiling water = 212°F = 100°C = 373°K

Atmospheric pressure is measured in (Earth) atmospheres, millimeters of
mercury (Hg), and bars.
1 atmosphere = 14.7 lbs/ft² = 760 mm of Hg = 1.013 bar
1 millibar = 10⁻³ bar = 0.75 mm of Hg = 0.00099 atmosphere

Wavelengths of light, x-rays, and gamma rays are measured in angstroms (A)
or microns (μ).
Frequency of radio waves is measured in hertz (Hz).
1 gigahertz (GHz) = 10⁹ hertz = 1 billion cycles per second (cps)
1 megahertz (MHz) = 10⁶ hertz = 1 million cycles per second (cps)
1 kilohertz (kHz) = 10³ Hz
1 hertz (Hz) = 1 cycle per second

Energy is measured in ergs and power in watts.
1 watt = 1 erg per second (erg/sec)

Energy of high-speed electrons and ions, and x-ray photons they create, is
measured in electron volts (eV), or kilo-electron volts (keV). Gamma-ray
energies are measured in mega-electron volts (MeV). See p. 356.

Magnetic field strength is measured in gauss (G) or oersteds. The Earth's
magnetic field at the surface is about 0.5 gauss.
1 milligauss (mG) = 10⁻³ gauss
1 gamma (γ) = 10⁻⁵ gauss

Gravitational field strength is measured in gals.
1 gal = 1 g (see p. 39)
1 g = 32.17 ft/sec² at lat 45° N or S on Earth
1 g = 980.7 cm/sec² (standard) on Earth's surface
0.362 g = 365 cm/sec² on Mercury's surface
0.915 g = 900 cm/sec² on Venus' surface
0.166 g = 162.5 cm/sec² on Moon's surface
0.370 g = 374 cm/sec² on Mars' surface
2.72 g = 2660 cm/sec² at Jupiter's cloud tops
1.18 g = 1160 cm/sec² at Saturn's cloud tops

Suggestions for
Further Reading

General

ABELL, G. O. *Exploration of the Universe.* New York: Holt, Rinehart and Winston, 2d ed., 1969—an informative introductory astronomy textbook. There is a paperback "Updated Brief Edition, 1975."

VON BRAUN, WERNHER, AND ORDWAY, F. I. III. *History of Rocketry and Space Travel.* New York: Crowell, 3d ed, 1975—very complete and well illustrated.

The Solar System. San Francisco: W. H. Freeman, 1976—a collection of *Scientific American* magazine articles by various authors, including Carl Sagan.

Chapter 2

MALLAN, LLOYD. *Suiting Up for Space.* New York: John Day, 1971—the full history of astronauts' space suits.

Sky and Telescope January issues for 1963 through 1968 carry *Rosters of Space Activity* similar to Table 4 for all announced launches.

New Catholic Encyclopedia, New York: Publishers Guild, 15 vols., 1974— under the entry "Space Exploration" contains a 15-year summary of space flight.

SHARPE, MITCHELL R. *It Is I, Sea Gull.* New York: Crowell, 1975—a revealing story of the training and political activity of Valentina Tereshkova, the only woman cosmonaut (1963, Vostok 66—see Vol. 1).

Chapter 3

COLLINS, MICHAEL. *Carrying the Fire.* New York: Farrar, Strauss and Giroux, 1974—an excellent account of astronaut training and the Apollo-11 mission to the Moon.

Apollo 14—Preliminary Science Report, NASA SP-272, Washington, D.C., U.S. Government Printing Office, 1972—well-illustrated articles about each experiment and the first results obtained. Similar 300-page paperbound books are SP-289 on Apollo 15, SP-325 on Apollo 16, and SP-330 on Apollo 17.

LEVINSON, A. A., ed. *Proceedings of the Second Lunar Science Conference,* 3 vols. Cambridge: Massachusetts Institute of Technology Press, 1971— detailed papers on results of Apollo experiments and the analyses of lunar samples. Paper-bound abstracts of later studies reported at the Third, Fourth, and Fifth Lunar Science Conferences are available from the Lunar Science Institute, 3303 NASA Road 1, Houston, Texas 77058.

LEWIS, RICHARD S. *The Voyages of Apollo.* New York: Quadrangle, 1975— a summary of the Apollo flights and findings, colorfully written and well illustrated.

FRONDEL, JUDITH. *Lunar Mineralogy.* New York: John Wiley, 1975—a summary of research findings on lunar rocks.

CORTRIGHT, EDGAR M., ed., *Apollo Expeditions to the Moon.* NASA Publication SP 350, 1975. A beautifully illustrated book; the 18 contributors include former key NASA officials, astronauts, engineers, and scientists. $8.90 from Government Printing Office, Washington, D.C. 20402.

Chapter 4

SAGAN, CARL, "The Solar System"; CAMERON, A. G. W., "The Origin and Evolution of the Solar System"; PARKER, E. N., "The Sun." *Scientific American* magazine, September 1975.

LEWIS, JOHN S. "The Chemistry of the Solar System," *Scientific American* magazine, March 1974—explains Table 8.

Mariner-10 Venus Encounter, Science 183, 1289–1321, March 29, 1974— detailed results written up by the investigators.

Mariner-10 Mercury Encounter, Science 185, 141–180, July 12, 1974—detailed results written up by the investigators.

Chapter 5

MURRAY, BRUCE C. "Mars from Mariner 9," *Scientific American* magazine, January 1973—further discussion, well illustrated.

SAGAN, CARL, and others. "Variable Features on Mars," *Icarus 17,* 346–372, 1972—more advanced treatment.

New Map of Mars (Fig. 84) is available at full size from Lowell Observatory, Flagstaff, Arizona 86001 for $1.50.

Mars as Viewed by Mariner 9, Mariner-9 Television Team and Planetary Program Principal Investigators. NASA, 1974.

HARTMANN, WILLIAM K. and RAPER, ODELL. *The New Mars: The Discoveries of Mariner 9.* NASA, 1974.

Chapter 6

"Pioneer-10 Mission: Jupiter Encounter," *Journal of Geophysical Research 79,* 3487–3694—detailed results of all experiments.

Science, 188, 445–477, May 2, 1975: Reports on Pioneer-10 and -11 results by the investigators.

Chapter 7

GATLAND, KENNETH. *Missiles and Rockets.* New York: Macmillan, 1975—a pocket encyclopedia with 84 color photos of military missiles and spacecraft, with details on their development and operation.

Skylab Explores the Earth, NASA SP 380, Washington, D.C., U.S. Government Printing Office, 1975—a broad summary of Skylab experiments.

Space World monthly magazine. Amherst, Wisconsin 54406—with semi-popular articles on specific equipment and missions.

Aviation Week and Space Technology weekly magazine. P. O. Box 430, Hightstown, New Jersey 08520—technical reports on aircraft, spacecraft, and new developments.

Chapter 9

GOLDBERG, LEO, ed. *Annual Review of Astronomy and Astrophysics*, vol. 5. Palo Alto, California, Annual Reviews, 1967—a chapter by G. G. Fazio summarizes the technique of gamma-ray astronomy.

GIACCONI, RICCARDO, and GURSKY, HERBERT. *X-Ray Astronomy*. Boston: D. Reidel, 1974—graduate-level summary of research advances through mid-1974. Largely devoted to Uhuru observations and their interpretation by several experts.

Chapter 10

MOORE, PATRICK. *The Next Fifty Years in Space*. New York: W. Luscombe, 1976. Realistic estimates and reasons for what can happen in space exploration.

Index

(Bold-face numbers indicate the more important pages on a subject.)